The Experience of Science
An Interdisciplinary Approach

The Experience of Science
An Interdisciplinary Approach

Martin Goldstein
Yeshiva University
New York, New York

and

Inge Goldstein
Columbia University
New York, New York

Plenum Press • *New York and London*

Library of Congress Cataloging in Publication Data

Goldstein, Martin, 1919–
 The experience of science.

 Includes bibliographies and index.
 1. Science. I. Goldstein, Inge F. II. Title.
 Q160.2.G65 1984   500   84-4755
 ISBN 0-306-41538-0

First Printing—April 1984
Second Printing—February 1986
Third Printing—June 1987

Cover: Raphael. "School of Athens." Fresco, Stanza della Segnatura, the Vatican, photographed by Peter Adelberg, European Art Color, New York, New York.

©1984 Plenum Press, New York
A Division of Plenum Publishing Corporation
233 Spring Street, New York, N.Y. 10013

All rights reserved

No part of this book may be reproduced, stored in a retrieval system, or transmitted, in any form or by any means, electronic, mechanical, photocopying, microfilming, recording, or otherwise, without written permission from the Publisher

Printed in the United States of America

For Eric, Michael, and Aviva

Preface

Our earlier book, *How We Know: An Exploration of the Scientific Process*, was written to give some conception of what the scientific approach is like, how to recognize it, how to distinguish it from other approaches to understanding the world, and to give some feeling for the intellectual excitement and aesthetic satisfactions of science. These goals represented our concept of the term "scientific literacy." Though the book was written for the general reader, to our surprise and gratification it was also used as a text in about forty colleges, and some high schools, for courses in science for the non-scientist, in methodology of science for social and behavioral sciences, and in the philosophy of science.

As a result we were encouraged to write a textbook with essentially the same purpose and basic approach, but at a level appropriate to college students.

We have drawn up problems for those chapters that would benefit from them, described laboratory experiments that illustrate important points discussed in the text, and made suggestions for additional readings, term papers, and other projects. Throughout the book we have introduced a number of chapters and appendices that provide examples of the uses of quantitative thinking in the sciences: logic, mathematics, probability, statistics, and graphical representation.

Recent experiences as teachers have strengthened our confidence in the case-history approach we have used. We felt and still feel that the methods scientists use arise out of problems and controversies within the fields in which they work: that no general scientific method exists that can be studied without reference to the different disciplines to which it may be applied. Not only do we believe this is how methods develop, we believe also that the student will be better able to grasp the essence of the scientific process by beginning with detailed case studies of discovery and research in particular fields.

We feel also that in spite of the independent development of different fields, a remarkable unity of approach emerges, and it is this unity on which we focus. In particular we feel that common features of scientific thinking are shared by most fields, whether they belong to the natural or to the social and behavioral sciences.

In this book we have made a special effort to introduce material that illustrates such thinking.

We conceive of two audiences for this text. The first and larger is composed of non-scientists, who in this era need an acquaintance with the outlook and cultural values of the sciences. The second is composed of students in science preparing to concentrate in one particular discipline, who ought to develop a broader perspective on the nature of science than is provided by that discipline alone.

Finally, we note that our view of science is an "internalist" one. We are describing what it is that scientists believe they are doing when they do science. We deal only briefly with important and interesting questions about the sociological and psychological setting of science: why people become scientists, who pays them and how it influences them, how the work they do influences society, and so on. To try to deal thoroughly with these issues would be to write another book.

If we can convey anything of the excitement and satisfactions of science, the conflict between its creative and critical sides, the beauty and frequent attention to tedious detail, we will feel we have done enough.

We would like to acknowledge the friends, colleagues, and family members who have helped us with criticisms, encouragement, information, or plain hard work: James Anderson, Gerhard Bry, Ora Fagan, Aviva Goldstein, Eric Goldstein, Michael Goldstein, Seymour S. Kety, John Kreniske, Thomas Lanigan, Arnold Lebow, Ernest Loebl, Suzanne Loebl, Zella Luria, Leonard Nash, Richard Neugebauer, Ruth Otmann, Martin Perl, Terri Perl, Neil Risch, and George Schmidt. We are grateful to all of them.

Note to the Student

You are using this text in a science course which is possibly the first you have taken at college. You may have begun this course with a mixture of trepidation and resentment—certainly this is how you will feel if you share the view of many that science is a difficult subject and a dry one.

We do not expect to convince you that this view is all wrong. Some sciences *are* difficult, others may seem dry. What we hope to convey in this book is an understanding of the scientific approach itself: the inherent logic (or lack of it) in scientific discovery, how theories come into being, how they are tested, why they are believed or discarded. Science is only one route by which human beings have tried to understand the world, but it is an important one that has helped to shape the intellectual climate of our time. Further, science is not a dry, orderly compilation of useful facts, although some of those who hold a negative view of science may think that it is. Science is an activity of creative and imaginative human beings, not of computers or other machines. The creativity and imagination must be controlled by discipline and self-criticism, but that is equally true of other kinds of creative activity such as the writing of poetry. And because science is a creative and imaginative activity, there are satisfactions in engaging in it no different from those felt by creative artists in their work, and there is a beauty in the results that can be enjoyed in the same way as poems, pictures, and symphonies.

In fact, the term *scientific method* is misleading. It may suggest that there is a precisely formulated set of procedures that if followed will lead automatically to scientific truth. There is no such "scientific method" in that sense at all, and one of the important things we want to convey in this book is the intuitive and unpredictable way scientists actually work. *The American Heritage Dictionary* gives as one sense of the word *art* the following: "A specific skill in adept performance, conceived as requiring the exercise of intuitive faculties that cannot be learned solely by study." Scientific research is, in this sense, an art.

Other incentives for gaining an understanding of scientific method are more practical. Many decisions that we as citizens—or those whom we allow to act for us—have to make require some specific scientific knowledge—some facts about chemistry, physics, biology, and so forth. What would be the risk to human beings of an accidental explosion in a nearby nuclear power plant? How should the government decide which type of research to support to find a cure for cancer or a new source of energy? Although this book is not meant to provide that kind of specific information, it can provide an understanding of how the knowledge in question has been acquired, and how sure we can be of its truth. Scientists called in as experts on matters like this often disagree profoundly. Whom should we believe and why? To a great extent we have no choice but to rely on experts in these matters, but we should understand something about the sources and limitations of even expert knowledge. Further, most people, in the course of making the various choices and decisions of daily life—whom to vote for, what to buy, where to live, what to eat—apply some features of scientific method in an intuitive way. They usually do not think of what they do as being an application of scientific method, nor do they use it to the maximum extent. But a clearer concept of some of the basic procedures of scientific thinking could be useful even in such ordinary activities. In turn, the fact that most people have some intuitive concept of scientific thinking makes it possible for them to acquire a more detailed understanding if it is explained properly.

THE CASE-HISTORY APPROACH

To illustrate the process of science we must also describe particular scientific discoveries. However the examples we use to illustrate the principles involved must fall within the comprehension and common experience of those for whom we are writing. Not all our examples therefore are on the frontiers of ongoing scientific research. Our primary purpose is not to teach current scientific knowledge, but the concept of science itself. What we hope to do is to impart the feeling of being present at the time these discoveries were being made, of understanding the conflicting views of the scientists involved, and seeing why the presently accepted view finally won out. Hence, those who read this book will not have to take this particular discovery on faith, and will, we hope, also come to understand the common features of the scientific process in general.

We wish we could have completely avoided asking our readers to take anything we say on faith, but unfortunately this proved impossible. As we indicated, our case histories of discovery were chosen because they illustrate principles of discovery, not necessarily because they describe the latest discoveries. But to create additional motivation to read them by relating them to current knowledge and current problems, we have found it occasionally necessary to describe certain scientific discoveries in a way that requires faith after all, though we hope these descriptions will still be both understandable and credible.

Our case histories have been chosen from three different fields of science: one is from physics, one from medical research, and one from abnormal psychology.

NOTE TO THE STUDENT

While the scientific method has shown its greatest success in the physical sciences, with the result that many scientists feel it can best be demonstrated by examples from physics, we chose to give only a single example from this field. One reason is that we wanted to show how the processes of research and discovery are similar in diverse fields of study. Another reason is that physics has a formidable reputation as a difficult subject, made worse by its heavy use of mathematics, a subject generally regarded as even more formidable. While we believe that science needs mathematics, and will spend some space trying to convince the reader of this, we are sadly aware that something about physics and mathematics frightens many people—so much so that if it were absolutely necessary for them to study physics and mathematics in order to understand science as a whole, they would give up entirely. This book is written for them, too.

Still another reason is that most people have more direct experience of and more intuitive feeling for the topics studied in psychology, medical research, and the social sciences than for topics in the physical sciences. Since we wish to build as much as possible on common sense, common experience, and the readers' interest, we have weighted our choice of examples accordingly.

The first case study (Chapter 4) describes the discovery by John Snow of the mode of transmission of cholera in nineteenth-century London. It provides us with a particularly beautiful example of how a controlled experiment is done, and how it can make the case for a theory overwhelming. We will also learn that even a correct theory does not agree with every experimental fact, and that even incorrect theories may explain many facts and have useful applications to real problems. We will learn also how great a variety of kinds of facts can be relevant to a theory: Snow's keen observation of the life-styles of people of different occupations and social classes, no less than his careful chemical and microscopic examination of various water supplies, helped him establish his theory of how cholera is spread. At the end of this chapter we show how the methods of research used by Snow—the comparison of rates of disease in different groups of people—are applicable to other problems as well, and we use research into the causes of cancer as an example.

The second case study (Chapter 6) is an example from the history of physics concerning the conflict between two theories: that heat is a substance, and that it is the motion of the atoms of matter. This example will teach us something about the value of replacing qualitative impressions with quantitative measurements, which is what has given the physical sciences their power and authority. It also presents another situation where for a time the wrong theory could explain things better than the correct one, and may thus give us a more sympathetic understanding of why scientists sometimes seem too conservative in their rejection of new ideas. As the subject of this chapter forms part of the early history of the concept of energy, and energy as a subject has a significance of its own, we also discuss briefly what energy is, and what the laws are that govern its transformations.

The third case study (Chapter 8) deals with mental disorders. The field is one where the study of the psychology of the individual and the study of people in groups both play a part. We will learn something too about the role of classification in science: that classification is not a simple mechanical arrangement of facts to suit our convenience, but rather is involved in dynamic interaction with the theories we hold, changing as theory changes. We will learn sometimes to be skeptical of "facts," and to check them when necessary. Most important, we will learn how difficult it is to be scientific and objective where human beings are concerned.

THE GENERAL PRINCIPLES

Chapters 11–18 deal with some of the general features of scientific method. There is some recapitulation of points made earlier in the case studies, and some discussion of subjects not dealt with adequately there.

We had originally hoped that the case histories would illustrate all the important features of the scientific process. Instead we found that no one of them, or even all three together, could cover everything. If we had included additional case histories to cover the features missed, it would have necessarily led to considerable repetition of some other features. We therefore decided to limit the number of case histories, at the price of having to make some points without the support of detailed examples.

The last chapter, called "The Cultural Roots of Science," deals with a difficult but important problem: the relation of scientific beliefs to the culture in which they occur. Contrasting the beliefs of a primitive African tribe, the Azande, with our own, we find their beliefs fulfilling the same function for them as our scientific understanding of the world fulfills for us. In this chapter we have attempted to make the reader consciously aware of the body of concepts and modes of thought—taken for granted and therefore never recognized or analyzed—that are shared by all members of a culture, and that limit and shape the beliefs that can be held.

QUANTITATIVE THINKING AND SCIENCE

Thinking quantitatively—learning how to describe things in numerical measures—is essential for all of science. We have therefore included chapters on logic and on mathematics, whose main purpose is not to teach these subjects, but rather to justify them—to explain why they are so essential for science. There are chapters on probability and statistics, subjects so generally useful that we felt it necessary to explain some of their basic concepts (although these chapters are not, of course, the equivalents of even elementary courses in either subject). There is also a chapter on graphical representation, and appendices that apply probability and statistics to particular problems in molecular motion, genetics, and errors of measurement. Our hope is that this material will make the subjects in question both comprehensible and challenging.

Contents

Chapter 1
What Is Science? 1

Why Learn Science? 1
A Definition of Science 2
    Understanding 3
    Generality 4
    Experimental Test 4
Science versus the Humanities 5
Reference Note 6
Suggested Reading 6

Chapter 2
Facts 7

Common Sense 7
Facts: What Are They? 9
    Seeing after Blindness 11
Facts Are "Theory Laden" 14
How Facts Are Used 15
How Science Begins 15
    Collecting All the Facts 16
The Facts about Motion 17
Which Facts Are Relevant? 17
    The Chicken or the Egg? 18
Science and Public Facts 18
Reference Notes 19
Suggested Reading 19
Problem 19

Chapter 3
Logic 21

Introduction 21
The Nature of Logic 21
The Logic of an Experimental Test 23
Logical Difficulties and Fallacies 23
An Example of Logical Reasoning 25
Probable Inference 27
Reference Notes 27
Suggested Reading 27
Problems 28

Chapter 4
Snow on Cholera 29

Introduction: The Man, the Background 29
The Disease 31
Introduction to the Study 32
    The History of Cholera 32
    Cholera Is Contagious 33
    How Does Cholera Spread? The "Effluvia" Theory 34
    Snow's Theory 36
The London Epidemics of 1853–1854 39
    The Broad Street Pump 39
A Controlled Experiment—Where Did They Get Their Water? 43
    Laboratory versus Natural Experiments 44
    Where Did They Get Their Water? 45
Being Critical 48
    Objections to Snow's Theory 48
    Other Theories: Effluvia, Elevation, Hard Water, and Soft Water 50
Applications to Other Problems 53
    What About Other Diseases? 53
    Prevention: Measures to Combat the Spread of Cholera 54
    What Snow Overlooked 55
The Causes of Cancer 56
    Biological and Chemical Approaches 57
    Geography 57
    Within a Country 59
    Genetics and Cancer 60
    Migration Studies 60

CONTENTS

    Environment and Life-Style 61
    Cancer and Smoking 64
    Nonepidemiological Evidence 66
    Unsolved Problems 67
    How Avoidable Is Cancer? 67
    Putting the Question Differently 69
Reference Notes 69
Suggested Reading 70
Problems 70

Chapter 5
Mathematics 73

The Nature of Mathematics 73
    Symbols and the Fear of Mathematics 74
    The Rules of the Game 74
The Truth of Mathematics 75
The Use of Mathematics in Science 76
The Reasons for Mathematics 77
    Economy of Effort 78
    Precision 78
    Another Kind of Precision 79
    Mathematics versus Words 80
    Quantitative Laws 81
How Many Prime Numbers Are There? An Example of Mathematical
Reasoning 82
Mathematics without Quantities—The Bridges of Koenigsberg 85
Calculation and Mathematics 88
Reference Notes 90
Suggested Reading 90
Problems 90

Chapter 6
Is Heat a Substance? 93

The Conflict between the Caloric and Kinetic Theories of Heat 93
    What Is Heat? 93
    The Caloric Theory 94
    The Kinetic Theory 94

The Usefulness of the Wrong Theory	95
Summary of What Is to Come	96
Measuring Hotness	97
Making Things Quantitative	97
Thermometers	99
Standardization and Calibration	100
The Equilibrium of Heat	102
Science and Quantification	103
Exact and Inexact Sciences	104
Heat and Heat Capacity	105
The Invention of "Caloric"	105
Conservation of Heat	106
Joseph Black	106
Heat versus Temperature	106
The Capacity for Heat	109
Latent Heat	113
Melting Ice	113
Latent Heat and Caloric	116
Other Triumphs of the Caloric Theory: The Flow of Heat	117
The Efficiency of Steam Engines	118
Rumford: Does Heat Have Weight?	119
Benjamin Thomson, Count Rumford	119
Rumford's War against the Caloric Theory	120
Does Heat Weigh Anything?	120
Heat from Friction	124
The Boring of Cannons	124
Why Rumford Did Not Win	128
Molecular Motion	129
Do Atoms and Molecules Move?	129
Demonstration of the Constant Motion of Molecules	130
Heat as Molecular Motion	133
The Conversion of Motion, through Friction, into Invisible Molecular Motion	135
Why Caloric Survived	136
How Does Heat Get through a Vacuum?	136
What Is Light?	137
The Objectivity of Scientists	138
The Laws of Energy	139
The Forms of Energy	139
The Transformations of Energy	140
Measuring Energy	141
The Laws of Thermodynamics	141
Reference Notes	143

CONTENTS

Suggested Reading 144
Problems 144
Experiments 145

Chapter 7
Probability 147

How to Deal with Uncertainty 147
How to Gamble and Win 147
Heads or Tails? 148
   Numerical Magnitudes 148
   Are Tosses Independent? 149
   The "Law of Averages" 149
   Sequences of Tosses 150
   A Proof of the Obvious 152
   The Bell-shaped Curve 152
   Another Paradox 156
   The Law of Averages Justified 157
   Uncertainty Remains 157
   Black Balls and White Balls 158
Another Meaning of Probability 160
Appendix 7-1: Computing Averages 161
   Introduction 161
   Tricks with Mirrors 163
   The Average of the Unaverage 164
   A New Headache: The Root-Mean-Square 165
   More Tricks with Mirrors 167
Appendix 7-2: Applications of Probability Theory to Molecular Diffusion .. 170
   Introduction 170
   Molecules in Motion 170
   The Tortuous Path: A Random Walk 171
   Making a Model 172
   Molecules, Dice, and Coins 173
   How Fast Does It Happen? 174
   The Random Walk in One Dimension 175
   Applying Our Mathematics 176
   The Experimental Test 176
   What We Do Not Know 178
   What Is Testable 179
Reference Notes 180
Suggested Reading 180
Problems 180
Experiments 181

Chapter 8
What Is Madness? The Scientific Study of Mental Disorders 183
Section I: Who Is Mad? 183
A Depressed Genius 184
History 185
   Treatment of the Insane 186
   Does the Treatment of the Mad Make Them Madder? 186
Classification as the Starting Point of Science 188
   Classification 188
   Facts and Their Classification 189
The Kinds of Mental Disorders 191
Schizophrenia and Depressive Disorders 194
   Description of Schizophrenia 194
   Description of Depressive Disorders 195
   Comparison of the Two Groups 196
Diagnosis 196
   Pattern Recognition—Art or Science? 197
The Experience of Madness 198
Theories of the Causes of Mental Disorders 202
   Psychogenic Theories 202
   Theories of Social Causation 203
   Biological Theories 205
   Interaction of Biological, Psychogenic, and Social Factors 206
An Epidemiological Study 207
   United States and British Rates of Mental Disorder—A Clue to Causes .. 207
   Explanations 207
   Caution! Discovery or Artifact? 208
   Previous Studies of Reliability of Diagnoses 210
   A Study of Diagnostic Practices 211
   A Thermometer for Mental Disorder? 212
   The Project Diagnosis 212
   The Results 213
   How Good Is the Project Diagnosis? 214
   Other Studies 215
   The Schizophrenia Epidemic in New York State 216
   The Psychiatrists Again 216
   What Have We Learned? 217

Section II: The Mental Hospital 218
Deterioration in Schizophrenia 218
   Institutional Neurosis or Schizophrenia? 223
   The Origin of Institutional Neurosis 224
   The Cure 226

CONTENTS

Generalizing a Concept 228
   Focus in Science 228
   A New Discovery? 230
Total Institutions 231
   What Do Convents and Concentration Camps Have in Common? 232
   The Institution and the Condition 234
   Belief and Evidence 235
   What Have We Learned So Far? 236
Labeling 236
   Eskimos and Yorubas 237

Section III: Genetic Studies 238

Why Does Schizophrenia Run in Families? 238
Inheritance 240
   Introduction 240
   Mendel's Work 240
   Difficulties 242
Adoption Studies 243
   The Danish Folkeregister 244
   "Blind" Classification 245
   The Control Group 245
   The Uterine Environment 246
   Cross-Fostering Studies 246
Twins 247
The Conclusions and the Controversies 249
   Another Classification of Schizophrenia 251
   The "Myth" of Schizophrenia 252
   Implications for the Psychogenic Approach 252
Postscript 253
Appendix 8-1: Genetics and Probability 254
   Four O'Clock Plants 254
   Mendel's Hypothesis 255
   Dominant and Recessive 256
   Human Genetics 258
Reference Notes 259
Suggested Reading 261
Problems 262

Chapter 9
Statistics 263

The Problem Turned Around 263
   How Tall Is the Average Person? 264

Is the Drug Effective? 264
　　Random versus Nonrandom 265
　　Another Meaning of "Statistics" 267
　　How Good Is an Average? 267
　　Again, the Average of the Unaverage 268
Statistical Tests in Science 270
　　Statistics and Science 272
Appendix 9-1: Statistics Experiment 273
　　The Chi-squared Test on a 2 × 2 Table 273
Reference Note 278
Suggested Reading 278

Chapter 10
Science—The Search for Understanding 279

Understanding as a Common Experience 279
The Flash of Insight—The Birth of an Idea 280
　　The Dancing Atoms 280
The Sense of Exhilaration 282
　　Religion, Poetry 283
　　. . . Alcohol 283
　　. . . and Insanity 284
Science Is a Consensus 284
Reference Notes 285

Chapter 11
Science—The Goal of Generality 287

Generality in Science: Examples 287
　　Einstein's Generalization 288
　　The Quantum Theory 288
　　Cholera and the Germ Theory 288
The Price of Generality 288
　　The Loss of Individuality 289
　　Science and Maps 290
Is History a Science? 290
Reference Notes 292

CONTENTS

Chapter 12
Science—The Experimental Test ... 293

Testing Theories ... 293
The Development of the Experimental Method ... 293
   The End of Authority ... 295
Repeatability ... 295
Quantity Rather Than Quality—The Faith in Mathematics ... 297
Testing by Experiment—Planned and Unplanned ... 298
   The Experiment Must Make a Difference ... 299
   An Awareness of Alternatives ... 299
   Refutability ... 302
   You Cannot Prove a Theory Right ... 302
   You Cannot Prove a Theory Wrong ... 303
Indirectness of Experimental Tests ... 304
   Generality and Indirectness ... 304
What Do We Test, and When? ... 305
Appendix 12-1: The Experimental Method in the Humanities ... 306
   Images as Facts ... 306
   A Controlled Experiment ... 307
   Results ... 307
Appendix 12-2: The Controlled Experiment Revisited ... 309
   Introduction ... 309
   Dichotomies ... 311
   A Common Mistake ... 311
   A More Subtle Mistake ... 312
Reference Notes ... 315
Suggested Reading ... 316
Problems ... 316

Chapter 13
The Experimenter and the Experiment ... 317

The Uncertainty Principle ... 317
   A Useful Metaphor—Interaction between Experiment and Observer ... 318
   The Smart Mice ... 318
   Placebo Pills in Drug Trials ... 319
Blind and Double-Blind Experiments ... 320
   The Lively Flatworms ... 321
   Mental Telepathy ... 321
   The Clever Horse ... 321

Interviewers and Interviewees 323
Rumford's Mistake 323
The Self-Fulfilling Prophecy 324
Reference Notes 325
Suggested Reading 325

Chapter 14
Measurement and Its Pitfalls 327

Measurement and Science 327
    Reliability and Validity 328
    Precision 328
    Random Errors 329
    Accuracy 329
    The Point of Diminishing Returns 331
    Counting 332
    How to Fool People 332
    How to Fool Oneself 333
Appendix 14-1: Significant Figures 336
Appendix 14-2: Describing Precision Quantitatively 338
Reference Notes 340
Suggested Reading 341
Problems 341

Chapter 15
Graphs and Sketches 343

Why Graphs? 343
    Comparison Graphs 343
    Discrete versus Continuous 346
    Mathematical Functions 346
    Graphing the Laws of Physics 348
    Noise 351
    Neglected Variables 352
    Useful Distortions 353
    Exponential Growth 355
Sketches 358
Reference Notes 358
Suggested Reading 358
Problems 358

Chapter 16
Where Do Hypotheses Come From? 361

Posing Hypotheses 361
   We All Make Them 361
The Moment of Insight 362
   Poetry Also 365
   Folk Wisdom 365
Chance 367
   The Lost Keys 368
The Collective Unconscious 368
The Tactics of Science 369
Reference Notes 370
Suggested Reading 370

Chapter 17
The Dispassionate Scientist 371

The Myths 371
The Reality 371
   For Example: Isaac Newton 372
   Freud Also 373
Why Scientists Care So Much 373
The Depersonalization of Discovery 374
Reference Notes 376
Suggested Reading 376

Chapter 18
The Cultural Roots of Science 377

The Subjective Element 377
The Tacit Component 378
The Belief in Witchcraft 379
   Arguing with the Azande 380
   Carelessness and Witchcraft 381
   The Poison Oracle 382
   The Confirmatory Test 385
   Dealing with Contradictory Results 385

Science versus Witchcraft 386
Cultures and Subcultures 387
Scientific Subcultures 388
Breaking Through 388
Reference Notes 390
Suggested Reading 390

Index 391

1
What Is Science?

WHY LEARN SCIENCE?

We have not yet said what we mean by the term *science,* and indeed it has many definitions. For some, the term applies only to the "exact" sciences, such as physics, which are characterized by laws of great generality and scope from which numerically precise predictions can be made. Isaac Newton, for example, discovered a way of describing motion in a few simple mathematical equations which could be used to describe all the different forms of motion in the then-known universe—the motion of the planets around the sun, the fall of an apple, the tides and waves of the ocean, and the vibration of a violin string. These motions can be described with great precision; we would not be able to design satellites, send them into predictable orbits around the earth, and predict when they will fall back to the earth as well, without Newton's Laws. If laws of great generality and accurate predictive power are taken as essential to what we define as science, then none of the social or behavioral sciences satisfies the criterion, nor for that matter do many branches of natural science, such as biology and geology. Such "laws" as have been found in psychology or sociology, unlike the laws of physics, are of very limited scope, are imprecise in prediction, and are often quite controversial within the field.

For others the term *science* implies the ability to perform experiments in which the properties or factors believed to cause some phenomenon are under the control of the experimenter, who can vary them at will and thus determine their influence on the outcome. An experiment in this sense creates a situation that does not exist in nature. The word *experiment* is used here in the narrow sense that distinguishes it from the process of discovering facts by *observation,* in which the observer does not change nature, but merely reports it.

Thus Thomson, studying the motion of electrons in a cathode ray tube under the influence of electrical and magnetic fields whose strength he could control, Pasteur, injecting sheep with a culture of anthrax bacilli, and Pavlov, conditioning a dog to salivate at the sound of a bell, were all doing *experiments*. In contrast, Tycho Brahe, using instruments of his own invention to obtain more accurate determinations of the positions of the stars and planets, Darwin, noting the varieties of finches on the Galapagos Islands, and Freud, examining the notebooks, paintings, and writings of Leonardo Da Vinci as clues to the deep structure of his personality, were making *observations*.

Limiting the definition of science to those disciplines which make use of experimental, as opposed to observational, methods would exclude many of what we customarily consider the greatest of scientific achievements.

In astronomy, one of the most exact of the exact sciences, we cannot move Mars nearer the sun to see how the length of the Martian year would change. In geology we interpret many of the geological features of northern latitudes as the result of the action of glaciers during an "ice age" 25,000 years ago, but we have no way of making glaciers retreat or advance at will to see if they really produce the features observed. The experiments possible in a biology laboratory provided only a small part of the initial evidence for the theory of evolution; most of the evidence was "out there" in nature, existing independently of human control.

Still another reason for rejecting this distinction as the boundary of science is that a close look at even the most "experimental" sciences, such as physics and chemistry, shows them to be based in reality upon an inextricable mixture of experiment and observation. The first great test of Newton's Laws came in fact from astronomical observation rather than laboratory experiments, though it was mainly through subsequent laboratory experiments that the power and generality of those laws could be appreciated.

But the most telling reason for refusing to accept this distinction as crucial to the definition of science was given by the philosopher of science, Ernest Nagel, who points out that even without the scientist's deliberate control of properties or factors, observational investigations identical in logical function with controlled experiments are possible, and can carry as much conviction.[1] Chapter 4 of this book, "Snow on Cholera," reports just such an observational investigation into the question of whether cholera is transmitted through pollution of the water supply.

In our view (and the view of most other scientists as well), the distinguishing feature of science is its willingness to test the truth of theories by examination of facts, either facts discovered by experiment or those found by observation. The term *empirical* is used to cover both experimental and observational methods, and we could have used it, but as a term it is less familiar and less evocative than *experimental*. As Nagel suggests, "experimental" need not be confined to the narrow sense given above, but can also be used more broadly as a synonym of "empirical." This is how we will use the term in this book.

A DEFINITION OF SCIENCE

We choose to define *science* very broadly, as an activity characterized by three features:

WHAT IS SCIENCE?

1. It is a search for understanding.
2. The understanding is achieved by means of statements of general laws or principles—laws applicable to the widest possible variety of phenomena.
3. The laws or principles can be tested experimentally.

Understanding

A search for understanding, for the revelation of an underlying pattern in some complex and confusing aspect of reality, is a major goal of science. But it is hard to specify precisely what constitutes understanding. It is clearly subjective: what satisfies one person does not satisfy another; different cultures have different standards of what is a good explanation; what satisfied people one hundred years ago may not work today. As vague and ill-defined as the concept is, however, the subjective sense of gratification which comes from understanding some aspect of reality is strong, and it is one of the important reasons for doing science in the first place.

That the goal of science is understanding is a statement all will welcome. But there is an unwelcome qualification that must be added: the people who engage in any particular field of scientific research are not like the rest of us, at least in their ways of viewing their field. While science is in principle open to anybody, each discipline requires a period of training, of getting used to its concepts, its specialized jargon, its criteria of what is and is not to be trusted. By the time the novice has finished his apprenticeship, he no longer thinks like the novice he once was, but like a member of a profession. The result is, among other things, that what will satisfy his need to understand, what will sound convincing and plausible as an explanation, is different from before.

Because scientists are members of a specialized subculture, it is often hard for the lay person—and for scientists from other disciplines—to understand and evaluate the scientific knowledge in a particular field. And as science itself has grown and ramified, the number of such specialized disciplines has increased tremendously. It is no longer a matter of nonscientists failing to understand science, or of psychologists having trouble with physics. It has reached the point where major disciplines such as chemistry are broken up into specialties such as physical chemistry and organic chemistry, in which few practitioners are able to speak a language common to both.

While this fragmentation of scientific knowledge is unfortunate, one should realize that it is an inevitable aspect not just of scientific expertise, but of any kind.

It is no different for lovers of baseball or music. By the two hundredth baseball game or symphony concert, one has acquired a complex body of knowledge, a set of shared assumptions, and a specialized jargon common to other fans, but unfamiliar to those who have not taken the trouble. One who is not an enthusiast for these activities can be bored beyond endurance if forced to listen to conversations between those who are. The values and judgments sound all wrong, the nuances of discrimination sound too trivial to think about, and so on.

Recognizing this, one may also recognize the necessity of patience.

Generality

The understanding we look for from science is expressed in the form of laws or principles that enable us to predict what will happen and to see why it has happened. By *generality* we mean the property of being applicable to the widest possible variety of phenomena. We want fewer laws, but we want them to cover more cases.

In subsequent chapters we will give some examples from the history of science to show that a science often begins with the discovery of a large number of seemingly unrelated laws or regularities, each applicable to a narrow range of phenomena, but that in time these laws are combined into a smaller number which apply over a much broader range. Ultimately the apparently independent laws which made up the original collection are seen to represent special cases of a single general law. The outstanding example of this is provided by the laws of motion discovered by Isaac Newton, which we referred to earlier in this chapter.

Science is a search for unity in diversity, for common patterns in what seem like quite unlike events. The more general our laws, the more unity we have uncovered.

Experimental Test

The requirement that we be willing to subject our explanations to experimental test is the distinguishing feature of science.

Ways of understanding the world other than the scientific way also have as their goal a sense of subjective satisfaction with the explanations found, and they too express the desire for generality. It is the possibility of experimental test, the recognition that we may have to change our minds if forced to by facts, that is unique to science.

In order for facts to force us to change our minds, there must first of all *be* facts: interested observers must be able to agree on what is and is not a fact (a problem which is not so simple as it sounds; in the next chapter we will spend some time on it).

Second, we ourselves must be open to the possibility that we may be wrong. After all, we do experiments because we expect them to tell us something. If we will not change our beliefs no matter what the outcome, then the experiment is hardly worth doing in the first place. Only experimentally testable theories are scientific theories.

There are other things we know that are of the utmost importance to us but are not experimentally testable: issues of values and esthetics, and religious and

WHAT IS SCIENCE?

ethical beliefs. Not everything that is important need be experimentally demonstrable. But scientific truth must be.

SCIENCE VERSUS THE HUMANITIES

The concept of an experimental test does seem to distinguish science sharply from other types of scholarly disciplines, such as literary criticism.

A new interpretation of *Hamlet* may or may not be convincing, but one cannot conceive of Shakespearean scholars agreeing on some precise experimental procedure for testing it. Rather than to the experimental test, one appeals to the consensus of informed practitioners in the field, who judge by subjective criteria: Is it a good explanation? Does it bring into a coherent picture a large number of what were previously thought to be unrelated facts? Is it fruitful in the sense of suggesting new directions of research that were not previously thought of?

We do not minimize the gap between the use of such criteria to judge a theory and the criterion of a precise experimental test, but we do want to point out three factors that make the differences less sharp than they might seem.

First, research in the humanities is as relentlessly grounded on facts as is research in the "exact" sciences. For example, no interpretation of a Shakespeare play is likely to be worth much if the critic proposing it does not really understand the precise meaning of the words of the text. Research is necessary: to know what a word means in one scene of *Hamlet,* one may have to examine carefully how it is used not only in the rest of Shakespeare's plays but also throughout Elizabethan literature. Understanding it may also depend on knowledge of some political crisis in the court of Queen Elizabeth that occurred while the play was being written.

Second, decisions between rival scientific theories, even in physics and chemistry, have not always been based on experiment alone, at least not in the idealized sense in which experimental testing is understood. Of course, if two theories agree in many areas but disagree in some, and if experiments show that where they disagree, one theory *always* gives the right answer and the other theory *always* the wrong answer, it is easy to decide in favor of the first one. In reality, however, no theories explain every possible experimental fact, and there is always considerable leeway in judging what experiments are relevant for testing the theories. Major scientific controversies have raged over competing theories, each of which had some area of application where it did better than its competitor. The disputes have been resolved by the same appeal to a consensus of informed practitioners we described as the court of last resort in the humanities, using criteria of explanatory power, coherence, and fruitfulness.

Third, both science and the humanities demand the constant operation of the critical faculty. The criterion of the experimental test in science is a reflection of a permanent obligation to be critical of one's beliefs, to be always asking, How do we know? Why are we sure? Could we be wrong? If we were wrong, how would we know? While in nonscientific disciplines the criterion for whether we are right or wrong is not an experimental test but rather a vaguer, less easily formulated standard, the same questions need always to be asked. We are committed in both science and the humanities to constant critical examination and to the search for better and deeper insights.

We are not making any exaggerated claim that historical or literary studies could be made into "sciences" if historians or professors of literature would only make the effort. We are saying that there are some things the physicist and historian do that are similar, similar enough so that each can develop some appreciation and respect for the work of the other, and the person who is neither can appreciate and understand the common features of the work of both.

We will give some brief examples later of research in literature and history that illustrate these common features.

REFERENCE NOTE

1. Ernest Nagel, *The Structure of Science: Problems in the Logic of Scientific Explanation*. (Indianapolis: Hackett, 1979).

SUGGESTED READING

The books listed below have helped us clarify our understanding of science and its methods.

Cohen, Morris R., and Ernest Nagel. *An Introduction to Logic and Scientific Method*. New York: Harcourt Brace, 1934.

Conant, James B., and Leonard K. Nash, et al., eds. *Harvard Case Histories in Experimental Science*. 2 vols. Cambridge: Harvard University Press, 1957.

Kuhn, Thomas S. *The Structure of Scientific Revolutions*. 2nd ed. Chicago: University of Chicago Press, 1970.

Nagel, Earnest. *The Structure of Science: Problems in the Logic of Scientific Explanation*. Indianapolis: Hackett, 1979.

Popper, Karl R. *The Logic of Scientific Discovery*. New York: Harper Torchbooks, Harper and Row, 1965.

Quine, W. V., and J. S. Ullian. *The Web of Belief*. 2nd ed. New York: Random House, 1978.

2

Facts

COMMON SENSE

Thomas Huxley, biologist, friend of Darwin and defender of and publicist for the theory of evolution, was a great popularizer of science. In a public lecture he defined science as follows: "Science is nothing but trained and organized common sense . . . its methods differ from those of common sense only as far as the guardsman's cut and thrust differs from the manner in which a savage wields his club."[1]

This is a claim that sounds reasonable at first, but a problem arises when we ask, What is common sense? A dictionary we consulted gives two meanings:[2]

1. Native good judgment, sound ordinary sense.
2. The set of general unexamined assumptions as distinguished from specially acquired concepts.

Huxley, presumably, had the first meaning in mind. It is interesting that the dictionary gives as an example of usage for the second meaning: "Common sense holds that heavier bodies fall faster than lighter ones," which puts science and common sense in direct opposition.

The world today and in the past has been the home of diverse cultures, each of which possesses a set of shared and unexamined general beliefs that form the "common sense" of each culture. What was common sense in Victorian London was not common sense in a village in India at that time, nor was it common sense in Elizabethan London. The reason Huxley could see parallels between science and common sense in his time was because the culture he and the other people of England shared had been shaped for several hundred years by discoveries in science, and by related advances in technology. This is not to say that scientific knowledge in a formal sense was widespread in England then, but rather that the ideas and concepts of science had for some time been in the air, and formed part of the intellectual climate of the period.

The point has been put very bluntly: Common sense is the collection of prejudices we have acquired by the time we are twenty years old.⁽³⁾ Indeed, as the dictionary example of usage suggests, science from time to time in its history has had to wage a resolute battle against common sense. One does not need to look at such abstruse examples of the conflict between the two as Einstein's theory of relativity, with its blurring of the distinctions between space and time, and between matter and energy. Would common sense have ever led us to believe that the world is round? That the earth is moving, rather than the sun? That human beings are descended from one-celled animals? That diseases are caused by living organisms, too small to see, that invade the body?

But again the dictionary example makes the point more sharply: *Common sense holds that heavier bodies fall faster than light ones*. This example, as noted earlier, derives its force from the idea that common sense is wrong, and in conflict with scientific knowledge. Common sense is often wrong, as well as in conflict with scientific knowledge, but in this particular case common sense is absolutely right: Heavy bodies do fall faster. If a leaf, a light pebble, and a large rock are dropped from a certain height at the same moment, the large rock will hit the ground first, and the leaf last.

What then does science say about this commonly observable experimental fact? Only that it is a result of air resistance, which has more of an effect the lighter the body is (or more exactly, the lighter it is in proportion to its size: a sheet of newspaper weighing an ounce falls more slowly than a one-ounce iron nail). If we were to do our experiment on the moon, where there is no atmosphere, all bodies would fall at the same rate, and if we create a vacuum in a container in the laboratory, they also fall at the same rate.

But science did not discover this by doing experiments in a vacuum (not experimentally possible until the eighteenth century) or by travelling to the moon (twentieth century). Galileo stated it as a law of nature in the sixteenth century, as a conjecture about what *must* happen if we *could* produce a vacuum. This conclusion of Galileo's was basic to the development of the laws of motion by Newton a generation later. It is clearly not based on common sense but rather is in direct contradiction to ordinary experience and verifiable knowledge.

This is not to say that scientific thinking begins by discarding common sense, or by discarding all preconceptions whatever. Scientists begin with the common sense, the shared assumptions, of the scientific community they belong to and of the larger culture as well. They have no choice. There is no way to rid our minds of everything we believe. We have to start somewhere, and alter our beliefs painfully, one at a time, as necessity compels.

In explaining scientific method, we will rely as much as possible on the common sense and common experiences of our readers. The reason we do so, as indicated earlier, is that our readers have grown up in a society in which many of the ideas of science have become absorbed into the common culture of its members.

From time to time, though, it will be necessary to point out divergences between the popular concepts and the reality of how science works. Among the most widespread misconceptions are those dealing with *facts* in science. What are they? How do we know them when we see them? What is their role in the making of theories and the testing of theories?

FACTS: WHAT ARE THEY?

There are some commonsense notions about the role of facts in science that are so mistaken that we need to confront them at the start.

Briefly the notions are the following:

1. Facts are the only reality, and they exist independently of the theories we hold about them.
2. Theories are discovered by first collecting and examining the facts.
3. Unless a theory agrees with *all* the facts, it must be discarded.

None of these is correct, though they are believed to be correct, even by many scientists. Scientific progress often depends on recognizing just how they can be wrong.

The word "fact" has a number of meanings. The *American Heritage Dictionary* lists, among others, two that are relevant to this discussion: "Something that has been objectively verified, something having real or demonstrable existence."[2]

We are surrounded by facts, by things about us that we can see, hear, feel, and smell. If science is to be a valid description of reality, it should begin with such basic experiences on which our knowledge of reality is founded, and return to them to test the theories proposed to explain them. Nothing could sound more sensible, or be more misleading.

What is misleading is the belief that facts exist prior to, and independently of, the theories we hold. Instead, as we look more closely at facts, we realize that they are inextricably entangled with theories and preconceptions. Unfortunately, this is not an easy point to demonstrate in two or three well chosen examples. Instead we will have to return to it repeatedly throughout this book, in the hope that gradually it will become clear.

We will begin, however, by trying to illustrate the point in a metaphorical way. There are parallels between the scientist's perception of the world and the way the ordinary person sees the world. This permits us to use ordinary sense-perception as a kind of model for scientific understanding.

It is natural to believe what we see: it is hard to imagine doing anything else. But the eye can be fooled. This happens often enough for the term "optical illusion" to have been devised. The existence of such illusions raises deep questions: if the eye can be fooled, when should we trust it? When is it giving us "facts," and when is it misleading us?

When we study these failures of perception, we realize that "seeing" is not a simple business. It is liable to error for two kinds of reasons. The first is physio-

FIGURE 2-1. R. L. Gregory suggests that the sense of movement given by this figure, and the nonexistent patterns the human eye perceives in it, may result from the way the figure stimulates movement detectors, and from after-images superposed on the figure itself.

logical, having to do with the structure of the eye and of the nervous system by which visual data is processed. The second is psychological, the result of our training, of our having developed, through a lifetime of experiences, methods of interpreting what we see.[4]

The eye does not work like a color camera or like a color television set, in which each point on a film or viewing screen records or duplicates the color and brightness of the light emitted from some point in the field of view. Instead it has receptors that actually detect features of what is present or changing in the field of view, such as motion, the angle a viewed object makes with the vertical direction,

FIGURE 2-2. The eye sees a circular white band completely covering the radial lines, continuing between them.

FACTS

FIGURE 2-3. R. L. Gregory states about this figure: "When this spiral is rotated, it appears to shrink or expand, depending on the direction of rotation. But when stopped, it continues to *appear* to shrink (or expand), in the *opposite direction*. This cannot be due to eye movement, since the apparent shrinkage or expansion occurs in all directions at once. The effect is paradoxical: there is movement, but no change in position or size."

and the contrast between light and dark objects, which cameras do not. Because of the existence of these physiological receptors in the eye and their mode of operation, it is possible to find stimuli which will disturb or saturate their capacity to respond. Examples of patterns that confuse the eye in this manner are given in Figures 2-1, 2-2, and 2-3.

The psychological sources of illusions are more germane to this discussion. They occur because seeing is not a passive process operating independently of the prior experience of the person doing the seeing. We see by *interpreting* a visual image, not just by *recording* it. Optical illusions that arise from the learned component of seeing are shown in Figures 2-4, 2-5, and 2-6.

Seeing after Blindness

The learned component of perception is demonstrated very strikingly by the experiences of people blind from birth who have undergone surgery that gives them vision in maturity. Surgery for removal of a cataract (a deterioration of the lens of the eye that makes it opaque) has been introduced in recent years. Infants are occasionally born with this condition and without the operation grow up blind. When the technique was first developed, there were many adults who had been blind from birth due to cataracts and to whom surgery gave the ability to see.

Their experiences on first seeing have been described by John Z. Young:[5]

> What would such a person see; what would he say, on first opening his eyes on a new world? During the present century the operation has been done often enough for systematic and accurate reports to be collected. The patient on opening his eyes for the first time gets little or no enjoyment; indeed, he finds

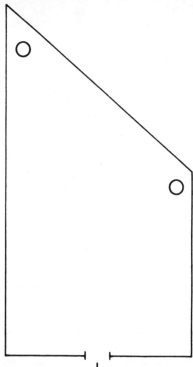

FIGURE 2-4. This is not "trick" photography: it is an undoctored photograph of two women standing in a "trick" room. The room has been constructed so that the left hand corner is twice as far from the camera as the right hand corner, but from the camera position looks like an ordinary rectangular room. The schematic diagram shows how the room was constructed. The two circles show the positions of the two women, and the opening at the bottom is the position of the camera.

FACTS

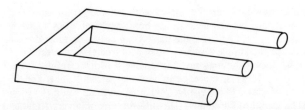

FIGURE 2-5. The discomfort produced by this figure arises from our persistent attempts to perceive it as a representation of a three-dimensional object.

the experience painful. He reports only a spinning mass of lights and colours. He proves to be quite unable to pick out objects by sight, to recognize what they are, or to name them. He has no conception of a space with objects in it, although he knows all about objects and their names by touch. "Of course," you will say, "he must take a little time to learn to recognize them by sight." Not a *little* time, but a very, very long time, in fact, years. His brain has not been trained in the rules of seeing. We are not conscious that there are any such rules; we think that we see, as we say, "naturally." But we have in fact learned a whole set of rules during childhood.

If our blind man is to make use of his eyes he, too, must train his brain. How can this be done? Unless he is quite clever and very persistent he may never learn to make use of his eyes at all. At first he only experiences a mass of colour, but gradually he learns to distinguish shapes. When shown a patch of one colour placed on another he will quickly see that there is a difference between the patch and its surroundings. What he will not do is to recognize that he has seen that particular shape before, nor will he be able to give it its proper name. For example, one man when shown an orange a week after beginning to see said that it was gold. When asked, "What shape is it?" he said, "Let me touch it and I will tell you!" After doing so, he said that it was an orange. Then he looked long at it and said, "Yes, I can see that it is round." Shown next a blue square, he said it was blue and round. A triangle he also

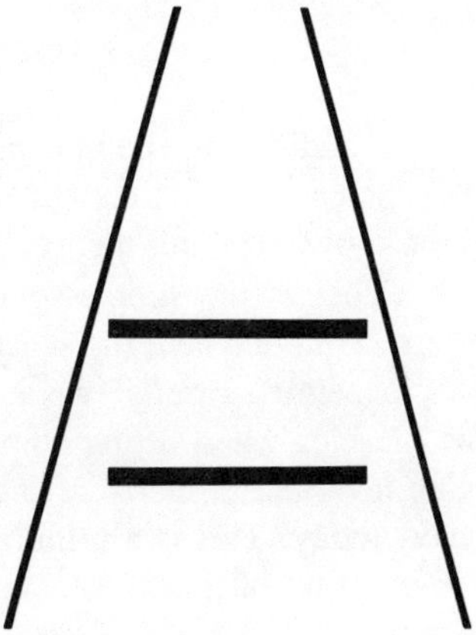

FIGURE 2-6. The Ponzo Illusion. The two horizontal black rectangles have the same width. The converging lines give an illusion of perspective. The upper rectangle is perceived as farther away and therefore larger.

described as round. When the angles were pointed out to him he said, "Ah. Yes, I understand now, one can *see* how they feel." For many weeks and months after beginning to see, the person can only with great difficulty distinguish between the simplest shapes, such as a triangle and a square. If you ask him how he does it, he may say, "Of course if I look carefully I see that there are three sharp turns at the edge of the one patch of light, and four on the other." But he may add peevishly, "What on earth do you mean by saying that it would be useful to know this? The difference is only very slight and it takes me a long time to work it out. I can do it much better with my fingers." And if you show him the two next day he will be quite unable to say which is a triangle and which a square.

The patient often finds that the new sense brings only a feeling of uncertainty and he may refuse to make any attempt to use it unless forced to do so. He does not spontaneously attend to the details of shapes. He has not learned the rules, does not know which features are significant and useful for naming objects and conducting life. Remember that for him previously shapes have been named only after feeling the disposition of their edges by touch. However, if you can convince him that it is worthwhile, then, after weeks of practice, he will name simple objects by sight. At first they must be seen always in the same colour and at the same angle. One man having learned to name an egg, a potato, and a cube of sugar when he saw them, could not do it when they were put in yellow light. The lump of sugar was named when on the table but not when hung up in the air with a thread. However, such people can gradually learn; if sufficiently encouraged they may after some years develop a full visual life and be able even to read. (pp. 61–63)

It is apparent that seeing—the sense we think of as most directly putting us in touch with facts—is learned rather than automatic. We see with our minds, not with our eyes, and we are subject to whatever unconscious biases and misconceptions are produced by the training that teaches us to see.

We are not arguing a case for disbelieving what we see. Most of the time we have no choice, really. However, being aware that perception is not passive observation but rather a learned use of our intellectual faculties, no matter how unconsciously it is done, should alert us to the possibility that things need not be what they seem, and that changes in our own thinking may change what we see.

FACTS ARE "THEORY LADEN"

In addition to the unconscious cultural component in our perception of fact, most scientific facts contain a consciously chosen and analyzable component of prior knowledge and theory.

Consider the simple-sounding statement, "This stone weighs three pounds." There is implied here the acceptance of a whole body of scientific laws and agreed-on procedures. Briefly, we start with the subjective sensation of heaviness felt by the muscles when lifting something. Then we learn by experience that heaviness is an invariant property of most solid things: a stone that was heavy yesterday is heavy today. This is a primitive statement of an important scientific principle: the conservation of mass. Next we establish a criterion of equality of "heaviness" by constructing a balance to compare weights. We are now introducing the concept of the lever. We note that equally "heavy" things (as judged by our muscles) placed at equal distances from the central pivot of our lever tend to balance. Then we choose one object as a standard, and call it, say, 1 pound. Anything that balances it on our set of scales also weighs one pound. Then we *define* a weight of two pounds as anything that balances two one-pound objects placed together on one side of the scale, and so forth. There is thus a lot of physics in the statement, "This stone weighs three pounds." That it really weighs three pounds is an objective fact, verifiable by any observer who wants to take the trouble to do so, but it would be misleading to call it a fact of nature: it is too man-made. It didn't exist until we invented it. This feature of facts is often described by saying that facts are "theory laden."

So we recognize that there is a difference between the commonsense view of "facts" as hard, inescapable, unchangeable things and the reality in science where the things we call facts are fuzzier. First of all, facts have a culturally conditioned component, and secondly, are partly created by the theories we hold, and thus are subject to change if the theories themselves are changed.

HOW FACTS ARE USED

Other misconceptions concern the way facts are used in the development of theories and in their testing. It is often said, even in scientific text books, that we arrive at theories by first examining facts, being led from these facts to generalizations. The trouble with this statement is that the world is composed of an unlimited number of facts, and if we tried to examine all of them we would never get beyond them. Every grain of sand on a beach is a separate fact, no two grains are likely to have the same shape, weight, or exact chemical composition.

But suppose we modify the statement to mean that we select a group of facts out of the infinite variety the world presents us with. Do we demand that our theory at least explain all of these? The answer is no. The facts we select are in large part already determined by some theory or preconception as to what facts are important and what facts are not. We will see in this book examples of conflicts between theories that lasted for long periods of time. If choosing between theories were a simple matter of finding which one agreed with a limited number of facts, the conflicts would have ended quickly. The reason they could drag on so long is that scientists disagreed about which facts were the important ones to explain.

HOW SCIENCE BEGINS

Science does not begin with facts; it begins with the perception of a problem and the belief in the possibility of an answer. Astronomy did not begin with the gathering of data on the motion of the sun, moon, and stars; it began with the belief that knowledge of such motions was worth having. Why the Babylonians of 5000 years ago wanted such information can only be a matter of conjecture, but it must have involved religious beliefs, astrological hypotheses about the influence of the stars on the course of history or on the lives of men, or the idea that knowledge of the motion of the heavenly bodies had some practical predictive value here on earth. Certainly knowledge of the phases of the moon is useful in predicting the height of the tides; the position of stars and the sun in the sky correlates with the season of the year and thus can be used for timing events that recur annually.

Once a problem is perceived and formulated, does the gathering of facts begin then? It was once believed that science could be reduced to a precise methodology that could be applied to the facts in a mechanical way by anyone wanting to determine the cause of a given phenomenon. This view was first expressed by Francis Bacon,[6] and formulated much later in a set of "methods of experimental inquiry" by John Stuart Mill.[7,8] The procedure Mill proposed was to consider as many instances of the occurrence of a phenomenon as one wishes to explain, and examine *all* the circumstances or facts attending each occurrence. Those circumstances that are absent from any occurrences of the phenomenon cannot be the cause of it. Ultimately, if the phenomenon has a single cause, only that circumstance that is present every time it occurs and absent every time it does not can be the cause.

The above statement is perfectly plausible, but useless. It is useless because the number of circumstances attending a phenomenon is infinite and no criteria of relevance are provided. Relevance requires some prior hypothesis, and the method proposed offers no directions for forming hypotheses. Yet no science is possible otherwise.

Collecting All the Facts

Let us take an example. To find the cause of lung cancer, one might try to locate a large number of cases of lung cancer and an equal number of people who do not have the disease, and start to list the circumstances attending each case: age, sex, employment, ethnic origin, smoking history, dietary habits, income, the number, age, and sex of children, marital status, neighborhood lived in, type of house lived in, number of rooms, type of furniture, street, age of parents and cause of their death, details of education, model of car driven, likes and dislikes in books and music, etc. One sees very quickly that the list has no end. Further, under each category there is a tremendous number of individual details. If one objects that most of the details are irrelevant, the answer must be, "how do we know?" It is only because we already have some feeling, even if hazy, as to possible causes of cancer, that we can rule out most of the detailed circumstances on the list. Needless to say, many of the great scientific discoveries resulted from recognizing the relevance of facts that had been previously overlooked, and putting aside facts previously considered important.

THE FACTS ABOUT MOTION

Consider the facts about motion known to the ancient world. Here on earth motion tends to come to a stop unless some agent acts to keep it going. A stone rolling downhill comes to rest some time after it has reached level ground. An arrow shot from a bow eventually falls to the earth. Living things die and cease moving. When the winds stop blowing, the ocean becomes calm. However, in the sky the heavenly bodies keep moving, most of them in regular paths, with no sign of slowing down. To Aristotle, these facts clearly showed that the heavenly bodies in their motion obey different laws from things here on earth, for on earth motion occurs only if there is an agent to cause it, and when that agent stops acting, the motion stops.

However, the laws that govern motion, discovered by Galileo and Newton, were found by making an imaginative leap that went beyond the facts and was not contained in them. It consisted of looking at the facts in a new way—of guessing that motion on earth stops not because of the *absence* of an agent to produce it but rather because of the active *presence* of an agent to stop it—namely, the friction that is always present to resist motion taking place on earth but is nearly absent in the emptiness of outer space, where the planets and stars move. Once this step was taken, laws that apply equally to the heavenly bodies and to objects on earth could be discovered.

WHICH FACTS ARE RELEVANT?

Let us consider another example. Suppose members of one family were all suddenly struck by a serious illness, followed by death within a day or two. What facts would we gather to explain this? In an earlier era, one might have looked for witches in the neighborhood, or suspected the victims of sacrilege or heresy. Today we have other hypotheses about disease, and we look for other circumstances, but the hypotheses with which we start still determine which facts we look for. Mill's methods of inquiry are useful only if we have rival hypotheses and must choose between them. The hypotheses come first.

This point can be made more concisely by some dialogue from a Sherlock Holmes story:[9]

> [Colonel Ross:] Is there any other point to which you wish to draw my attention?
> [Holmes:] To the curious incident of the dog in the night-time.
> [Colonel Ross:] The dog did nothing in the night-time.
> [Holmes:] That was the curious incident.

To Colonel Ross, the circumstance noted by Holmes was not a circumstance at all, yet this non-fact was the key to the solution: the crime was committed by someone known to the dog.

Thus we conclude that although facts are indeed stubborn things, they are inextricably interwoven with our prior hypotheses and our cultural prejudices. It is best to think of them as having a man-made component, rather than as purely objective facets of an already existing nature (although they may be no less tangible and inescapable than such other man-made objects as ten-ton trucks).

The Chicken or the Egg?

The requirement that we begin with hypotheses rather than facts may surprise and puzzle the reader, who may reasonably ask: How do we invent hypotheses before we look at facts? Can we make any fruitful speculations about how the world works before we know something about the world? Do we first have to invent the theory of relativity before we can measure the velocity of light or the energies given off in nuclear transformations?

The answer is that, yes, we must be aware of some facts before we look for hypotheses to explain them, but there is nothing systematic or logical about the process, no one right way to select or arrange them.

Of course we usually do not begin our research by selecting exactly the one correct hypothesis. We start rather with some vaguely formulated intuitive guess. Any hypothesis, even a wrong one, is better than none, though of course the closer our first effort is to the truth the easier our task. The great theories of science were not born fully grown, like Athena from the brow of Zeus.

From the time a hypothesis is proposed, it is subjected to a constant dynamic confrontation with those facts deemed relevant to testing its truth. Most hypotheses die young. Even the survivors do not emerge unscathed by the confrontation, but are altered, refined, made more precise and more general. By the time a hypothesis has reached this point it will have earned considerable credibility, and will now be called a "theory." The term "hypothesis," with its implication of uncertainty and tentativeness, is dropped.

More about the origins of hypotheses is given in Chapter 16.

SCIENCE AND PUBLIC FACTS

So facts are not really independent of the observer and his theories and preconceptions. However, at any one time, in any one culture, it is usually possible for most observers to agree on them. To put it better, *facts are what all observers agree on*.

This statement implies something crucial about those facts with which science is concerned: they must have more than one observer. There must be a group of observers that anyone can join—it cannot be a private club. Of course, joining the club imposes duties. Often, one cannot judge the truth of some claimed observation without going to the trouble of learning a lot of things that most people do not automatically know. Is the sparkplug removed from the motor of this car burnt out or not? This is a question of fact, but not everyone knows off-hand how to verify it. One must be not merely an observer, but an informed and interested observer.

This allows for the possibility that the informed and interested observers could all be wrong. It has happened in the past and will happen again. But it is the best we can do, and it is what makes science possible.

REFERENCE NOTES

1. Thomas Henry Huxley, *On the Educational Value of the Natural History of Science,* 1854.
2. William Morris, ed., *The American Heritage Dictionary of the English Language* (Boston: American Heritage, 1969).
3. We have seen this remark attributed to Einstein, but have been unable to verify it or to identify another source.
4. R. L. Gregory, *Eye and Brain: The Psychology of Seeing,* 3rd ed. (New York: McGraw-Hill, 1978).
5. John Z. Young, *Doubt and Certainty in Science: A Biologist's Reflections on the Brain* (New York: Oxford University Press, 1960). Reprinted by permission of Oxford University Press.
6. Francis Bacon, *The New Organon and Related Writings,* ed. Fulton H. Anderson (Indianapolis: Bobbs-Merrill, 1960).
7. Morris R. Cohen and Ernest Nagel, *An Introduction to Logic and Scientific Method* (New York: Harcourt Brace, 1934).
8. Ernest Nagel, ed., *John Stuart Mill's Philosophy of Scientific Method* (New York: Hafner, 1950).
9. Sir Arthur Conan Doyle, "Silver Blaze" in *The Memoirs of Sherlock Holmes* (Harmondsworth: Penguin, 1971), p. 25.

SUGGESTED READING

Gombrich, Ernst H. *Art and Illusion: A Study in the Psychology of Pictorial Representation.* Princeton, N.J.: Princeton University Press, 1969.
Gregory, R. L. *Eye and Brain: The Psychology of Seeing.* 3rd ed. New York: McGraw-Hill, 1978.
Gregory, R. L., and Ernst H. Gombrich, eds. *Illusion in Nature and Art.* New York: Charles Scribner's Sons, 1973.
Harris, Errol E. *Hypothesis and Perception: The Roots of Scientific Method.* New York: Humanities Press, 1970.
Nash, Leonard K. *The Nature of the Natural Sciences.* Boston: Little, Brown, 1963.
Scientific American. *Image, Object, and Illusion: Readings from Scientific American.* San Francisco: W. H. Freeman, 1974.
Ziman, John M. *Public Knowledge: An Essay Concerning the Social Dimension of Science.* Cambridge: Cambridge University Press, 1968.

PROBLEM

1. Study Figure 2-5 for one minute (time yourself with a watch). Then close the book and try to draw the figure from memory. Did you have any special difficulty? To what do you attribute it?

3
Logic

INTRODUCTION

In this chapter and a later one we will try, without teaching much logic or mathematics, to explain a little about how these disciplines work, why they are important in science, and how they are used.

Logical is a word with positive connotations even for people who are not well grounded in the formal principles of logic. Most of us are flattered if some chain of reasoning we use is acknowledged by others to be logical. The term *mathematical* does not always have positive connotations. Many of us are afraid of mathematics and try to avoid situations where it seems to be needed. Whether this is because one needs a "mathematical mind" to understand mathematics and few people have one, or whether there is something wrong with the way mathematics is taught, we do not know, but present-day science is inconceivable without both logic and mathematics. This is not to say that there are no branches of science that do not use mathematics; there are, and they are important. For those who are firmly convinced that mathematics is not for them, this is probably encouraging. However, there is a hitch. It may seem a truism to state that, for those problems that *can* be handled by mathematics, mathematics is the best and often the only way to handle them. The hitch is in deciding when a particular problem can be handled by mathematics. This can be done only by those who have some grasp of mathematics to start with, and some appreciation of its power and versatility. We will return to this question in a later chapter.

THE NATURE OF LOGIC

Logic has been defined as "the science of valid inference." As such, it is concerned with what *must* follow from a given set of starting assumptions. Let us look at some simple examples of valid inference. Consider the following classic group of three statements (the example is from Aristotle):

1. All men are mortal.
2. All heroes are men.
3. Therefore all heroes are mortal.

A group of statements in this form is an example of what is called a *syllogism*.

The above statements will be accepted by everyone as being true, just as everyone will accept as true the statement *All heroines are women,* but their truth is not the important issue at the moment. Consider the following group of three *false* statements:

1. All ostriches are citizens of Guatemala.
2. All cows are ostriches.
3. Therefore all cows are citizens of Guatemala.

This syllogism, in spite of the falsity of each of its statements, has something very important in common with the first syllogism, as does the following:

1. All nachtigalls are vogels.
2. All pferds are nachtigalls.
3. Therefore all pferds are vogels.

We can say nothing about the truth or falsity of these statements unless we know a little German.

What these three syllogisms have in common is that whether or not the first two statements of each are true, the third follows from them. In other words, *if* the first two statements were true, the third *must be*. This is what *valid inference* means.

Notice the distinction that has been made between truth and validity: logic is concerned with the latter. Only after we have established the logical validity of a chain of reasoning need we worry about truth. Then we know that *if* the initial statements (premises) are true, so is the conclusion.

A corollary of the above is the following: Suppose we do not know if the premises of some argument are true or not, and direct knowledge of their truth or falsity is not easily come by. Suppose, however, we can deduce from these premises by valid logical reasoning a new statement, which by direct observation is found to be false. Then we can be sure that at least one of the premises is false. If a hero were proven to be immortal, then either he is not a man or else not all men are mortal after all. It is this aspect of valid inference that makes it useful in science, because scientific hypotheses are often framed in such a way that a direct experimental test of their truth is difficult or impossible, but other statements logically deducible from the hypotheses are testable. If they are false, at least one of the hypotheses is false also.

The argument does not work both ways. If the logical consequences of some set of hypotheses are found to be true, the hypotheses may still be false. This is often a trap for those who have not had some training in formal logic, but a simple example will show it.

1. All Mexicans are Virginians.
2. George Washington was a Mexican.
3. Therefore George Washington was a Virginian.

Thus a true conclusion can follow from two false premises. It is easy to construct other examples of syllogisms where one of the premises is true and the other is false. The conclusion can be true or false. The fact that a true conclusion can follow from false premises but a false conclusion cannot follow from true ones means that from direct knowledge that a conclusion is false we can prove some of the starting premises wrong, but we cannot from knowledge that the conclusion is true prove them right.

THE LOGIC OF AN EXPERIMENTAL TEST

The lack of symmetry in logical arguments—that true conclusions can follow from false premises, but false conclusions cannot follow from true ones—has an important bearing on the logic of an experimental test of a scientific theory. Scientific theories are usually more complicated in nature than simple statements like "all heroes are mortal," and experimentally testing them is correspondingly more complicated than looking around to see if we can locate an immortal hero anywhere. But just as a failure to find any examples of immortal heroes does not *prove* the truth of either "all men are mortal" or "all heroes are men," so the fact that an experiment agrees with the prediction of a theory does not *prove* the truth of the theory. Plausibility is the best we can hope for.

But if the experiment *disagrees* with the theory, we can be sure that *something* we assumed must be false. This sounds as though it should mean that if the experiment disagrees, the theory is *proven* false. Unfortunately things are not so simple. The reason is that when we look at all the assumptions we make to predict the outcome of the experiment we find that they include more than just the truth of the scientific theory. They always include assumptions about the application of the theory to this particular experiment; assumptions that the instruments used really measure, with sufficient accuracy, the quantities the theory refers to; assumptions that the effects of various extraneous conditions on the outcome are negligible, or can be corrected for; and, for that matter, tacit assumptions about the integrity and the carefulness of the scientists doing the work.

So while a false conclusion assures us of a false premise, the false premise need not necessarily be the truth of the theory that the experiment was designed to test.

LOGICAL DIFFICULTIES AND FALLACIES

We hope the above discussion has conveyed the concept of a valid chain of inference, and given some idea why it is important. However, the examples taken, such as the proof of the mortality of heroes, are fairly transparent, and may give the reader the impression that it is a straightforward matter to decide what the logical consequences of premises are. This is far from the case. Some simple-seeming sequences of statements may sound valid, but are not.

An example of such a sequence is the following:

1. All animals that have hair and suckle their young are mammals.
2. Goldfish neither have hair nor suckle their young.
3. Therefore goldfish are not mammals.

Indeed, goldfish are not mammals. All the statements are true, but the appearance of validity is deceptive. The falseness of the logic is revealed if we change the subject matter but not the form of the argument:

1. All citizens of New York are citizens of the United States.
2. The citizens of Nevada are not citizens of New York.
3. Therefore, the citizens of Nevada are not citizens of the United States.

If the chain of reasoning about mammals had begun with a different statement, *(A) Only animals that have hair and suckle their young are mammals,* then the reasoning would have been valid. Statement (A) is not logically equivalent to the previous statement (1). We tend automatically to trust arguments that lead to conclusions we believe to be true. We should learn not to.

Other difficulties arise when we have to derive the consequences of not two starting premises but a whole host of them. Under these conditions, the consequences are often far from obvious. Further, there may be an enormous number of consequences if the starting premises are "rich" enough.

An example of a set of starting premises the logical consequences of which are not obvious is provided by the author of *Alice in Wonderland*:[1]

1. No kitten, that loves fish, is unteachable.
2. No kitten without a tail will play with a gorilla.
3. Kittens with whiskers always love fish.
4. No teachable kitten has green eyes.
5. No kittens have tails unless they have whiskers.

These premises lead to more than one conclusion. Among them are:

1. Green-eyed kittens do not have tails.
2. No kittens that hate fish have whiskers.
3. Kittens that do not love fish have no tails.
4. Green-eyed kittens will not play with a gorilla.

These are not immediately apparent, nor do they exhaust the possibilities of the premises.

Examples of more complicated premises arise in quite real, practical situations. The rules of a game like chess, a complicated legal contract such as a lease or an

LOGIC

insurance policy, the entire body of the laws of the United States or of some state government, and the regulations of the Internal Revenue Service all can be looked upon as sets of premises that have large numbers of far-from-obvious consequences, as anyone who has experience of them will attest.

It sometimes happens that we start with premises that look as though there is nothing wrong with them, but after deductions of their logical consequences are made, contradictions among them are revealed.

An example of a set of starting assumptions that contains a contradiction, demonstrable only after careful analysis, is given by Cohen and Nagel;[2] it is a table purporting to be a statistical study of 1000 students at a university:

Freshmen	525
Male	312
Married	470
Male freshmen	42
Married freshmen	147
Married males	86
Married male freshmen	25

There is nothing obviously wrong with the above figures, yet they imply that the number of unmarried female nonfreshmen is -57. Since one of the tacit premises of any tabulation of this type is that no negative numbers are possible, there is a contradiction.

AN EXAMPLE OF LOGICAL REASONING

The logical consequences of even a relatively simple set of premises can sometimes be surprising and unexpected. The following is an example from Cohen and Nagel:[2]

> Let us consider the proposition: *There are at least two persons in New York City who have the same number of hairs on their heads,* and let us symbolize it by q. How could its truth be established? An obvious way would be to find two individuals who actually do have the same number of hairs. But this would require an extremely laborious process of examining the scalps of perhaps six million people. It is not a feasible method practically. We may be able to show, however, that the proposition q follows from or is necessitated by other propositions whose truth can be established more easily. In that event, we could argue for the truth of the proposition q, in virtue of its being implied by the others, and in virtue of the established truth of the propositions offered as evidence. Let us try this method.
>
> Suppose it were known by an actual count that there are five thousand barber shops in New York City. Would the proposition *There are five thousand barber shops in New York City* be satisfactory evidence for q? The reader will doubtless reply, "Nonsense! What has the number of barber shops to do with

there being two persons with an identical number of scalp hairs?" In this way the reader expresses the judgment (based on previous knowledge) that the number of barber shops is no evidence at all for the equality in the number of hairs. Not all propositions are relevant, even if true, to the truth of a proposition in question.

Let us now consider the proposition *The number of inhabitants in New York City is greater than the number of hairs that any one of its inhabitants has on his head.* We shall denote this proposition by p. Is the truth of p sufficient to establish the truth of q? The reader might be inclined to dismiss p, just as he dismissed the information about the number of barber shops, as irrelevant. But this would be a mistake. We can show that if p is true, q must be true also. Thus suppose, taking small numbers for purposes of illustration, that the greatest number of hairs that any inhabitant of New York City has is fifty, and that there are fifty-one people living in New York City, no one of whom is completely bald. Let us assign a number to each inhabitant corresponding to the number of hairs that he has. Then the first person will have one hair, the second person two hairs, and so on, until we reach the fiftieth person, who will have, at most, fifty hairs. There is one inhabitant left and, since we have assumed that no person has more than fifty hairs, he will necessarily have a number of hairs that is the same as that possessed by one of the other fifty persons. The argument is perfectly general, as a little reflection shows, and does not depend on the number fifty we have selected as the maximum number of hairs. We may, therefore, conclude that our proposition p, the number of inhabitants in New York City is greater than the number of hairs that any one of its inhabitants has on his head, implies proposition q, there are at least two persons in New York who have the same number of hairs on their heads. The two propositions have been shown to be so related that it is impossible for the first (called the *evidence* or premise) to be true, and the second (called the *conclusion* or *that which is to be proved*) to be false. (pp. 6–7)

In the above example, Cohen and Nagel, being philosophers rather than experimental scientists, have not told us how to prove statement p, that the number of hairs on any person's head is less than the population of New York City. Certainly if the only way to prove p is first to count the hairs on every inhabitant's head, we have gained nothing from our elegant logical argument. Indeed, the statement p could be proved rigorously only by just such a counting procedure. But we can make p plausible enough to accept as true by various simpler procedures. We can measure the diameters of hairs from the heads of a large number of individuals. We will find that the diameters vary, depending on the individual and on the color of the hair (blonde hair is thinner). Then we can measure the area of people's heads that is covered with hair. This too varies from individual to individual. Now let us choose the smallest diameter of any hair we observed and calculate its cross-sectional area. Also, let us choose the individual with the largest scalp area that we have found. The area of this scalp, divided by the cross-sectional area of the thinnest hair we have found, will give us a number that is probably equal to or larger than the number of hairs on any one person's head. We can have a lot of confidence in this without having tediously counted the hairs on the head of any one individual. But of course we could be wrong and not know it. (The number of hairs on the average person's head has been estimated to be 150,000, and at the time Cohen and Nagel's book was published New York City had a population of 6 million.)

PROBABLE INFERENCE

We have stated that logic is concerned with valid inference. Consider the following chain of reasoning.

1. Most Republicans are conservative in fiscal matters.
2. Mr. Robinson is a Republican.
3. Therefore Mr. Robinson is conservative in fiscal matters.

Note the word "most" in the first statement. The three statements constitute an example of what is called probable inference, and it is a type of reasoning we cannot do without, either in daily life or in science. Yet it is very different from the type of inference we have called *valid:* it is perfectly possible for the first two statements to be true, yet the third might be false. While direct observation of Mr. Robinson's fiscal carelessness might surprise us, it does not refute either his Republicanism or the fiscal conservatism of most Republicans.

REFERENCE NOTES

1. Lewis Carroll, *Symbolic Logic and the Game of Logic* (New York: Dover Publications, 1958), p. 119.
2. Morris R. Cohen and Ernest Nagel, *An Introduction to Logic and Scientific Method* (New York: Harcourt Brace, 1934), p. 322. Copyright 1934 by Harcourt Brace Jovanovich, Inc; renewed 1962 by Ernest Nagel and Leonora Cohen Rosenfield. Reprinted by permission of the publisher.

SUGGESTED READING

Cohen, Morris R., and Ernest Nagel, *An Introduction to Logic and Scientific Method.* New York: Harcourt Brace, 1934.
Flew, Anthony. *Thinking Straight*. Buffalo: Prometheus Books, 1977.
Kahane, Howard. *Logic and Contemporary Rhetoric: The Use of Reason in Everyday Life.* 3rd ed. Wadsworth, Calif.: Wadsworth, 1980.

The books by Flew and Kahane deal not only with logic, but with clear reasoning generally, including questions of evaluating evidence, detecting deliberate fallacies, and so on. It is interesting that Flew finds most of his examples of erroneous or fallacious reasoning from arguments supporting politically liberal or radical viewpoints, while Kahane finds his from conservative sources. The idea that careless or misleading arguments are found mainly on one end of the political spectrum is itself deserving of careful and rational examination.

PROBLEMS

1. Are the following syllogisms valid? Explain.
   (1) All mammals suckle their young.
   (2) Humans suckle their young.
   (3) Hence, humans are mammals.

   (1) Animals that have cloven hooves and chew their cud are kosher.
   (2) Rabbits do not have cloven hooves.
   (3) Hence, rabbits are not kosher.
2. Without referring to the text, reproduce the proof by Cohen and Nagel that there are at least two people in New York City with the same number of hairs on their heads.
3. Can you show that the number of unmarried female freshman in the example on p. 25 is -57? Can you find any other contradictions in the table?

4
Snow on Cholera

INTRODUCTION: THE MAN, THE BACKGROUND

John Snow (1813–1858) was the son of a farmer in York, England. At the age of 14 he was apprenticed to a surgeon in Newcastle, who sent him when he was 18 to attend the victims of a major outbreak of cholera in the vicinity. In 1838 Snow passed his examination in London and became a member of the Royal College of Surgeons. He quickly made significant contributions to medical research: he participated in the development of an air pump for administering artificial respiration to newborn children unable to breathe and invented an instrument for performing thoracic surgery. He made major contributions to the new technique of anesthesia, becoming the leading specialist in London in the administration of ether, but switching to the easier-to-use chloroform when his own experimental studies convinced him of its practicality. He administered chloroform to Queen Victoria on the birth of her children, Prince Leopold and Princess Beatrice. His greatest achievement was his study of cholera, which he described in his monograph "On the Mode of Communication of Cholera," one of the classics of scientific method and a fascinating story fascinatingly written. Snow died in 1858, a relatively young man, while at work on a book entitled *On Chloroform and Other Anaesthetics*.

The concept of communicable diseases—that some diseases are transmitted by close contact from the sick to the well—came into being in the Middle Ages.[1] The ancient Greeks were the first to attempt to look at disease scientifically. They rejected the idea of disease as a punishment for sin or as a consequence of witchcraft, and studied instead the relation of diseases to aspects of the natural environment or the way men and women live, eat, and work. They noted, for example, that it was unhealthy to live near swamps. But in spite of the fact that they suffered from epidemics of various sorts, they somehow missed recognizing that some diseases are contagious.

The prescriptions for isolation and purification described in the Hebrew Bible for physiological processes such as menstruation and for diseases characterized by discharges or skin lesions apparently are based on the idea of the contagiousness of spiritual uncleanliness, of which the physical disease was merely an external symptom. In the Middle Ages, the Church, confronted with a major epidemic of leprosy, revived the biblical practice of isolating the sick, and the same methods were applied during the outbreak of the Black Death (bubonic plague) in the fourteenth century. By this time the concept of contagion was well established.

It is interesting that the belief that disease was a consequence of evil behavior coexisted with the recognition of contagion for hundreds of years afterwards. Attempts to develop treatments for syphilis were opposed on the grounds that syphilis was a deterrent to sexual immorality. Cholera was most prevalent among the poor for reasons that will become apparent, and there were many who regarded it also as a just punishment for the undeserving and vicious classes of society. A governor of New York once stated during a cholera epidemic, ". . . an infinitely wise and just God has seen fit to employ pestilence as one means of scourging the human race for their sins, and it seems to be an appropriate one for the sins of uncleanliness and intemperance. . . ." The president of New York's Special Medical Council declared at the onset of an epidemic in 1832 that "the disease had been confined to the intemperate and the dissolute with but few exceptions." A newspaper report noted,

> Every day's experience gives us increased assurance of the safety of the temperate and prudent, who are in circumstances of comfort. . . . The disease is now, more than before, rioting in the haunts of infamy and pollution. A prostitute at 62 Mott Street, who was decking herself before the glass at 1 o'clock yesterday, was carried away in a hearse at half past three o'clock. The broken down constitutions of these miserable creatures perish almost instantly on the attack. . . . But the business part of our population, in general, appear to be in perfect health and security.

A Sunday school newspaper for children explained, *"Drunkards and filthy wicked people of all descriptions* are swept away in *heaps,* as if the Holy God could no longer bear their wickedness, just as we sweep away a mass of filth when it has become so corrupt that we cannot bear it. . . . The Cholera is not caused by intemperance and filth in themselves, but is a *scourge,* a rod in the hand of God."[2]

By the middle of the nineteenth century some of the major communicable diseases had been identified and well differentiated from each other. This was no easy task in itself. For example, a large number of children's diseases have fever and a sore throat as symptoms; these diseases are still not easy to distinguish today. There are many conditions associated with severe diarrhea, for example cholera, typhoid, dysentery, bacterial food poisoning, noninfectious diseases of the lower intestine such as colitis, and poisoning with certain drugs. In the nineteenth century it had also been demonstrated that some of the contagious diseases could be artificially transmitted by inoculation of small amounts of "morbid matter" taken from the sick. The modes of transmission of particular diseases such as syphilis, intestinal worms, and skin diseases were known. Further, certain types of living organisms had been shown to cause disease directly: the itch mite in scabies, and various types of fungus in a disease of silkworms, in ringworm, and in other conditions. Bacteria and protozoa were discovered with the invention and further development of the microscope. These were often observed in the bodies of victims of certain diseases, and various scientists were beginning to speculate that they might be the cause of communicable diseases. This idea in various forms was in the air by the time of Snow, and he made use of it. Solid proof of the germ theory of disease came only in the 1860s and 1870s, after Snow's study, in the work of Pasteur and Koch.

THE DISEASE

Cholera is a bacterial disease characterized by severe diarrhea, vomiting, and muscular cramps. The diarrhea can produce extreme dehydration and collapse; death is frequent, and often occurs within hours after the onset of sickness.

The disease had been known to exist in India since the eighteenth century, and occurs there and in other parts of the world today. In the nineteenth century, as travel between Asia and the West became more common and as the crowding of people in urban centers increased as a result of the industrial revolution, major epidemics occurred in Europe and America. England had epidemics in 1831–1832, 1848–1849, and 1853–1854.

The question of how cholera is transmitted was especially difficult. On the one hand there was good evidence that it could be transmitted by close personal contact. Yet there was equally good evidence that some who had close personal contact with the sick, such as physicians, rarely got it, and that outbreaks could occur at places away from the immediate neighborhood of existing cases of the disease.

A number of theories were proposed, some of which were too vague to be rationally examined but some of which had solid experimental support. A number of people, including both physicians and uneducated laymen, had blamed the water supply. Snow adopted this theory, but refined it by specifically implicating the excretions of the cholera victims.

Snow's genius lay not so much in hitting on the correct mechanism for the spread of the disease as in providing an elegant and convincing experimental proof of it; he recognized the importance of the circumstance that, by chance, in a single district of London where an outbreak had occurred, some houses got their drinking water from one water company, and others from a different one. The two companies drew their water from two different parts of the Thames River which were not equally contaminated by the sewage of the city.

That Snow made the case for his own theory so convincing did not relieve him of the obligation to test alternate theories and show that they did not explain the experimental observations as well as his own. We have collected in one section his discussions of these theories and his arguments against them.

The reader should be aware that Snow's theory did not, at the time he proposed it, explain every single experimental fact. There were some facts that did not fit, and others that were explained as well or better by other theories. Most successful scientific theories, especially when new, are in this position, and those who propose them must have the courage and the judgment to put discordant facts aside at times. There are obvious risks in doing this, but no advance would be possible without it. We will point out from time to time places where Snow's explanations of discordant facts were shaky.

INTRODUCTION TO THE STUDY

We have chosen to tell the story as much as possible in Snow's own words, partly because it conveys more of the direct personal experience of making a major scientific discovery, and partly because Snow tells it so well. Page references are to *Snow on Cholera,* a reprint of two of Snow's major monographs, published by the Commonwealth Fund, New York, 1936. All quotations are from "On the Mode of Communication of Cholera," originally published in 1854. All italics are ours. In the selections from the monograph we have mostly followed Snow's order of presentation except in a few cases where logical clarity is achieved by deviating from it.

Snow begins with a brief historical review, following which he cites evidence to show that cholera can be transmitted by close personal contact with the sick:

The History of Cholera

> The existence of Asiatic Cholera cannot be distinctly traced back further than the year 1769. Previous to that time the greater part of India was unknown to European medical men; and this is probably the reason why the history of cholera does not extend to a more remote period. It has been proved by various documents, quoted by Mr. Scot, that cholera was prevalent at Madras in the year above mentioned, and that it carried off many thousands of persons in the peninsula of India from that time to 1790. From this period we have very little account of the disease till 1814, although, of course, it might exist in many parts of Asia without coming under the notice of Europeans. . . .
>
> In 1817, the cholera prevailed with unusual virulence at several places in the Delta of the Ganges; and, as it had not been previously seen by the medical men practising in that part of India, it was thought by them to be a new disease.

At this time the cholera began to spread to an extent not before known; and, in the course of seven years, it reached, eastward, to China and the Philippine Islands; southward, to the Mauritius and Bourbon; and to the north-west, as far as Persia and Turkey. Its approach towards our own country, after it entered Europe, was watched with more intense anxiety than its progress in other directions.

It would occupy a long time to give an account of the progress of cholera over different parts of the world, with the devastation it has caused in some places, whilst it has passed lightly over others, or left them untouched; and unless this account could be accompanied with a description of the physical condition of the places, and the habits of the people, which I am unable to give, it would be of little use.

General Observations on Cholera

There are certain circumstances, however, connected with the progress of cholera, which may be stated in a general way. It travels along the great tracks of human intercourse, never going faster than people travel, and generally much more slowly. In extending to a fresh island or continent, it always appears first at a sea-port. It never attacks the crews of ships going from a country free from cholera, to one where the disease is prevailing, till they have entered a port, or had intercourse with the shore. Its exact progress from town to town cannot always be traced; but it has never appeared except where there has been ample opportunity for it to be conveyed by human intercourse. (pp. 1–2)

Cholera Is Contagious

It was important to demonstrate that cholera was indeed a contagious disease, transmittable by contact with victims of the disease. Snow cites the following examples to prove this. In each of them, the occurrence of several cases among members of a single family or among people living in close proximity at a time when no other cases of the disease existed in the vicinity would be hard to explain unless contagion or a common source of infection were at work.

> There are also innumerable instances which prove the communication of cholera, by individual cases of the disease, in the most convincing manner. Instances such as the following seem free from every source of fallacy. (pp. 1–2)
>
> I called lately to inquire respecting the death of Mrs. Gore, the wife of a labourer, from cholera, at New Leigham Road, Streatham. I found that a son of the deceased had been living and working at Chelsea. He came home ill with a bowel complaint, of which he died in a day or two. His death took place on August 18th. His mother, who attended on him, was taken ill on the next day, and died the day following (August 20th). There were no other deaths from cholera registered in any of the metropolitan districts, down to the 26th August, within two or three miles of the above place. (p. 3)
>
> John Barnes, aged 39, an agricultural labourer, became severely indisposed on the 28th of December 1832; he had been suffering from diarrhea and cramps

for two days previously. He was visited by Mr. George Hopps, a respectable surgeon at Redhouse, who, finding him sinking into collapse, requested in an interview with his brother, Mr. J. Hopps, of York. This experienced practitioner at once recognized the case as one of Asiatic cholera; and, having bestowed considerable attention on the investigation of that disease, immediately enquired for some probable source of contagion, but in vain: no such source could be discovered. . . .

Whilst the surgeons were vainly endeavouring to discover whence the disease could possibly have arisen, the mystery was all at once, and most unexpectedly, unravelled by the arrival in the village of the son of the deceased John Barnes. This young man was apprentice to his uncle, a shoemaker, living at Leeds. *He informed the surgeons that his uncle's wife (his father's sister) had died of cholera a fortnight before that time, and that, as she had no children, her wearing apparel had been sent to Monkton by a common carrier. The clothes had not been washed; Barnes had opened the box in the evening; on the next day he had fallen sick of the disease.*

During the illness of Mrs. Barnes [the wife of John Barnes: she and two friends who visited Barnes during his illness also got cholera] her mother, who was living at Tockwith, a healthy village 5 miles distant from Moor Monkton, was requested to attend her. She went to Monkton accordingly, remained with her daughter for 2 days, washed her daughter's linen, and set out on her return home, apparently in good health. Whilst in the act of walking home she was seized with the malady, and fell down in collapse on the road. She was conveyed home to her cottage, and placed by the side of her bedridden husband. He, and also the daughter who resided with them, took the malady. All the three died within 2 days. Only one other case occurred in the village of Tockwith, and it was not a fatal case. . . .

It would be easy, by going through the medical journals and works which have been published on cholera, to quote as many cases similar to the above as would fill a large volume. *But the above instances are quite sufficient to show that cholera can be communicated from the sick to the healthy; for it is quite impossible that even a tenth part of these cases of consecutive illness could have followed each other by mere coincidence, without being connected as cause and effect.* (p. 9)

How Does Cholera Spread? The "Effluvia" Theory

The transmission of contagious diseases was frequently imputed to "effluvia" given off in the exhalations of the patient or from bodies of the dead, and subsequently inhaled into the lungs of a healthy person.

It would be a mistake to look down on the effluvia theory as so much unenlightened superstition. The fact is that the effluvia theory does not really contradict the germ theory. Some germ (and virus) diseases really are transmitted through the air from victims of the disease to others (the common cold is one example). The effluvia theory erred only in not identifying the active agent or "poison" as a microorganism. One should recognize that the germ theory then was highly speculative, had very little evidence in its favor, and was not widely accepted by physicians. The idea that disease could be spread by foul odors or other poisonous emanations represented a great advance over views attributing disease to witchcraft or sin, and, in the absence of any knowledge of microorganisms, was a plausible explanation of contagion.

Further, the effluvia theory led to justified concern over the crowded and unsanitary living and working conditions of the poor. Interested readers should consult the report prepared for Parliament by E. Chadwick in 1842 for a description of these conditions.[3] Chadwick's report led to the first serious public health measures taken by the British government, and in fact these measures resulted in improved health of the population of England.

This illustrates a truism of scientific research: an incorrect theory is better than no theory at all, or, in the words of an English logician, Augustus De Morgan, "Wrong hypotheses, rightly worked, have produced more useful results than unguided observations."[4]

Snow now points out two arguments against this theory: (1) not everyone in close contact with a patient gets the disease, even though anyone having close contact breathes the "effluvia" given off in the exhalations of the patient, and (2) sometimes cholera breaks out during an epidemic in new areas remote from other cases, where there has been no opportunity for exposure to "effluvia."

He further criticizes this theory for restricting too much our examinations of possible ways by which a disease can be spread. He points out that there are various diseases whose mode of transmission is known and which are transmitted in different ways—syphilis, certain diseases of the skin, intestinal worms—none of them by "effluvia."

Snow also gives another argument against the effluvia theory: cholera begins without any prior evidence of a systemic infection, but rather with intestinal symptoms directly, whereas if it resulted from breathing in a poison there should be some evidence of general illness first. With cholera, however, the very first symptom is the severe diarrhea—there is no fever or other symptoms preceding it. This strongly suggests that the disease begins when some "morbid material" is introduced into the alimentary canal, with only a brief period of incubation—24 to 48 hours—which allows the "morbid material" to reproduce. Snow infers from this that the morbid material is a microorganism:

> For the morbid matter of cholera having the property of reproducing its own kind, must necessarily have some sort of structure, most likely that of a cell. It is no objection to this view that the structure of the cholera poison cannot be recognized by the microscope, for the matter of small-pox and of chancre can only be recognized by their effects, and not by their physical properties. (p. 15)

The theory that contagious diseases are caused by microorganisms had been proposed, as noted, by others before Snow, using the same argument. Convincing confirmation did not come for another 20–30 years.

In the first sentence of the next quotation, Snow states the basic hypothesis of his work.

Snow's Theory

> The instances in which minute quantities of the ejections and dejections of cholera patients must be swallowed are sufficiently numerous to account for the spread of the disease; and on examination it is found to spread most where the facilities for this mode of communication are greatest.

In the following, Snow points out that people belonging to different social classes perform different functions around the sick, live in different kinds of houses, and have different personal habits and lifestyles. The result is that they have different risks of catching diseases.

Why Doctors Did Not Get Cholera and Those Laying Out the Body Did

> Nothing has been found to favour the extension of cholera more than want of personal cleanliness, whether arising from habit or scarcity of water, although the circumstance till lately remained unexplained. The bed linen nearly always becomes wetted by the cholera evacuations, and as these are devoid of the usual colour and odour, the hands of persons waiting on the patient become soiled without their knowing it; and unless these persons are scrupulously cleanly in their habits, and wash their hands before taking food, they must accidentally swallow some of the excretion, and leave some on the food they handle or prepare, which has to be eaten by the rest of the family, who, amongst the working classes, often have to take their meals in the sick room: hence the thousands of instances in which, amongst their class of the population, a case of cholera in one member of the family is followed by other cases; whilst medical men and others, who merely visit the patients, generally escape. The post mortem inspection of the bodies of cholera patients has hardly ever been followed by the disease that I am aware, this being a duty that is necessarily followed by careful washing of the hands; and it is not the habit of medical men to be taking food on such an occasion. On the other hand, the duties performed about the body, such as laying it out, when done by women of the working class, who make the occasion one of eating and drinking, are often followed by an attack of cholera; and persons who merely attend the funeral, and have no connexion with the body, frequently contract the disease, in consequence, apparently, of partaking of food which has been prepared or handled by those having duties about the cholera patient, or his linen and bedding. (pp. 16–17)

Why the Rich Did Not Get Cholera So Often

> The involuntary passage of the evacuations in most bad cases of cholera, must also aid in spreading the disease. Mr. Baker, of Staines, who attended 260 cases of cholera and diarrhea in 1849, chiefly among the poor, informed

> me . . . that "when the patients passed their stools involuntarily the disease evidently spread." It is amongst the poor, where a whole family live, sleep, cook, eat and wash in a single room, that cholera has been found to spread when once introduced, and still more in those places termed common lodging-houses, in which several families were crowded into a single room. It was amongst the vagrant class, who lived in this crowded state, that cholera was most fatal in 1832; but the Act of Parliament for the regulation of common lodging-houses, has caused the disease to be much less fatal amongst these people in the late epidemics. When, on the other hand, cholera is introduced into the better kind of houses, as it often is, by means that will be afterwards pointed out, it hardly ever spreads from one member of the family to another. The constant use of the hand-basin and towel, and the fact of the apartments for cooking and eating being distinct from the sick room, are the cause of this. (p. 18)

We may recall the two observations cited by Snow against the "effluvia" hypothesis: (1) not everyone (such as doctors) in close contact with a cholera victim gets it, and (2) sometimes it appears at great distances from the nearest case. Note that Snow's hypothesis (communication through evacuation) explains the first observation plausibly. Now he deals with the second, by making an additional subsidiary hypothesis.

How Did Cholera Get to the Rich?

> If the cholera had no other means of communication than those which we have been considering, it would be constrained to confine itself chiefly to the crowded dwellings of the poor, and would be continually liable to die out accidentally in a place, for want of the opportunity to reach fresh victims; but there is often a way open for it to extend itself more widely, and to reach the well-to-do classes of the community; I allude to the mixture of the cholera evacuations with the water used for drinking and culinary purposes, either by permeating the ground, and getting into wells, or by running along channels and sewers into the rivers from which entire towns are sometimes supplied with water. (pp. 22–23)

In the following quotations Snow gives evidence for his second hypothesis: *that cholera spreads through the water supply.* It should be noted that the idea that water is responsible was suggested by many others quoted by Snow: Mr. Grant, Dr. Chambers, Mr. Cruikshanks, and various other named and unnamed individuals. Snow's hypothesis, although it must have owed much to these people, is more specific in that it identifies the excretions of the victims as the source of the contamination of the water supply, and explains transmission by direct contact as well.

> In 1849 there were in Thomas Street, Horsleydown, two courts close together, consisting of a number of small houses or cottages, inhabited by poor people.

FIGURE 4-1. "Monster soup, commonly called Thames Water." Etching by William Heath, ca. 1828. The pollution of the London water supply was not a discovery of John Snow. This etching was apparently a response to a report of a London commission that investigated the water from the Thames, and reported it was "charged with the contents of the great common sewers, the drainings of the dunghills and laystalls, the refuse of the hospitals, slaughterhouses, and manufacturers." (Given by Mrs. William H. Horstmann to the Philadelphia Museum of Art, and reproduced with permission of the museum.)

The houses occupied one side of each court or alley—the south side of Trusscott's Court, and the north side of the other, which was called Surrey Buildings, being placed back to back, with an intervening space, divided into small back areas, in which were situated the privies of both the courts, communicating with the same drain, and there was an open sewer which passed the further end of both courts. *Now, in Surrey Buildings the cholera committed fearful devastation, whilst in the adjoining court there was but one fatal case, and another case that ended in recovery. In the former court, the slops of dirty water, poured down by the inhabitants into a channel in front of the houses, got into the well from which they obtained their water; this being the only difference that Mr. Grant, the Assistant-Surveyor for the Commissioners of Sewers, could find between the circumstances of the two courts as he stated in a report that he made to the Commissioners.* (p. 23)

In Manchester, a sudden and violent outbreak of cholera occurred in Hope Street, Salford. The inhabitants used water from a particular pump-well. This well had been repaired, and a sewer which passes within 9 inches of the edge of it became accidentally stopped up, and leaked into the well. The inhabitants of 30 houses used the water from this well; among them there occurred 19 cases of diarrhea, 26 cases of cholera, and 25 deaths. The inhabitants of 60 houses in the same immediate neighbourhood used other water; among these there occurred 11 cases of diarrhea, but not a single case of cholera, nor one death. It is remarkable, that, in this instance, out of the 26 persons attacked with cholera, the whole perished except one. (pp. 31–32)

The Washerwoman Was Spared

Dr. Thomas King Chambers informed me, that at Ilford, in Essex, in the summer of 1849, the cholera prevailed very severely in a row of houses a little way from the main part of the town. It had visited every house in the row but one. The refuse which overflowed from the privies and a pigsty could be seen running into the well over the surface of the ground, and the water was very fetid; yet it was used by the people in all the houses except that which had escaped cholera. That house was inhabited by a woman who took linen to wash, and she, finding that the water gave the linen an offensive smell, paid a person to fetch water for her from the pump in the town, and this water she used for culinary purposes, as well as for washing.

How the Landlord Got It

The following circumstance was related to me, at the time it occurred, by a gentleman well acquainted with all the particulars. The drainage from the cesspools found its way into the well attached to some houses at Locksbrook, near Bath, and the cholera making its appearance there in the autumn of 1849, became very fatal. The people complained of the water to the gentleman belonging to the property, who lived at Weston, in Bath, and he sent a surveyor, who reported that nothing was the matter. The tenants still complaining, *the owner went himself, and on looking at the water and smelling it, he said that he could perceive nothing the matter with it.* He was asked if he would taste it, *and he drank a glass of it. This occurred on a Wednesday; he went home, was taken ill with the cholera, and died on the Saturday following, there being no cholera in his own neighbourhood at the time.* (pp. 31–32)

THE LONDON EPIDEMICS OF 1853–1854

The Broad Street Pump

In the years 1853 and 1854 there were severe epidemics of cholera in London. During their course, Snow performed a number of studies which are described in his monograph. The most famous, although not the most conclusive, concerned an outbreak in the vicinity of a pump on Broad Street, Soho. By the time of the outbreak, Snow already was convinced that cholera is spread through the water supply, but the data he was able to gather by close observation guided by his hypothesis made the case for it much more convincing. Because of his hypothesis, he asked certain questions and noticed certain things, for example, the high rate of the disease among the customers of a certain coffee shop, and the low rate among the inhabitants of a workhouse and the employees at a brewery.

He also took the first public health measure based on his ideas. He told the Board of Guardians of the parish to remove the handle of the Broad Street pump to prevent any further use of the contaminated water and thus any further cases of cholera arising from this source. He hoped that this would provide experimental proof of his theory. It would have done so if there had been a sudden drop in the number of new cases of the disease after the pump handle was removed. But in this he was disappointed. The epidemic had already passed its peak and the number of new cases was already falling rapidly.

"The Mortality in This Limited Area Equals Any That Was Ever Caused in This Country by the Plague"

> The most terrible outbreak of cholera which ever occurred in this kingdom is probably that which took place in Broad Street, Golden Square, and the adjoining streets, a few weeks ago. Within 250 yards of the spot where Cambridge Street joins Broad Street, there were upwards of 500 fatal attacks of cholera in 10 days. The mortality in this limited area probably equals any that was ever caused in this country, even by the plague; and it was much more sudden, as the greater number of cases terminated in a few hours. The mortality would undoubtedly have been much greater had it not been for the flight of the population. Persons in furnished lodgings left first, then other lodgers went away, leaving the furniture to be sent for when they could meet with a place to put it in. Many houses were closed altogether, owing to the death of the proprietors; and, in a great number of instances, the tradesmen who remained had sent away their families so that in less than six days from the commencement of the outbreak, the most afflicted streets were deserted by more than three-quarters of their inhabitants.
>
> There were a few cases of cholera in the neighbourhood of Broad Street, Golden Square, in the latter part of August; and the so-called outbreak which commenced in the night between the 31st August and the 1st September, was, as in all similar instances, only a violent increase of the malady. As soon as I became acquainted with the situation and extent of this irruption of cholera, I suspected some contamination of the water of the much-frequented street-pump in Broad Street, near the end of Cambridge Street; but on examining the water, on the evening of the 3rd September, I found so little impurity in it of an organic nature, that I hesitated to come to a conclusion. Further inquiry, however showed me that there was no other circumstance or agent common to the circumscribed locality in which this sudden increase of cholera occurred, and not extending beyond it, except the water of the above mentioned pump. (pp. 38–39)

Snow began his study by obtaining from the London General Register Office a list of the deaths from cholera in the area occurring each day. These figures showed a dramatic increase in cases on August 31, which he therefore identified

Table 4-1
Results of Snow's Investigation

83 Deaths[a]						
73 Living near Broad Street pump			10 Not living near pump			
61 Known to have drunk pump water	6 Believed not to have drunk pump water	6 No information	5 In families sending to Broad St. pump for water	3 Children attending school near pump	2 No information	

[a] Of the 83 individuals who had died of the disease, 69 were known definitely or could be assumed to have drunk the pump water, 6 were believed not to have drunk it, and for 8 there was no information.

as the starting date of the outbreak. He found 83 deaths that took place from August 31 to September 1 (see Table 4-1), and made a personal investigation of these cases. He found, first, that nearly all the deaths had taken place within a short distance of the pump, only ten of the deaths having occurred in houses nearer to another pump. Of these ten deaths, five were in families that went to the Broad Street pump for their water by choice, and three of the others were among children who attended school near the Broad Street pump and were known to drink the water from it. Snow was able to ascertain that 61 of the people who died were known to have drunk the pump-water, and only 6 were known not to have done so. In another six cases, no information was available. (pp. 39–40)

Who Drank the Pump Water?

For reasons of clarity we summarize the results of Snow's investigation of these 83 deaths in Table 4-1, which shows that there were deaths among people not known to have drunk water from the Broad Street pump. These deaths therefore are facts that seem to contradict Snow's hypothesis. A scientist faced with facts contradictory to a hypothesis has many alternatives, only one of which is to discard the hypothesis. Another alternative is to make a closer examination of these facts, to see whether in some plausible way they can be shown either not really to contradict the hypothesis or actually to support it. It occurred to Snow to look for ways *the individuals in question might have drunk the water without being aware of it.*

> The additional facts that I have been able to ascertain are in accordance with those above related; and as regards the small number of those attacked, who were believed not to have drunk the water from Broad Street pump, it must be obvious that there are various ways in which the deceased persons may have taken it without the knowledge of their friends. The water was used for mixing with spirits in all the public houses around. It was used likewise at dining-

rooms and coffee-shops. The keeper of a coffee-shop in the neighbourhood, which was frequented by mechanics, and where the pump-water was supplied at dinner time, informed me (on 6th September) that she was already aware of nine of her customers who were dead. The pump-water was also sold in various little shops, with a teaspoonful of effervescing powder in it, under the name of sherbet; and it may have been distributed in various other ways with which I am unacquainted. The pump was frequented much more than is usual, even for a London pump in a populous neighbourhood. (pp. 41–42)

Snow next gives two striking observations that confirm the role of the pump. There were two large groups of people living near the Broad Street pump who had very few cases of cholera: the inhabitants of a workhouse and the employees of a brewery.

Why Were the Workhouse and the Brewery Spared?

In the workhouse, which had its own water supply, only 5 out of 535 inmates died. If the death rate had been the same as in the surrounding neighborhood, over 100 would have died.

There is a brewery in Broad Street, near to the pump, and on perceiving that no brewer's men were registered as having died of cholera, I called on Mr. Huggins, the proprietor. He informed me that there were about 70 workmen employed in the brewery, and that none of them had suffered from cholera—at least in a severe form—only two having been indisposed, and that not seriously, at the time the disease prevailed. The men are allowed a certain quantity of malt liquor, and Mr. Huggins believes that they do not drink water at all; and he is quite certain that the *workmen never obtained water from the pump in the street. There is a deep well in the brewery,* in addition to the New River water. (pp. 41–42)

The Pump Handle

On September 7, Snow met with the Board of Guardians of the parish and informed them of his evidence as to the role of the pump in the outbreak. On September 8, the handle of the pump was removed, but, as Snow notes, by this time the epidemic had subsided, perhaps because many inhabitants had fled the neighborhood. So the removal of the pump handle did not produce any dramatic effect on the number of new cases (Figure 4-2).

Following the epidemic, the pump was opened and examined. No direct evidence of leakage from nearby privies was found, but Snow states his belief that it must have occurred, perhaps by seepage through the soil, as on microscopic examination "oval animalcules" were found, which Snow points out are evidence of organic contamination. (They were not the bacteria causing cholera, which were not detectable by the microscopic techniques of the time, nor did Snow take them seriously as a causative agent—rather, he knew that "animalcules" were very common in natural waters contaminated with sewage or other organic matter, even when no cholera was present.)

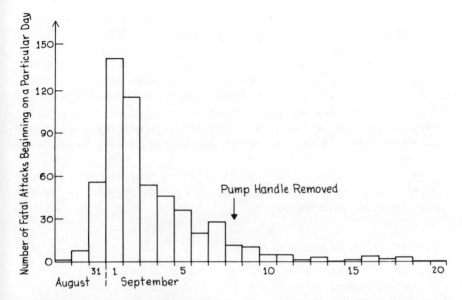

FIGURE 4-2. The Broad Street pump outbreak. The figure shows the number of fatal cases that began on a given date, plotted against that date. The arrow indicates when the pump handle was removed.

Additional evidence for the contamination of the pump water with sewage was provided by inhabitants of the neighborhood who had noticed a disagreeable taste in the water just prior to the outbreak and a tendency for the water to form a scum on the surface when it was left to stand a few days. Further, chemical tests showed the presence of large amounts of chlorides, consistent with contamination by sewage, but, like the animalcules, not constituting overwhelming proof. The question of chlorides in the drinking water will come up again more dramatically later on.

Snow's conclusion on the Broad Street pump outbreak is as follows:

> Whilst the presumed contamination of the water of the Broad Street pump with the evacuations of cholera patients affords an exact explanation of the fearful outbreak of cholera in St. James's parish, there is no other circumstance which offers any explanation at all, whatever hypothesis of the nature and cause of the malady be adopted. (pp. 51–54)

A CONTROLLED EXPERIMENT—WHERE DID THEY GET THEIR WATER?

In the next section of Snow's monograph he describes another study that he made during the 1853–1854 epidemics, which provided far more convincing evidence for his hypothesis. He was able to perform, in effect, a controlled experiment.

The concept of a controlled experiment is one of the most crucial in gaining an understanding of the nature of the scientific process. The basic idea is simple, perhaps deceptively so.

When a television repairman wants to find out why a set is not functioning properly, he can try replacing the suspect parts—transistors, condensers, tubes—with new ones, one at a time. A psychologist testing whether the race of the teacher makes a difference to how well black children learn may compare the performance of black children in classes with white teachers to their performance in classes with black teachers. Both the repairman and the psychologist are doing controlled experiments.

Not all experiments in science are "controlled" in this sense: some are more exacting than this and some are less so, as we will point out when we come to discuss appropriate examples later on.

Laboratory versus Natural Experiments

Questions about the causes of a disease that afflicts human beings are easier to answer if animals can get the same disease. A controlled experiment can be done to test any hypothesis we have. Suppose mice could get cholera, and one wished to prove that water containing the excretions of cholera victims could produce the disease. One would take two large groups of mice similar in every relevant respect, put cholera excretions in the drinking water of one group (the test group), and leave them out of the water of the second group (the control group). If a large number of cholera cases were found in the test group and none in the control group, the case would have been made. Unfortunately, mice do not get cholera, nor did any other laboratory animal known to Snow. Thus there was no choice but to experiment on human beings.

Human beings are more difficult to experiment on than mice. First, there is an ethical question—if you are inclined to believe that contaminated drinking water produces cholera, but you are not yet sure, do you have the right to let people drink it? Even if they would be drinking it anyway, do you not have an obligation to stop them?

The ethical problems can be avoided if by chance a "natural" experiment is available: it may happen that a group in the population has been exposed fortuitously to what is believed to be the cause of a disease. A controlled experiment* is then possible if another group in the population can be found, similar in every relevant respect to the first one, except that it has not been exposed to the suspected cause. If the disease occurs in the first group and not the second, we have confirming evidence that the suspected cause really is the cause. But in such a "natural" situation it may be hard to prove the two groups similar in "every relevant respect."

---

* Some scientists would prefer the term "controlled observation" here. (See the discussion in Chapter 1.)

SNOW ON CHOLERA

For example, different districts of London had different water supplies and different cholera rates. Unfortunately, from the point of view of testing Snow's hypothesis that cholera is caused by contaminated water, the people in the different districts were different in other ways also. The rich lived in different neighborhoods from the poor and suffered less from cholera. Was it because they had uncontaminated water supplies or because they ate better food, worked shorter hours at easier jobs, lived in newer, cleaner houses?

Also different groups of equally "poor" people might differ in other significant ways. In London at that time there was a greater tendency for people of the same occupation to live in a single neighborhood than is the case today, so that one neighborhood might have a lot of butchers, another might have tailors, and a third drivers of carts. Might susceptibility to cholera depend on occupation? Snow himself was aware that some occupational groups such as doctors were less likely to get cholera, and some, such as coal miners, were more likely. Perhaps there was some overlooked causative factor related to one's work.

Since we know that Snow's theory about the water supply was correct, we can feel that all these other differences are irrelevant and can be disregarded. But at the time this was not yet clear, and of course the whole purpose of the experiment was to test this theory. If the control and test groups differed in three or four other possibly significant ways besides getting their water supplies from a different source, we would not feel safe in blaming the water supply alone; any of these other differences between the groups might be responsible for the differences in cholera rates.

Where Did They Get Their Water?

It was Snow's genius to recognize the importance of the fortuitous circumstance that two different water companies supplied a single neighborhood in an intermingled way.

The two water companies in question had, previous to the 1849 epidemic, drawn their water from the Thames, from places that could be expected to be contaminated with the sewage of the city. But in 1852 one of these companies, the Lambeth Company, moved its waterworks upstream to a place free of London sewage. The other, the Southwark and Vauxhall Company, remained where it was. Both companies delivered drinking water to a single district of the city:

> The pipes of each Company go down all the streets, and into nearly all the courts and alleys. A few houses are supplied by one Company and a few by the other, according to the decision of the owner or occupier at that time when the Water Companies were in active competition. In many cases a single house has a supply different from that on either side. Each Company supplies both rich and poor, both large houses and small; *there is no difference either in the condition or occupation of the persons receiving the water of the different Companies.*

In the next sentence, Snow summarizes the basic idea of the experiment:

> As there is no difference whatever, either in the houses or the people receiving the supply of the two Water Companies, or in any of the physical conditions with which they are surrounded, it is obvious that no experiment could have been devised which would more thoroughly test the effect of water supply on the progress of cholera than this, which circumstances placed ready-made before the observer.
>
> The experiment, too, was on the grandest scale. No fewer than 300,000 people of both sexes, of every age and occupation, and of every rank and station, from gentlefolks down to the very poor, were divided into two groups without their choice, and, in most cases, without their knowledge; one group being supplied with water containing the sewage of London, and, amongst it, whatever might have come from the cholera patients, the other group having water quite free from such impurity.
>
> To turn this grand experiment to account, all that was required was to learn the supply of water to each individual house where a fatal attack of cholera might occur. I regret that, in the short days at the latter part of last year, I could not spare the time to make the inquiry; and, indeed, I was not fully aware, at that time, of the very intimate mixture of the supply of the two Water Companies, and the consequently important nature of the desired inquiry. (pp. 75–76)

Carrying out the idea required putting together two kinds of data: cholera cases and water supply. The first was easier to come by then the second.

> When the cholera returned to London in July of the present year, however, I resolved to spare no exertion which might be necessary to ascertain the exact effect of the water supply on the progress of the epidemic, in the places where all the circumstances were so happily adapted for the inquiry. I was desirous of making the investigation myself, in order that I might have the most satisfactory proof of the truth or fallacy of the doctrine which I had been advocating for 5 years. I had no reason to doubt the correctness of the conclusions I had drawn from the great number of facts already in my possession, but I felt that the circumstance of the cholera-poison passing down the sewers into a great river, and being distributed through miles of pipes, and yet producing its specific effects, was a fact of so startling a nature, and of so vast importance to the community, that it could not be too rigidly examined or established on too firm a basis. (p. 76)

Snow began to gather data on cholera deaths in the district. The very first results supported his conjecture: of 44 deaths in the district in question, 38 occurred in houses supplied by the Southwark and Vauxhall Company.

> As soon as I had ascertained these particulars I communicated them to Dr. Farr, who was much struck with the result, and at his suggestion the Registrars of all the south districts of London were requested to make a return of the water supply of the house in which the attack took place, in all cases of death from cholera. This order was to take place after the 26th August, and I resolved to carry my inquiry down to that date, so that the facts might be ascertained for the whole course of the epidemic. (p. 77)

Determining which water company supplied a given house was not always straightforward. The residents of houses often did not know the name of their supplier, and did not always keep receipts. In houses in which rooms were rented by working people by the week, the landlord or his agent paid the water bills, and did not always live in the house or the neighborhood. Fortunately Snow developed a simple chemical test, based on the high sodium chloride content of the water supplied by one of the two companies: he added a solution of silver nitrate to a sample of the water: when sodium chloride was present it immediately produced a white cloud of insoluble silver chloride.

Snow thus had both his chemical test and the receipts when available to identify the source of the water. His painstaking attention to detail, to be as sure as possible of his experimental data, illustrates one of the more important and more tedious aspects of scientific research.

Deaths and Death Rates

Snow now expresses the result of his study in quantitative terms. He notes that the Southwark and Vauxhall Company supplied about 40,000 houses in London during 1853 and the Lambeth Company (drawing its water upstream) about 26,000. In the rest of London, where there were over 250,000 houses, there were *more* deaths than in the houses supplied by Southwark and Vauxhall—1422 compared with 1263—but there were 6 times as many houses, also. What matters here is not the total number of deaths, but the rate of deaths per house or per family. Put another way, if you live in a house supplied by Southwark and Vauxhall, what are your chances of dying, compared with your chances if you live in a house supplied by the Lambeth Company? Snow expressed the rate in deaths per 10,000 houses, according to the following formula:

$$\text{Rate} = \frac{\text{deaths}}{\text{number of houses}} \times 10,000$$

Table 4-2 gives the results.

The death rate in the houses supplied by the Southwark and Vauxhall Company

Table 4-2
Death Rates from Cholera in Houses Supplied by the Two Water Companies

	No. of houses	Deaths from Cholera	Deaths in each 10,000 houses
Southwark & Vauxhall Company	40,046	1,263	315
Lambeth Company	26,107	98	37
Rest of London	256,423	1,422	59

was therefore more than eight times greater than in the houses supplied by the Lambeth Company.

BEING CRITICAL

Objections to Snow's Theory

Snow next considers an objection to his hypothesis: not everyone who drinks the polluted water gets sick. Note that he had used a similar objection against the "effluvia" hypothesis: not everyone exposed to the effluvia of cholera patients gets sick. However, he dealt differently with the two cases. He was able to find a relevant factor consistent with his own hypothesis to separate those who became ill from those who did not: they were members of different social groups with different sanitary practices that caused them to have different chances of ingesting excreta. On the other hand, Snow did not find, among those with equal chances of ingesting excreta, any factor that distinguished those who became ill from those who did not. Those who did not accept Snow's hypothesis, and had cited as evidence against it the fact that not all known to ingest excreta got the disease, would have been able to make a better case if they had been able to identify a factor distinguishing those who became ill from those who did not that was consistent with an alternative hypothesis.

Here is Snow's discussion of this problem:

Why Some Who Should Get Cholera Don't

There are one or two objections to the mode of communication of cholera which I am endeavouring to establish, that deserve to be noticed. Messrs. Pearse and Marston state, in their account of the cases of cholera treated at the Newcastle Dispensary in 1853, that one of the dispensers drank by mistake some rice-water evacuation* without any effect whatever. In rejoinder to this negative incident, it may be remarked, that several conditions may be requisite to the communication of cholera with which we are as yet unacquainted. Certain conditions we know to be requisite to the communication of other diseases. Syphilis we know is only communicable in its primary stage, and vaccine lymph must be removed at a particular time to produce its proper effects. In the incident above mentioned, the large quantity of the evacuation taken might even prevent its action. It must be remembered that the effects of a morbid poison are never due to what first enters the system, but to the crop or progeny produced from this during a period of reproduction, termed the period of incubation; and if a whole sack of grain, or seed of any kind, were put into a hole in the ground, it is very doubtful whether any crop whatever would be produced.

An objection that has repeatedly been made to the propagation of cholera

---

* The evacuations of victims of cholera are so watery that they do not have the usual color and odor of feces. The term "rice-water evacuation" has been used to characterize the appearance of these evacuations.

through the medium of water, is, that every one who drinks of the water ought to have the disease at once. This objection arises from mistaking the department of science to which the communication of cholera belongs, and looking on it as a question of chemistry, instead of one of natural history, as it undoubtedly is. It cannot be supposed that a morbid poison, which has the property, under suitable circumstances, of reproducing its kind, should be capable of being diluted indefinitely in water, like a chemical salt; and therefore it is not to be presumed that the cholera-poison would be equally diffused through every particle of the water. The eggs of the tape-worm must undoubtedly pass down the sewers into the Thames, but it by no means follows that everybody who drinks a glass of the water should swallow one of the eggs. As regards the morbid matter of cholera, many other circumstances, besides the quantity of it which is present in a river at different periods of the epidemic must influence the chances of its being swallowed, such as its remaining in a butt or other vessel till it is decomposed or devoured by animalcules, or its merely settling to the bottom and remaining there. In the case of the pump-well in Broad Street, Golden Square, if the cholera-poison was contained in the minute whitish flocculi visible on close inspection to the naked eye, some persons might drink of the water without taking any, as they soon settled to the bottom of the vessel. (pp. 111–113)

In some respects Snow's defense against this objection would be accepted as valid today, in the light of knowledge gained in the century that has elapsed since the germ theory of disease was accepted. We know, for example, that individual susceptibilities to a given disease vary widely, often for reasons that even now are not well understood. Also, some individuals may suffer an attack of a disease in a mild and clinically unrecognized form and may subsequently be immune for a longer or shorter period of time. It is a very rare epidemic in which everyone gets sick.

Snow's explanation of why the individual who drank cholera evacuation by mistake did not contract the disease is not plausible today, nor can we believe that it would have been plausible at the time, especially to anyone skeptical of Snow's theory. It would have been more admirable, but less human, had Snow acknowledged that this was one experimental fact he could not explain, and let it go at that.

Scotland Is Different

The next section discusses one oddity of the behavior of cholera: it was mainly a summer disease in England and would not spread in winter even when introduced then, but it seemed not to be seasonal in Scotland, running through its epidemic course as soon as it appeared, even in winter. Snow's explanation in terms of his theory is charming, and it shows the kinds of things a scientist has to be alert to.

It also may help to demolish a myth about the scientific method we have referred to in an earlier chapter: that scientific hypotheses are obtained by first examining the facts. In reality, the hypothesis comes first, and tells us which facts are worth examining. It is easy to see how Snow was led to compare the drinking habits of the English with those of the Scots, given his theory, but if one had only the facts about the seasonal differences in cholera between the two countries, would one have inferred a theory blaming the water supply?

> The English people, as a general rule, do not drink much unboiled water, except in warm weather. They generally take tea, coffee, malt liquor, or some other artificial beverage at their meals, and do not require to drink between meals, except when the weather is warm. In summer, however, a much greater quantity of drink is required, and it is much more usual to drink water at that season than in cold weather. Consequently, whilst the cholera is chiefly confined in winter to the crowded families of the poor, and to the mining population, who, as was before explained, eat each other's excrement at all times, it gains access as summer advances to the population of the towns, where there is a river which receives the sewers and supplies the drinking water at the same time; and, where pump-wells and other limited supplies of water happen to be contaminated with the contents of the drains and cesspools, there is a greater opportunity for the disease to spread at a time when unboiled water is more freely used.
>
> In Scotland, on the other hand, unboiled water is somewhat freely used at all times to mix with spirits; I am told that when two or three people enter a tavern in Scotland and ask for a gill of whiskey, a jug of water and tumbler-glasses are brought with it. Malt liquors are only consumed to a limited extent in Scotland, and when persons drink spirit without water, as they often do, it occasions thirst and obliges them to drink water afterwards. (pp. 117–118)

Other Theories: Effluvia, Elevation, Hard Water, and Soft Water

We have collected below Snow's discussion of alternate theories and his reasons for rejecting them. Giving fair consideration to theories opposed to one's own is something all scientists should try to do; but not all scientists are really capable of it, and are not necessarily bad scientists because of this shortcoming. Science proceeds by a consensus of scientists: one person's failure to be objective about a theory he doesn't like is made up for by the opposite bias of his opponents and the fairness of the less emotionally involved. Snow was better at it than most.

> Whilst the presumed contamination of the water of the Broad Street pump with the evacuations of cholera patients affords an exact explanation of the fearful outbreak of cholera in St. James's parish, there is no other circumstance which offers any explanation at all, whatever hypothesis of the nature and cause of the malady be adopted. . . . Many of the non-medical public were disposed to attribute the outbreak of cholera to the supposed existence of a pit in which persons dying of the plague had been buried about two centuries ago; and, if the alleged plague-pit had been nearer to Broad Street, they would no doubt still cling to the idea. The situation of the supposed pit is, however, said to be Little Marlborough Street, just out of the area in which the chief mortality occurred. With regard to effluvia from the sewers passing into the streets and houses, that is a fault common to most parts of London and other towns. There is nothing peculiar in the sewers or drainage of the limited spot in which this outbreak occurred; and Saffron Hill and other localities, which suffer much more from ill odours, have been very lightly visited by cholera. (pp. 54–55)

The low rate of mortality amongst medical men and undertakers is worthy of notice. If cholera were propagated by effluvia given off from the patient, or the dead body, as used to be the opinion of those who believed in its communicability; or, if it depended on effluvia lurking about what are by others called infected localities, in either case medical men and undertakers would be peculiarly liable to the disease; but, according to the principles explained in this treatise, there is no reason why these callings should particularly expose persons to the malady. (p. 122)

Dr. William Farr, a colleague of Snow and himself a distinguished scientist, had discovered an interesting relation between the cholera rates in various districts of London during the 1847–48 epidemic and their elevations. London was divided into 38 districts; Farr tabulated the death rate from cholera in each district (deaths per 10,000 of population) and the mean elevation of the district above an arbitrarily chosen reference point, the high water mark of the Thames River at one point of its course through London. He found a close relation: the greater the elevation the lower the death rate, ranging from 102 deaths per 10,000 in the districts at elevations less than 20 feet above high water, to 8 deaths per 10,000 people for one district at an elevation of 350 feet above high water. His conclusion was that the "elevation of habitation reduces the effects of cholera to insignificance."[5]

Height above Sea Level

Dr. Farr gave a number of reasons why elevation should be important: among them are the tendencies of regions of lower elevation to collect deposits of refuse and sewage, and to be more poorly drained. The result in his view was a greater pollution of the atmosphere in such places by the vapors emitted from decomposing organic matter—in other words, effluvia. While Snow did not particularly respect the effluvia theory, he was respectful in treating the ideas of his friend and colleague. He pointed out, however, that the limited correlation between the disease and elevation noted by Farr was actually better explained by his own theory: the low-lying districts of London were also those more likely to have water supplies contaminated by sewage.

Farr's observations can be thought of as a controlled experiment, the control and test groups being inhabitants of London living at different elevations above sea level (Figure 4-3). Indeed, the people living at the lower elevations suffered more cholera, but the two groups differed, as pointed out by Snow, in other significant ways, even though elevation was directly connected to the differences in the significant factor.

Dr. Farr was inclined to think that the level of the soil had some direct influence over the prevalence of cholera, but the fact of the most elevated towns in this kingdom, as Wolverhampton, Dowlais, Merthyr Tydfil, and Newcastle-upon-Tyne, having suffered excessively from this disease on several occasions, is opposed to this view, as is also the circumstance of Bethlehem Hospital, the Queen's Prison, Horsemonger Lane Gaol, and several other large buildings,

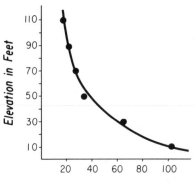

FIGURE 4-3. Cholera rates and elevation of residence. Plotted from the data of Dr. William Farr.

which are supplied with water from deep wells on the premises, having nearly or altogether escaped cholera, though situated on a very low level, and surrounded by the disease. The fact of Brixton, at an elevation 56 feet above Trinity high-water mark, having suffered a mortality of 55 in 10,000, whilst many districts on the north of the Thames, at less than half the elevation, did not suffer one-third as much, also points to the same conclusion.

I expressed the opinion in 1849, that the increased prevalence of cholera in the low-lying districts of London depended entirely on the greater contamination of the water in these districts, and the comparative immunity from this disease of the population receiving the improved water from Thames Ditton, during the epidemics of last year and the present, as shown in the previous pages, entirely confirms this view of the subject; for the great bulk of this population live in the lowest districts of the metropolis. (pp. 97–98)

Limestone and Sandstone

Another hypothesis, which agreed with at least some of the experimental facts, was proposed by John Lea of Cincinnati. Lea had found that districts in which the underlying rock formations were limestone had much more cholera than districts overlying sandstone. He conjectured that the calcium and magnesium salts, which were present in water in limestone districts, were somehow necessary for the cholera "poison" to have its effect. He noted as supporting evidence for this hypothesis the fact that towns that relied on river water, in which there was much calcium and magnesium, suffered more than towns that used rain water.

Snow's criticisms of Lea's hypothesis is in part specious. He attributed the difference in cholera rates between sandstone and limestone districts observed by Lea to a greater oxidizing power of sandstone on organic substances. This explanation is not very plausible, as Snow himself was aware—he had no evidence that limestone might not be equally oxidizing. We can be even more sure today that the correlation between cholera and rock formation found by Lea was entirely fortuitous. Snow of course attributed the higher cholera rates in towns using river water to the greater likelihood of river water being contaminated with sewage.

APPLICATIONS TO OTHER PROBLEMS

What About Other Diseases?

Having established convincingly that cholera could be communicated through the water supply, Snow then extended his theory beyond its original area of applicability: was it possible that other infectious diseases were also transmitted in the same way? He considered four other epidemic diseases: yellow fever, malaria (intermittent fever, ague), dysentery, and typhoid fever. He proved to be wrong about the first two and right about the second two. His reasoning in the cases where he was wrong is interesting to quote, because he makes a plausible case:

> Yellow fever, which has been clearly proved by Dr. M'William and others to be a communicable disease, resembles cholera and the plague in flourishing best, as a general rule, on low alluvial soil, and also in spreading greatly where there is a want of personal cleanliness. This disease has more than once appeared in ships sailing up the river Plate, before they have had any communication with the shore. The most probable cause of this circumstance is, that the fresh water of this river, taken up from alongside the ship, contained the evacuations of patients with yellow fever in La Plata or other towns. . . . (p. 127)

> Intermittent fevers are so fixed to particular places that they have deservedly obtained the name of endemics. They spread occasionally, however, much beyond their ordinary localities, and become epidemic. Intermittent fevers are undoubtedly often connected with a marshy state of the soil; for draining the land frequently causes their disappearance. They sometimes, however, exist as endemics, where there is no marshy land or stagnant water within scores of miles. Towards the end of the seventeenth century, intermittent fevers were, for the first time, attributed by Lancisi to noxious effluvia arising from marshes. These supposed effluvia, or marsh miasmata, as they were afterwards called, were thought to arise from decomposing vegetable and animal matter; but, as intermittent fevers have prevailed in many places where there was no decomposing vegetable or animal matter, this opinion has been given up in a great measure; still the belief in miasmata or malaria* of some kind, as a cause of intermittents, is very general. It must be acknowledged, however, that there is no direct proof of the existence of malaria or miasmata, much less of their nature.
> The preventive of ague, draining the land, must affect the water of a district quite as much as it affects the air, and there is direct evidence to prove that intermittent fever has, at all events in some cases, been caused by drinking the water of marshes. (pp. 129–130)

In the following paragraph, Snow, to explain the apparent absence of direct person-to-person contagion in malaria, makes an inspired guess: the malaria parasite, he speculates, must spend part of its life cycle outside the human body. Indeed it does, in the body of the *Anopheles* mosquito.

---

* For Snow, *malaria* meant "bad air" and did not describe the disease that we now call by that term. (His terms for that disease were "intermittent fever" and "ague.")

> The communication of ague from person to person has not been observed, and supposing this disease to be communicable, it may be so only indirectly, for the materies morbi [disease-producing substance] eliminated from one patient may require to undergo a process of development or procreation out of the body before it enters another patient, like certain flukes infesting some of the lower animals, and procreating by alternate generations. (p. 133)

Snow's explanation of why yellow fever breaks out on ships arriving on the River Plate before they even land is quite reasonable, but is of course wrong. Similarly, the close identification of malaria with marshes had been known since the time of Hippocrates, and Snow's conjecture that it comes from drinking marsh water is also plausible but wrong. It is interesting that the Italian physician Lancisi (1654–1720), whose "effluvia" theory of malaria is quoted by Snow, also suggested that mosquitoes might spread malaria.[1] Snow does not refer to this idea, and we do not know what he thought of it.

Prevention: Measures to Combat the Spread of Cholera

The last part of Snow's monograph gives his list of recommended measures for preventing the spread of cholera. His ideas did not win immediate acceptance from his medical contemporaries, who, although they felt he had made a good case for some influence of polluted water in cholera, continued to believe in "effluvia" theories as an alternate or contributing cause. In any event, his recommendations regarding the water supply were adopted, which spared London further cholera epidemics.

> The measures which are required for the prevention of cholera, and all diseases which are communicated in the same way as cholera, are of a very simple kind. They may be divided into those which may be carried out in the presence of an epidemic, and those which, as they require time, should be taken beforehand.
> The measures which should be adopted during the presence of cholera may be enumerated as follows:
> 1st. The strictest cleanliness should be observed by those about the sick. There should be a hand-basin, water, and towel, in every room where there is a cholera patient, and care should be taken that they are frequently used by the nurse and other attendants, more particularly before touching any food.
> 2nd. The soiled bed linen and body linen of the patient should be immersed in water as soon as they are removed, until such time as they can be washed, lest the evacuations should become dry, and be wafted about as a fine dust. Articles of bedding and clothing which cannot be washed, should be exposed for some time to a temperature of 212° or upwards.
> 3rd. Care should be taken that the water employed for drinking and preparing food (whether it come from a pump-well, or be conveyed in pipes) is not contaminated with the contents of cesspools, house-drains, or sewers; or, in the event that water free from suspicion cannot be obtained, it should be well boiled, and if possible, also filtered. . . .
> 4th. When cholera prevails very much in the neighbourhood, all the provisions which are brought into the house should be well washed with clean water and exposed to a temperature of 212° Fahr.; or at least they should undergo one of these processes, and be purified either by water or by fire. By being

careful to wash the hands, and taking due precautions with regard to food, I consider that a person may spend his time amongst cholera patients without exposing himself to any danger.

5th. When a case of cholera or other communicable disease appears among persons living in a crowded room, the healthy should be removed to another apartment, where it is practicable, leaving only those who are useful to wait on the sick.

6th. As it would be impossible to clean out coal-pits, and establish privies and lavatories in them, or even to provide the means of eating a meal with anything like common decency, the time of working should be divided into periods of four hours instead of eight, so that the pitmen might go home to their meals, and be prevented from taking food into the mines.

7th. The communicability of cholera ought not to be disguised from the people, under the idea that the knowledge of it would cause a panic, or occasion the sick to be deserted.

The measures which can be taken beforehand to provide against cholera and other epidemic diseases, which are communicated in a similar way, are:

8th. To effect good and perfect drainage.

9th. To provide an ample supply of water quite free from contamination with the contents of sewers, cesspools, and house-drains, or the refuse of people who navigate the rivers.

10th. To provide model lodging-houses for the vagrant class, and sufficient house room for the poor generally. . . .

11th. To inculcate habits of personal and domestic cleanliness among the people everywhere.

12th. Some attention should be undoubtedly directed to persons, and especially ships, arriving from infected places, in order to segregate the sick from the healthy. In the instance of cholera, the supervision would generally not require to be of long duration. . . .

I feel confident, however, that by attending to the above-mentioned precautions, which I consider to be based on a correct knowledge of the cause of cholera, this disease may be rendered extremely rare, if indeed it may not be altogether banished from civilized countries. And the diminution of mortality ought not to stop with cholera. (pp. 133–137)

What Snow Overlooked

Snow's monograph ends with a paragraph stating that typhoid fever, which killed many more in England than did cholera, might also be controlled by the measures he proposed. This was right, and both diseases were soon brought under control.

We would like to close Snow's story with one more quotation, because it tells us something important about one aspect of scientific research. In order to make the problems we want to solve tractable, we need to limit the range of what we study. Yet by doing so we risk overlooking important possibilities not included within the narrowed scope of our inquiry.

Early in the monograph, Snow gives his reasons for believing that the "morbid matter" causing cholera reaches the digestive tract directly by ingestion, rather than through a preliminary systemic infection.

> If any further proof were wanting than those above stated, that all the symptoms attending cholera, except those connected with the alimentary canal, depend simply on the physical alteration of the blood, and not on any cholera poison circulating in the system, it would only be necessary to allude to the effects of a weak saline solution injected into the veins in the stage of collapse. The shrunken skin becomes filled out, and loses its coldness and lividity; the countenance assumes a natural aspect; the patient is able to sit up, and for a time seems well. If the symptoms were caused by a poison circulating in the blood, and depressing the action of the heart, it is impossible that they should thus be suspended by an injection of warm water, holding a little carbonate of soda in solution. (p. 13)

Today it is recognized that cholera kills by dehydration and that if victims receive sufficient fluid either orally or intravenously the disease is rarely fatal. It is ironic that it should not have occurred to Snow that the observation he reported suggests a way of treating the disease. But Snow was not looking for a treatment of cholera, he was trying to establish how it is transmitted, and in that he succeeded.

THE CAUSES OF CANCER

In many ways Snow's approach to his problem may seem out of date now. We have the advantage of the germ theory of disease, the benefits of more than a century of research into the behavior and life cycles of bacteria and other infective organisms, powerful microscopes and other laboratory apparatus to search for and identify them, knowledge of viral diseases, and statistics, a highly developed branch of mathematics, to help us analyze our experiments.

But the basic process by which Snow made his discovery—the recognition of a problem, the formation of a hypothesis, the design of an experiment to test that hypothesis, the critical evaluation of the results of that experiment, the consideration of alternative hypotheses—are the underlying processes of all scientific research. They are as applicable to other fields of science as they are to the study of epidemics of contagious diseases, and as applicable in medical research today as they were in Snow's time.

The specific approach used by Snow—the search for the causes of diseases or clues to their origins by comparison of the naturally occurring distribution of diseases in different populations—forms a branch of medical research known as epidemiology. Although historically, as the name suggests, it began as a means of studying epidemics, the methods are applicable to a great variety of diseases or conditions, contagious or otherwise. In Chapter 8 we will discuss attempts to understand the causes of mental disorders by similar methods. In the remainder of this chapter we will discuss briefly the application of epidemiological methods to a current and still unsolved problem, that of cancer, a major cause of death and disability. We do so partly for its intrinsic interest and partly to demonstrate some things about the interdependence of different fields of science.

The term "cancer" describes not a single disease but a collection of diseases characterized by the wild and uncontrollable proliferation of abnormal cells produced by the body of the victim.[6] The different forms of cancer are distinguished not only by the organ affected, but by the precise type of abnormal cell produced, which can be determined by microscopic examination. The U.S. Surgeon-General's 1979 Report, *Smoking and Health*,[7] refers to six distinct types of "lung cancer," of which three are associated with cigarette smoking and three are not.

Biological and Chemical Approaches

The biology and chemistry of cancer have been studied for several decades, and extensive research is being carried out today. Some of the questions asked are: What are the factors that produce the abnormal cell in the first place? How does the abnormal cell differ from the normal cells of the body in its metabolism? Why do abnormal cells multiply so rapidly? Can their behavior be related to molecular processes taking place within the cells, processes involving the proteins present or the nucleic acids which are the carriers of the genetic heritage of the cell? If these questions can be answered, perhaps some knowledge of the cause of cancer will be obtained that will suggest ways to prevent or treat it.

As an alternative to attempts to explain cancer in terms of the biological and biochemical functioning of normal and abnormal cells, the epidemiologist studies the distribution of cancer among populations. If differences in the cancer rates in different populations can be found, clues to the cause may emerge.

To decide which populations to compare, one must start with some hypotheses about which factors may make a difference to cancer rates. Often the initial hypotheses are vaguely felt impressions, but as the data gathered to test them are examined, sharper, more precise hypotheses are formulated. Examples of factors that have been considered are as follows:

1. Age
2. Occupation
3. Ethnic origin
4. Country or area of residence
5. Factors of "life-style," including smoking, diet, and sexual practices

In addition, changes in cancer rates in a given population over long periods of time and changes in rates when a group of people migrate from one country to another have been studied.

Geography

Differences in cancer rates between different countries of the world are striking, and often puzzling as well.

Table 4-3
Examples of Extreme Differences in Worldwide Cancer Rates

Site of cancer	Area of high incidence	Area of low incidence	Ratio of highest to lowest rate
Esophagus	Northeast Iran	Nigeria	300
Liver	Mozambique	England	100
Prostate	U.S. Blacks	Japan	40
Breast	British Columbia (Canada)	Israel (Non-Jewish)	7
Colon	Connecticut, U.S.	Nigeria	10
Buccal Cavity	Bombay, India	Denmark	25
Nasopharynx	Singapore (Ethnic Chinese)	England	40
Penis	Parts of Uganda	Israel (Jewish)	300

Table 4-3 gives some examples of these differences.* The table lists, for certain specific kinds of cancer, the country with the largest *known* rate of incidence (number of new cases per year; in this table the age range 35–64 years is used), the country with the smallest *known* rate, and the *ratio* of the largest rate to the smallest rate. The word *known* appears in italics because in many countries the level of medical care available to large portions of the population is not adequate to provide reliable figures. The table includes only countries with reliable cancer mortality data.

Why is esophagus cancer 300 times as common in Iran as it is in Nigeria, and penis cancer 300 times as common in Uganda as among Israeli Jews? Why so much more cancer of the nose and pharynx among ethnic Chinese in Singapore? Why is liver cancer 100 times more common in one African country than in England? The 7 to 1 ratio of breast cancer when British Columbians are compared with the Arab inhabitants of Israel does not seem so dramatic, but breast cancer is a leading cause of death among women in Western countries, and if we could discover some means

---

* The older a person is, the more likely he or she is to get cancer. This is true for almost all forms of cancer, and it is a very significant effect. The chance of getting cancer at age 50 is about 5 times what it is at age 30, at age 70 it is about 15 times as great. This well-known fact is not well understood, but it requires us to be careful in making comparisons. The death rate from cancer in Western industrialized countries is much greater than in less developed or undeveloped countries, but it is not primarily because the environment is poisoned by industry—it is because people in the industrialized countries live longer. They do not die of infectious diseases or malnutrition at early ages, and so live long enough to have high probabilities of dying from cancer instead. Life expectancy in Sweden is 75 years, and in Bangladesh 49 years. Thus, a much higher fraction of the Swedish population than of the Bangladesh population is over 50 years of age. This being the age group most susceptible to cancer, there will be many more cancer deaths per 100,000 people in Sweden, even if, as is probably the case, 60-year-old men and women in both countries have about the same chance of dying of cancer in a given year.

In comparing cancer rates in different countries, it is therefore necessary to take into account the possibility that the age distributions of the populations are different. One way to do it is to compare death rates of specific age-groups in the two countries, for example, 50-year-olds.

of reducing the Canadian breast cancer death rates to the Israeli Arab level, we would save many lives and prevent much suffering.

Most of the questions raised by these figures cannot as yet be answered; in a few cases we have fairly plausible hypotheses, to be discussed later.

Within a Country

Even within one country, rates can differ significantly.

Figure 4-4 is one of many maps of cancer distribution in the U.S., prepared by the U.S. National Cancer Institute.[9] This particular map shows counties in the United States in which melanoma, an often fatal form of skin cancer, is unusually common. (Most other forms of skin cancer are readily treatable.) Melanoma is clearly more common in the southern and southeastern parts of the country—the so-called "sun-belt"—and is attributed to excessive exposure to sunlight (and other forms of radiation as well).

Thirty different types of cancer, classified by sex and race of the victim, have been studied in this fashion. Certain interesting patterns have emerged which have suggested new lines of research. For example, rectal cancer is more common among the well-to-do than the poor; other types, such as cancer of the uterine cervix, are

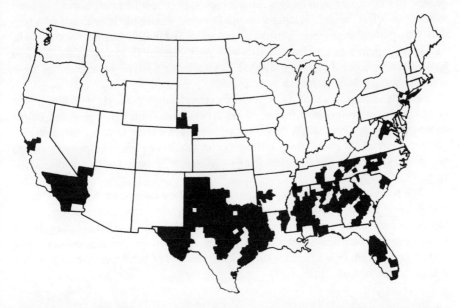

FIGURE 4-4. Distribution of melanoma (a form of skin cancer) among white males in the United States. The black areas shows counties in which the rates are above the national average. (From T. J. Mason *et al., Atlas of Cancer Mortality for U.S. Counties 1950–1969.* Reproduced with the permission of the U.S. Department of Health and Human Services.)

more common among the poor. In counties with high concentrations of chemical industries there is excess mortality from cancer of the bladder. This excess is greater among employees of chemical industries but is not confined to them: a recent study has found excess bladder cancer among women who reside in these areas, who do not normally work in chemical industries in large numbers, and among pet dogs as well.[10] Stomach cancer is high in certain rural areas of the Midwest where people of Russian, Austrian, Scandinavian and German descent live. The countries from which these people or their forebears emigrated are known to have high rates of stomach cancer also.

Genetics and Cancer

As can be seen in Figure 4-4, the differences in cancer rates between different geographical areas are striking. It is tempting to conclude that there must be differences in environmental factors to which the different populations are exposed, but before doing so, we must rule out another possibility: the differences may reflect differences in an inherited susceptibility to cancer.

Indeed, there is evidence that suggests that genetic factors do play a role in cancer. Breast cancer at an early age does appear to have a high frequency in certain families; so does a rather rare form of kidney cancer, called Wilm's Tumor. Further, genetic and environmental factors can interact: skin cancer from excessive exposure to sunlight affects mainly people who are fair-complexioned. Inheritance of a tendency to particular diseases can occur not only in families but in races as well. Sickle-cell anemia is a disease that affects only blacks. Only Jews from Eastern Europe suffer from Tay-Sachs disease. Thalessemia, a type of anemia, is most common among Mediterranean peoples.

Finally, stomach cancer is 20% more common among people with blood type A. As the distribution of different blood types among different races and ethnic groups does vary, very slight differences in stomach cancer rates in different geographic areas could result from differences in ethnic or racial composition of the populations, even in the absence of any difference in exposure to environmental factors.

We have already mentioned the excess rate of stomach cancer in both Central and Eastern Europe, and in descendants of immigrants from these regions living in certain rural areas of the midwestern United States. Do these midwestern Americans have an inherited susceptibility to this form of the disease, or have they maintained certain customs, including dietary habits, of their forebears in the "old country" that increase their risk?

Migration Studies

Although studies of the living habits of these particular groups are underway, definitive answers to these questions have not yet been obtained. But other immigrant

Table 4-4
Comparison of Cancer-Incidence Rates in Japan and for
Japanese and Caucasians in Hawaii[a]

Primary site of cancer	Patients' sex	Annual incidence/million people		
		Japan	Hawaii, 1968–1972	
			Japanese	Caucasians
Esophagus	M	112	46	75
Stomach	M	1,291	397	217
Colon	M	87	371	368
Rectum	M	90	297	204
Lung	M	299	379	962
Prostate	M	13	154	343
Breast	F	295	1,221	1,869
Cervix uteri	F	398	149	243
Corpus uteri	F	20	407	714
Ovary	F	55	160	274

[a] The table compares the age-standardized rates of new cases of cancer per million people in Japan and in Hawaii. The data given for Japan are from Osaka prefecture only, and for the years 1970–1971.

groups have been studied, and the results indicate that environment or life-style plays a larger role than genetic susceptibility in cancer rates. Breast cancer is much rarer (1/6 as great) and stomach cancer much more common (six times as great) in Japan than in the United States. Japanese began immigrating to the U.S. at the end of the nineteenth century and large numbers of them settled in Hawaii and on the West Coast. The first and second generations of immigrants continued to follow Japanese customs of diet, clothing, and other aspects of life-style, but later generations began to adopt American practices to a greater and greater extent. If the cancer rates had reflected genetic differences they would not have changed so rapidly with the change in life-style: changes in the genetic makeup of a human population do take place with changes in external environment, but they require many hundreds of generations. In fact, however, stomach cancer has declined and breast cancer increased with each successive generation, and among present day American descendants of Japanese immigrants, particularly those that have adopted American life-styles, the rates have approached those prevailing among Caucasian Americans (Table 4-4). Only a change in an environmental factor could act so rapidly.[11]

Environment and Life-Style

What factors in the environment or in the way people live could be responsible for the extraordinary differences between different populations?

Before we look for possible environmental causes, we must recognize that the word "environment" does not cover industrial pollutants only, but includes agents present in the natural environment, and equally important, aspects of the living habits—life-style—of various peoples. Smoking cigarettes, for example, is an aspect of life-style. There is almost unanimous agreement among physicians and epidemiologists that it is the single most important *known* environmental factor causing deaths from cancer.

The term *life-style* covers not only pleasant habits that can be sacrificed with more or less discomfort once their risks are recognized, but also practices that are inextricable features of the whole culture of a people.

A poignant example is the problem of breast cancer, the leading cause of cancer deaths among women in most Western countries. In the year 1700, Bernardo Ramazzini, an Italian physician, noted that nuns have a much higher rate of breast cancer than other women.[12] He attributed the greater rate to celibacy. He was not far wrong—the modern view is that early child-bearing, the sooner after the age of first menstruation the more so, decreases the risk of breast cancer significantly. How many women in Western societies, having this knowledge, would be willing to have babies before the age of 15?

Environmental Hazards (of Natural Origin)

As noted above, the term "environmental hazard" does not imply only industrially produced chemicals that pollute the natural environment, but includes some substances that are direct products of natural processes.

There is for example a certain amount of background high-energy radiation from radioactive minerals and cosmic rays all over the world which, if current views on the relation of radiation to cancer are correct, must be responsible for a small fraction of cancer cases. Aflatoxin, a substance produced by a fungus that grows on certain foods stored in tropical countries, is a potent carcinogen and has been thought responsible for the fact that liver cancer, rare in industrialized countries, is common in parts of Africa and China.

Denis Burkitt, an English physician practicing in East Africa, noted that a type of cancer of the lymph system (called a lymphoma), was unusually common among native children in East Africa, in whom it produced tumors of the jaw. Burkitt travelled around Africa, visiting hospitals in many different areas, and found that the lymphoma was found only in certain specific climatic zones: those where the rainfall was at least 20 inches (50.8 cm) per year and the temperature regularly above 60°F (15.6°C). In certain mountainous regions, such as the region east of Lake Victoria, the boundary between affected areas and tumor-free areas is strictly according to altitude: closely related tribes with similar cultures and life-styles were affected or not according to whether they lived above or below a 4400-foot (1350-meter) elevation. Further research showed that the areas of high incidence were also areas of high incidence of certain mosquito-borne diseases such as yellow fever and malaria. It has since been shown that this type of cancer is caused by a virus, most probably the Epstein-Barr virus, which first of all is not spread by mosquitos, but most often by oral transmission, and secondly is not uncommon in Western countries, where it produces infectious mononucleosis and some other diseases, but only rarely any form of cancer. The current theory is that intense and persistent malarial infection damages the body's immune system, somehow enabling the virus to produce the particular form of cancer observed.[13]

This discovery not only led to adoption of measures to combat the mosquito, and thus the disease, but also was an important contribution to basic biological research into the causes of cancer: only a few human cancers have been demonstrably related to virus infection; other examples include cervical cancer and forms of cancer of the liver. Some researchers believe that many other human cancers will ultimately be shown to have a viral origin.

Sunlight, as described earlier, is still another environmental hazard of natural origin.

Environmental Hazards (of Industrial Origin)

Cancers caused by chemical products of industrial activity have been discovered primarily by studies of particular occupational groups.

The earliest such example was the discovery, by Dr. Percivall Pott of London in 1775, that cancer of the scrotum was much commoner among chimney sweeps than among other men. He attributed it to exposure to the soot and dust in the chimneys.[14] 150 years later it was shown in the laboratory that soot and coal tar painted on the skin of animals produced cancers at the site.

A more recent example is the discovery that mesothelioma, a cancer of the lining of the chest cavity generally of extreme rarity, is quite common among people who have been employed in the asbestos industry. The observation was made almost simultaneously by a number of physicians in the United States, England, and South Africa, most notably by Dr. Irving Selikoff, then a physician in private practice in Passaic, New Jersey, a town in which there was an asbestos processing plant.[15]

Not all exposure to asbestos takes place in industrial settings. Workers in the asbestos industry bring the fibers home from their places of employment; members of their families are known to have a higher-than-normal rate of contracting mesothelioma. Zeolites, minerals chemically similar to asbestos, are used as building materials in certain rural villages in Turkey. Mesothelioma is common among these villagers.[16]

In addition to a greatly increased risk of mesothelioma, workers exposed to high asbestos concentrations are also at much higher risk for lung cancer if they are also smokers: their risk is much greater (eight-fold in one study) than among smokers not employed in the asbestos industry.[17] A similarly enhanced risk for lung cancer is found among uranium miners who smoke.[18]

Life-Style

An example of a form of cancer attributable to "life-style" is lung cancer, which is believed by most scientists who have studied the problem to be in the great majority of cases a direct result of cigarette smoking. If their views are correct, the elimination of cigarette smoking would cut the overall cancer death rate in the United States by about 30% and the lung-cancer death rate by 80%.*(18) Because of the importance of this question, we would like to review the evidence briefly.

Cancer and Smoking

Since about 1920 there has been a dramatic increase in the rate of epidermoid carcinoma of the lung in men (see Table 4-5). This also corresponds with a great increase in cigarette smoking in men. Is cigarette-smoking the cause of this increase in lung cancer? What additional evidence can be cited either to support this hypothesis or to make it less likely? Are there alternative hypotheses that can explain the data better? One must be careful not to conclude too readily that if a disease and some hypothesized causal agent have increased (or decreased) together, the hypothesized agent really is the cause. The similar variation may be an accidental consequence of an indirect relationship, as was the close relation between elevation and cholera in London found by William Farr (Figure 4-3), or even entirely fortuitous.

Specifically, while both cigarette-smoking and lung cancer have increased dramatically in recent years, we could not argue that there is a causal relation unless those individuals who smoke more are just the ones more likely to get lung cancer. It has been shown that people who are known to smoke heavily are indeed just the

Table 4-5
Annual Rates of Respiratory Cancer Deaths per 100 Million People under Age 65 in the United States[a]

	1933–37	1938–42	1943–47	1948–52	1953–57
Men	5,812	8,517	11,818	15,498	19,495
Women	2,102	2,450	2,889	2,785	2,820
	1958–62	1963–67	1968–72	1973–77	1978
Men	23,230	26,766	30,683	32,356	33,816
Women	3,539	4,922	7,568	10,266	12,064

[a] From Doll and Peto, 1981, p. 1282.

* There is good reason to believe that the death rate from heart attacks would be cut also, saving even more lives than this decrease in the cancer death rate.

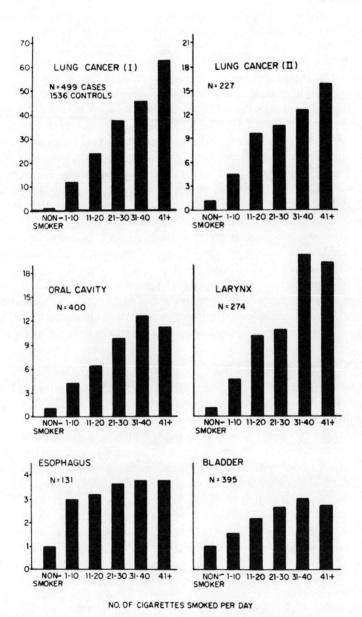

FIGURE 4-5. Relative risk of tobacco-related cancer for present (>10 years) male smokers.

ones who are more likely to develop lung cancer (Fig. 4-5), but a small number of victims have never smoked. Therefore, we cannot continue to maintain that smoking is the *sole* cause of the disease, but the hypothesis that it greatly increases the risk is still tenable.

Even though it has been established that smokers are much more likely to get lung cancer than nonsmokers are, this again may not necessarily imply a causal relation. People who smoke may differ from people who do not smoke in ways other than their smoking habits. In fact, smokers do drink more alcohol and coffee than nonsmokers do. But controlled experiments have shown that neither coffee nor alcohol is associated directly with an increased risk of lung cancer.

Another factor of difference between smokers and nonsmokers is that on the average people who live in cities smoke more than people who live in rural areas. Again, it could be that there is something in the city environment that causes lung cancer, and the correlation with smoking is only a coincidence. Here the problem is more complicated. Comparison of urban-rural differences in lung cancer shows that city-dwelling non-smokers are more likely to get lung cancer than rural nonsmokers, but the differences in rates are small compared to the differences in the rates associated with smoking habits.

Further confirmatory evidence comes from the recent rising rate of lung cancer among women together with the knowledge that women have also begun to smoke cigarettes heavily (see Table 4-5), and the fact that risk of lung cancer among former smokers decreases in proportion to the time elapsed since they stopped.

Nonepidemiological Evidence

Most scientists familiar with the epidemiological evidence for the hypothesis that cigarette smoking causes lung cancer find it fairly convincing, but it would be much more convincing if the precise biological mechanism by which the cancer is produced could be known. There are some scientists who use the word "proof" in this connection—who feel that only when a biological mechanism has been established can the causal relation be considered proven. In their view the high degree of correlation of smoking with lung cancer does not constitute proof. The position we take in this book is that no scientific hypotheses are "provable" if the word implies the certainty of mathematical proof. However, the subjective degree of confidence we can have in different hypotheses can differ enormously depending on the strength and kind of evidence. We prefer to put the matter this way: knowledge of a biological mechanism greatly increases our confidence in a hypothesis about disease that was originally established as plausible by epidemiological methods. Convincing as Snow's hypothesis was, it became more so with the development of a germ theory of disease and the identification of the specific organism responsible for cholera. It is to be expected, therefore, that the smoking–lung cancer relation so far supported by epidemiological evidence would seem even more probable if a biological basis could be established. Indeed, there is already some confirmatory evidence of this kind.

Clinical examination of the lungs and bronchial tissue of heavy smokers who do not have lung cancer shows changes that are considered pre-cancerous. Similar changes have been produced in the bronchial tissues of dogs exposed to tobacco smoke. Tobacco smoke condensate painted on the skin of mice and other animals produces skin cancers. Chemical analysis of tobacco smoke reveals that it contains a number of chemicals known individually to produce cancer in laboratory animals.

Unsolved Problems

It must be conceded that there are many unanswered questions. While the hypothesis that cigarette smoking is a direct cause of lung cancer is a highly probable one, it would be even more so if answers to these questions could be found.

For example, what is the precise mechanism by which lung cancer is produced? Is it the action of one of the known chemical carcinogens in tobacco smoke? Is it several of them acting cooperatively? Is it some other constituent of tobacco smoke not yet identified as a carcinogen? Some scientists have speculated, for example, that traces of certain radioactive elements known to be present in tobacco smoke might be responsible. Others have blamed constituents of the cigarette paper. Smoking has been shown to damage the body's mechanism for removing foreign particles deposited on the surfaces of the lung. It has been proposed that dust particles commonly found in the ambient air and bearing cancer-causing substances that all of us are exposed to are retained by the lungs of smokers for longer periods of time and thus have a greater chance of producing the disease.

Nonsmokers also get lung cancer, though their chances are much less. Why do they? Not all heavy smokers get it—the majority, in fact, do not. How do they avoid it?

Conclusions reached by epidemiological studies often depend on biological or biochemical research before they are considered to be firmly established. However, the traffic is not all one way. Epidemiology in turn suggests lines of research for biologists and biochemists to follow. This has been the case with cancer research, where the identification by epidemiological methods of substances in the environment that are associated with a high risk of cancer has led to their study in the laboratory. It has also been the case in other fields of medical research. For example, in the discovery of vitamins, only after the relation between certain diseases and the absence of certain foods from the diet was established epidemiologically, could the search for the specific disease-preventing substances in foods and their chemical identification be undertaken.

How Avoidable Is Cancer?

We have illustrated striking differences in cancer rates between different geographical areas, between different ethnic groups, between people working in different occupations, and between peoples following different life-styles. We have also indicated that most of these differences are not due to genetic factors. These data imply that it may be within our power to cut cancer death rates dramatically by changing living habits or eliminating environmental carcinogens. Cancer rates *everywhere* could thus be lowered at least to the lowest value they can have *anywhere*. It has been estimated that world cancer rates would be lowered 80% if this could be done—if the rate of esophagus cancer all over the world were as low as it is in Nigeria, and the rate of liver cancer as low as it is in England, and so on.

But this is not so simple as it sounds. We have already pointed out the difficulty of altering life-styles, and the fact that many environmental carcinogens are of natural rather than of industrial origin. Very few identifications of causal agents have been made which are sufficiently clear-cut to justify preventive measures. The most significant so far is cigarette smoking, which is associated not only with lung cancer but with several other forms of cancer as well. As mentioned, this factor alone is believed responsible for 30% of cancer deaths in the United States. Specific industrial chemicals have been identified as carcinogenic: 2-naphthylamine and benzidine in bladder cancer, asbestos in mesothelioma, vinyl chloride in a rare form of liver cancer. Measures to protect workers in industries that use these chemicals have been adopted—some as long as eighty years ago—and rates have diminished among them, but there is still controversy over where to set the limits of acceptable exposure. Part of the problem in setting standards is the difficulty of evaluating their effectiveness once they have been adopted—cancer is a disease that is slow to develop after exposure to a causative agent, often taking 10 to 50 years from initial exposure to the diagnosis of the disease. Another difficulty is that some cancers are caused by more than one agent. Bladder cancer in humans, for example, is known from epidemiological studies to be asssociated with smoking and also to follow exposure to any one of four or five different chemicals, and in laboratory animals it has been produced by many more. The maps prepared by the National Cancer Institute show excesses of bladder cancer among men in industrialized areas of the United States, and in fact several other cancers—mouth, throat, and esophagus cancer in men, and intestinal and rectal cancers in both sexes—show a similar pattern. But which of the enormous numbers of chemical substances present in these environments might be responsible, and to what extent? Areas of industrial development are also areas of high population density, which means more cars and trucks producing exhaust fumes, more lawns on which weed-killers and insecticides are sprayed, more power companies burning oil and coal to produce electricity, as well as more industrial pollution.

Last but not least is the economic cost of control measures. Setting tough standards for industries can be carried to a point that makes them uncompetitive. Factories close, and both profits and jobs are lost. We are forced into the painful position of setting a dollar value on human life. How many deaths from cancer per year can be tolerated to provide employment to 10,000 people?

The decisions are not easy, and it is not likely that we shall have the knowledge, the power, or the will to eliminate all causes of cancer from the environment in the immediate future. But we must continue research, in the clinic, the hospital, the laboratory, and among populations everywhere, in the hope that solutions to these problems will be found.

Putting the Question Differently

We would like to close this very brief overview of cancer epidemiology on a hopeful and instructive note.

The questions we have asked about life-style and environment have been focused entirely on what we are doing wrong: What are we eating or drinking or breathing in that is hurting us? These are important questions to ask, and some useful answers have already been obtained. But we could ask a different kind of question, and sometimes asking the right question is as important as getting the right answer.

Are there things present in the food which some of us eat and others do not, that *prevent* cancer rather than causing it, thus accounting for some of the differences in cancer rates? Some investigators have found evidence that vitamin A in the diet may play a role in the prevention of cancer—smokers with high vitamin A intake have been shown in one study to have less lung cancer than expected from their smoking histories. This observation has raised new possibilities for cancer prevention, and is being intensively studied.[19]

REFERENCE NOTES

1. George Rosen, *A History of Public Health* (New York: MD Publications, 1958).
2. Charles E. Rosenberg, *The Cholera Years: The United States in 1832, 1849, and 1866* (Chicago: University of Chicago Press, 1968).
3. Edwin Chadwick, *Report on the Sanitary Condition of the Labouring Population of Great Britain*, ed. M. W. Flinn (1842; Edinburgh: Edinburgh University Press, 1965).
4. Augustus De Morgan, *Budget of Paradoxes*, 2 vols., ed. David E. Smith (1915; facsimile edition, New York: Arno, 1977).
5. William Farr, *Vital Statistics: A Memorial Volume of Selections from the Reports and Writings of William Farr* (Metuchen, N.J.: Scarecrow Press, 1975), 343–351.
6. Richard Doll and Richard Peto, *The Causes of Cancer* (Oxford: Oxford University Press, 1981).
7. U.S. Department of Health, Education, and Welfare, *Smoking and Health. A Report of the Surgeon General* (Washington, 1979).
8. Doll and Peto, *Causes of Cancer*, table abstracted from Table 2, p. 1199.
9. Thomas J. Mason, et al., *Atlas of Cancer Mortality for U.S. Counties: 1950–1969* (Washington: U.S. Department of Health, Education, and Welfare, 1975).
10. Howard M. Hayes, Jr., Robert Hoover, and Robert E. Tarone, "Bladder Cancer in Pet Dogs: A Sentinel for Environmental Cancer?" *American Journal of Epidemiology* 114 (1981):229–233.
11. Doll and Peto, *Causes of Cancer*, 1201.
12. Bernardino Ramazzini, *Diseases of Workers*, trans. Wilmer Cave Wright (New York: Hafner, 1964).

13. Denis Burkitt, "The Discovery of Burkitt's Lymphoma," *Cancer* 51 (1983):1777–1786.
14. Harry Wain, *A History of Preventive Medicine* (Springfield, Ill.: Charles C Thomas, 1970).
15. William N. Rom, "Asbestos and Related Fibers," in *Environmental and Occupational Medicine*, ed. William N. Rom (Boston: Little, Brown, 1983).
16. Doll and Peto, *Causes of Cancer*, 1223.
17. U.S. DHEW, *Smoking and Health*, p. 7-11.
18. E. L. Wynder, and S. D. Stellman, "Impact of long-term filter cigarette usage on lung and larynx cancer risk. A case-control study," *Journal of the National Cancer Institute* 62 (1979):471–477.
19. Doll and Peto, *Causes of Cancer*, 1230–32.

SUGGESTED READING

Doll, Richard, and Richard Peto. *The Causes of Cancer: Quantitative Estimates of Avoidable Risks of Cancer in the United States Today*. Oxford: Oxford University Press, 1981.
Roueché, Burton. *The Incurable Wound*. Boston: Little, Brown, 1958.
———. *A Man Named Hoffman*. Boston: Little, Brown, 1965.
———. *The Orange Man and Other Narratives of Medical Detection*. Boston: Little, Brown, 1971.
———. *The Medical Detectives*. New York: Washington Square Press, 1982.
Snow, John. "On the Mode of Communication of Cholera." In *Snow on Cholera: Being a Reprint of Two Papers by John Snow, M.D.* New York: The Commonwealth Fund, 1936.
Terris, Milton, ed. *Goldberger on Pellagra*. Baton Rouge: Louisiana State University Press, 1964.

PROBLEMS

1. Give John Snow's explanation for the following observations:
   (a) In England cholera epidemics occurred primarily in the summer months while in Scotland they tended to occur any time of year.
   (b) Malaria tends to decrease when swamps are drained.
   (c) Removing the handle of the Broad Street pump had very little effect on the epidemic in its neighborhood.
2. Give two examples from the story of cholera of an experimental fact or observation successfully explained by a theory we now regard as incorrect.
3. Give two examples from the story of cholera of an experimental fact or observation that seemed to contradict a theory we now regard as correct.
4. What was Farr's observation on the relation between cholera rates in districts in London and their elevations above the Thames River? How did he explain it? How did Snow explain it?
5. State three different pieces of experimental evidence that support the hypothesis that cigarette smoking causes cancer.
6. State one example of an alternative explanation of one of the above pieces of evidence, which, if true, would have shown that the relation between cigarettes and lung cancer is fortuitous.
7. What kind of experiment is used to answer the question of whether differences in cancer rates among different groups of peoples such as ethnic groups, people living in different countries, etc., are genetic in origin?

8. Is the cancer death rate (deaths per year per million people) higher in India than in the United States or is it lower? Why?
9. Give examples of at least one form of cancer attributed to:
   (a) Life-style (other than lung cancer from smoking).
   (b) Exposure to industrial chemicals.
   (c) Exposure to a *natural* environmental hazard.

5
Mathematics

THE NATURE OF MATHEMATICS

As indicated in Chapter 3, logic is concerned with making valid inferences from a limited set of starting premises. The premises themselves are statements about certain entities—heroes, Virginians, mammals, football players, and so on—and relations between these entities. Any set of premises is associated with a particular subject matter; for example, the premises about mammals and goldfish are associated with a particular problem in biological classification.

If the starting premises are *mathematical* in nature, they, together with their logical consequences, constitute a *mathematical* system. The premises are then called axioms or postulates, and the statements logically deduced from them are called theorems. The concept "mathematical" is not easy to define, and is best illustrated by examples.

The essence of it is abstraction. Instead of dealing with individuals, we try to gain generality by dealing with classes. In going from individuals to classes we must ignore most of the features that the individuals possess, and pay attention to only a few that they have in common. The loss of detailed information is made acceptable by the greater generality of the relations we can demonstrate among them.

We can illustrate the procedure of abstraction by recalling the syllogisms we considered earlier. We recognized at that time that the three we initially discussed, the first about the mortality of heroes, the second about the citizenship of cows, and the third about the vogelness of pferds, were all of the same form and were all equally valid. We could have written the syllogisms more abstractly, using symbols:

1. All x's are y's.
2. All z's are x's.
3. Therefore all z's are y's.

We can see that any substitution for x, y, and z in the above gives a valid syllogism, so we now have reduced all syllogisms of this type to a single form. We have lost sight of heroes, cows, and pferds, but in turn, we have obtained a relation that includes these as special cases, and includes multitudes of other things as well.

Symbols and the Fear of Mathematics

We have mentioned in the preceding chapter the inordinate fear of mathematics felt by many. Teachers of mathematics and psychologists have concerned themselves with this problem, and written textbooks of mathematics designed precisely for the fearful. Our own feeling is that at least part of the problem is the use of symbols like x and y and $=$ for things and relations. It is intimidating partly because one is forced to deal with abstract entities instead of concrete realities like apples and shoes, and partly because the language of symbols is unfamiliar. To encounter the following collection of signs in the course of one's ordinary reading can also be intimidating:

העט של דודתי על השולחן

but it is only the Hebrew sentence, "The pen of my aunt is on the table."

The use of a symbol, x, to refer to a number whose magnitude is not yet known should not really be more intimidating than the use of the term "The Party of the First Part" in a contract.

The Rules of the Game

Mathematics began as a highly practical subject needed for counting, weighing, changing money, observing the stars, measuring areas of land. As time went on, it was found useful to deal not just with the measurements of this or that farmer's plot of land, but rather to invent such abstract entities as "straight line," "right angle," "rectangle," "circle," and so on, and relationships such as "equal to," "parallel," "greater than," and so on. These entities and relationships could be studied independently of the particular plots of land they referred to. Eventually, this line of development was codified in Euclid's treatise on geometry, in which an enormous number of theorems were deduced by valid logical procedures from a set of about a dozen postulates about the entities and the relationships among them.

We have noted previously the distinction between *validity* and *truth* in a chain of reasoning. Validity is concerned only with what is implied by the postulates or premises, whether or not they are true. This should suggest that one can construct mathematical systems starting with sets of postulates that either are untrue or have no connection with the real world at all. We have given earlier examples of valid syllogisms composed of false statements or statements about entities that seemed to have no meaning whatever: nachtigalls, vogels.

MATHEMATICS

For 2000 years after geometry was formalized in Euclid's treatise, no one doubted that it was both valid and true, and it was accepted that mathematics had to be both. But in the nineteenth century a change in viewpoint occurred. It was recognized that for a mathematical system, the real world need not matter. We are free to invent mathematical systems that are untrue to the world we live in, ones that do not relate to anything in the real world at all, or even ones that lie about it. For example, it was obvious to Euclid and the inventors of geometry that at a point off a given straight line one and only one straight line parallel to the first one can be drawn, and this statement was taken as a postulate of geometry, playing a crucial role in it. What if we pretend that there is a kind of universe where no line parallel to a given line can be drawn? We may believe that the statement is a false one, but it is a statement that has logical consequences, and a whole new "geometry" can be deduced from it. This geometry may not be a good one for measuring the areas of plots of land, but might be interesting to study just for fun, or even useful for some purpose other than land measurement. Whatever use we wish to put it to, or even if we do not put it to use at all, it can be as logically consistent as what we think of as the geometry of the real world. From this point of view, mathematical systems are games, like chess. People invent the game by specifying the rules. Within the rules, certain strategies are recognized as good, certain principles of play are accepted. But one can change the rules if one wants, or play a new game. One could limit the moves of the queen to a maximum of three squares in any direction, use a rectangular instead of a square board, or play checkers or bridge instead.

One may rightly ask what makes mathematics useful to science then, if it is just a game without any necessary connection to reality? Ironically, this is precisely what makes it useful. After all, in science, we are trying to discover which hypotheses about the real world work and which do not. If we needed a guarantee that some hypothesis is true before we explored its consequences logically, we would be paralyzed at the start. It is precisely the ability of mathematics to explore any world we choose to invent that gives it its power.

THE TRUTH OF MATHEMATICS

But is not Euclid's geometry true? Does it not describe exactly the relations of shapes, lengths, angles, areas, and so on, in the real world? The answer can be yes or no depending on what type of things in the universe we apply it to. If we measure the areas of small plots of land, it works within the accuracy of our ability to measure. If we look at large land areas, continents, for example, it fails. The reason, of course, is that the earth is round. If one objects that plane geometry was never meant to apply to a round earth, it might be answered that the original inventors of geometry were not aware of any limitation on its accuracy when applied to land areas. The experimental fact of the earth's roundness refutes at least their original belief in geometry's absolute truth. One might still object that one should apply plane geometry only to a plane; once we know that the earth is round we no longer expect geometry to work. But then the question arises, how do we know we have a plane? The answer is, unfortunately, that we often know this only if we find that plane geometry applies. The argument is circular. The "truth" of mathematics in the real world means no more than that we have found it to be applicable in some limited area, to within some acceptable error of measurement.

THE USE OF MATHEMATICS IN SCIENCE

To use mathematics in science, it is not strictly necessary that we frame our hypotheses initially in mathematical terms. Verbal statements have logical consequences also. But to apply mathematics the possibility must exist of reformulating the statements mathematically, so that the objects and their relations in the proposed hypotheses correspond to the entities and relations in a set of mathematical axioms. If the correspondence is precise, the logical consequences of our hypotheses are equivalent to the theorems validly deducible in the mathematical system. If experimentally we find that the equivalents of the mathematical theorems are not true in the real world, we know that one or more of the hypotheses we made are wrong. If, on the other hand, they are found to agree with experiment, we have made our hypotheses more probable (but have not proved them!).

Let us consider a simple example: counting. We begin by recognizing that many different kinds of objects can be counted. The realization that seven apples have something in common with seven deer and seven stones, and that we can abstract this common feature and study it by itself, was a great advance in human thought. Once it was made, the development of the whole number system was possible. It should be realized that not everything can be counted. If we count seven apples into a basket, then empty out the basket and count the apples again, there are still seven: they retain their identity. However, this does not happen to seven drops of water counted into a cup. Like plane geometry, counting has a limited range of applicability to the real world. Only certain types of objects can be counted in a way that satisfies the rules of arithmetic. For those objects that do, the system offers a tremendous economy. Modern life would be very difficult if we needed separate words and concepts for counting apples and for counting stones.

Simple arithmetic as used for counting does not inspire the fear that the rest of mathematics does; most people are used to ordinary numbers and feel they understand them.

We will show later that there are properties of the whole numbers of arithmetic that are surprising and not at all obvious: the reason for doing this is not to confirm the belief that mathematics must be difficult, but rather to suggest the richness of mathematical systems.

THE REASONS FOR MATHEMATICS

Granted that mathematics provides an alternate way of deducing the logical consequences of a scientific hypothesis, the reader still has a right to ask, Is it the only way, or the best way? Why must we use it, especially if we are not good in mathematics?

The reasons for preferring the mathematical route, *when it can be used at all*, can be summarized in a few words: *simplicity, economy of effort, generality, precision,* and *richness.* We will try to illustrate these properties by examples.

We have already used the whole number system as an example of simplicity and economy, and asked the reader to imagine the confusion and difficulties that would result if we needed different number words for each type of object we wished to count. The notion of mathematics as "simple" may come as a surprise to some. The fact is, however, that once you know it, mathematics is simpler than any other way of discussing its subject matter. In a very real sense, mathematics is a language, with its own vocabulary and grammar. Like a foreign language encountered for the first time, it sounds like gibberish, but it can be learned, and can become as familiar to the eye and ear as one's mother tongue.

Let us give an example of the *simplifying power of mathematics*. There is a well-known mind-reading trick that runs as follows: a "mind reader" asks you to pick a number between 1 and 10. Without telling him which you have picked, you are to double it, add 14, divide by 2, then subtract the number you started with. He then informs you, perhaps to your surprise, that the number you are left with is 7. You may confirm this easily for any number from 1 to 10. The various operations may be represented in the form of a table. Suppose you had picked 5 as your starting number:

Operation	Result
Pick a number	5
Double it	10
Add 14	24
Divide by 2	12
Take away the number you started with (5)	7

It is easy to confirm that the result is the same if we had started with 1, 2, 3, 4, 6, 7, 8, 9, or 10.

To understand how this trick works, it is helpful to use algebra. Algebra begins with the realization that while we cannot really double a number or add 14 to it and obtain a numerical result unless we know the number, certain sequences of arithmetical operations have results that are independent of whatever number we perform them on. For example, if we add 10 to a number, then subtract 3 from the result, it is the same as if we added 7. Similarly, if we multiply a starting number by 2, then multiply the result by 3, the result is the same if we had originally multiplied by 6. So there is a lot of arithmetic we can do even when we do not know what number we are talking about. Let us agree to call our chosen number y. Using this notation, we can analyze the mind-reading act.

Operation	Result
Take a number	y
Double it	$2 \times y$
Add 14	$2 \times y + 14$
Divide by 2	$y + 7$
Take away the number you started with (y)	7

The algebraic analysis of our problem shows how mathematics, in this case the use of algebraic notation, makes things simpler. Compare this explanation of how the trick works with how you would explain the trick using only ordinary language.

It also shows the feature of *generality*. We realize first that the restriction to starting numbers between 1 to 10 is unnecessary; it will work for any number whatever. But the generality provided by mathematics is even greater than this. Once we have caught on to the use of algebra, we can invent far more mystifying mind-reading tricks. Once we introduce y, we can carry on any series of operations, no matter how complicated, as long as we make sure at the end to eliminate y and end with a known number.

Economy of Effort

The number of mathematical systems so far invented is much greater than the number that have been used in the sciences. Further, many of these systems, especially those that have found some scientific use, have been extensively studied, and large numbers of theorems—valid logical consequences of the axioms—have been found. As science develops and new hypotheses are proposed, we find in the great majority of cases that if these can be expressed in mathematical form, the mathematics already exists. Only rarely is it necessary to invent a new mathematics in answer to a scientific need. This means in turn that once we find a way to express a hypothesis in mathematical language, the work of deducing its logical consequences is already done for us, and we need only look it up. This is what is meant by *economy of effort:* there is no need to do the work twice.

Precision

Our scientific theories, since we want to use them to predict what will happen in the future, are more useful to us to the extent that they are *precise*. The prediction that a 10-kilogram stone dropped from a height of 25 meters will develop a speed great enough to seriously injure anyone on whom it falls is an important and useful one, but the prediction that it will have a speed of 22.1 meters per second after falling 25 meters is even more useful. Mathematics, although not exclusively concerned with numerical magnitudes, provides the best language for talking about them. The above figure for the speed of fall was obtained from the physical law that the speed of fall, s, in meters per second, at any point along the path of fall of a body dropped near the earth's surface, is given approximately by

$$s = 4.43 \sqrt{h}$$

where h is the distance that the body has already fallen, measured in meters. This equation displays also the simplicity and generality of mathematical formulas.

However, as is true with most laws of science, its range of applicability is limited. It will not apply to any body whose size, shape, or speed of fall is such that air resistance affects its motion appreciably. It does not describe the fall of a feather, or even of a heavy body like a stone dropped from a great height, as eventually the stone will be falling so fast that air resistance becomes important. Another limitation is that it applies only near the earth's surface. Farther away, the force of the earth's gravity is less, and hence the speed of fall is less than that given by the formula. Most scientific laws have such limits on the conditions under which they apply: if the conditions are not met, the predictions of the formula become inaccurate to a lesser or greater degree. The formula can be regarded as an idealization which no real falling body obeys exactly, but to which many falling bodies conform within reasonable limits of accuracy. Within those limits, the formula describes the speed of anything, dropped from any height.

It would be hard to imagine any non-mathematical way to predict the enormously large number of possible velocities resulting from dropping objects from various heights.

Another Kind of Precision

Another facet of this feature of precision is that it requires us to formulate our hypotheses with more care. To convey this idea we would like to provide a partly imaginary example of the difference between qualitative, verbally formulated laws and quantitative, mathematically formulated laws. It again concerns the laws obeyed by falling bodies.

We are all familiar with the fact that when an object is dropped, its speed increases as it falls. If a rock must be dropped on our foot, we prefer it to be from 2 centimeters rather than 25 meters: the difference is dramatic. We can express the increase of speed in a number of ways, including the following two:

Statement (A): The farther the object has fallen, the faster it is moving.
Statement (B): The longer the time the object has fallen, the faster it is moving.

These statements are qualitative rather than quantitative. Neither says anything about *how much*. Both are in fact true, and there is little to choose between them as laws descriptive of falling bodies.

Now if we want to express our laws mathematically we will find that we are forced to be more precise. From the mathematical point of view, phrases like "the farther," "the faster," and "the longer" are not easy to deal with. Our very desire to use a mathematical formulation demands that we say *how much* faster? *How much* farther? and so on.

Mathematics versus Words

We will try to express statements (A) and (B) in mathematical language, using a little algebra.

We will use equations that describe the relation between two quantities (or variables), and will deal mostly with equations of simple proportionality. When we say that *y is proportional to x*, we mean simply that if x doubles, y doubles, and that if x is tripled, so is y, and so on. A simple example is the relation of distance traveled to time, for a car going at a constant speed. The distance is *proportional* to the time driven: we go twice as far in 2 hours as in 1 hour. Note that this is equally true for cars traveling rapidly as for cars traveling slowly. The speed does not matter, so long as it is constant: the proportionality holds for any constant speed.

To express this result mathematically, we write

$$d = st \qquad (1)$$

where d is the distance travelled, and t is the time. The constant speed of the car is s, which can be different in different cases. It is easy to see from the equation that doubling t doubles d, and so on. So the equation describes simple proportionality.

In any relation of simple proportionality between two quantities a constant always appears. We call s in equation (1) a "constant" even though in different situations its value may be different. At times we go 60 miles an hour for a while, at other times we go 10 miles an hour for a while. In either case the distance is proportional to the time travelled. The equation $y = bx$ represents proportionality just as does $d = st$, and $y = bx$ implies that as long as b remains the same, doubling x doubles y and so on. Of course if we are interested in a time period in which the speed of the car varies, it will no longer be true that the car will travel twice as far in twice the time.

There can be many mathematical relationships between d and t other than simple proportionality. We can list a few:

MATHEMATICS

$$d = \frac{a}{t} \quad (2)$$

$$d = at^2 \quad (3)$$

$$d = at^3 \quad (4)$$

In relation (2), doubling t cuts d in half, and tripling t cuts d to one-third of its value. In relation (3), doubling t quadruples d, tripling t increases d 9 times. In relation (4), doubling t increases d 8-fold, tripling t increases d 27-fold. Relations (2), (3), and (4) are thus not relationships of simple proportionality between d and t.

Quantitative Laws

Now let us return to the speed of our falling body, and try to reformulate our laws mathematically.

The statement (A), that the body falls faster the farther it has fallen, asserts that the speed s depends on the distance d fallen, and is no longer constant: it increases as d increases. But there are obviously an enormous number of possible relationships that describe this. For example,

$$s = ad$$
$$s = ad^2$$
$$s = ad^3$$
$$\ldots$$
$$\ldots$$
$$\ldots$$
$$\text{etc.}$$

All of these equations, and many more, are consistent with the qualitative statement (A). We cannot proceed unless we make one choice or another. Our wish to use mathematics forces us to be *precise* about the hypotheses we make.

Let us try simple proportionality as a start:

$$s = ad \quad (5)$$

If we wish to formulate a mathematical statement analogous to statement (B), we are in the same difficulty. Again we will begin with simple proportionality:

$$s = bt \quad (6)$$

So we have replaced the qualitative statements (1) and (2) with equations (5) and (6), which were suggested by statements (1) and (2) but are clearly not equivalent to them. Equations (5) and (6) are only special, possible cases of (A) and (B). In the above equations, a and b are constants. Their values do not change while the variables s, d, and t are changing all the time.

We noted that both (A) and (B) are found by observation to be true. We have not yet said whether (5) and (6) are true, but we can show easily that (5) and (6) are *inconsistent:* they cannot *both* be true. They might, as far as we know, both turn out to be false.

To show the inconsistency we note very simply, that if

$$s = ad \quad \text{and} \quad s = bt$$

simultaneously, then

$$ad = bt.$$

This is an equation that relates distance to time. Since a and b are constants, it is a relation of simple proportionality. Dividing both sides by a, we have

$$d = \frac{b}{a} t \qquad (7)$$

where the ratio of the two constants b and a is obviously also a constant. Equation (7) is thus one of simple proportionality between distance and time. But if distance is proportional to time, the speed of fall must be a *constant*. In fact, equation (7) could have been written

$$d = st$$

where the constant speed, s, takes the place of the ratio b/a. Since the direct observation we started with was that the speed is *not* constant, there is a contradiction. *Equations (5) and (6) cannot both be true.*

Whether either is true can only be answered by experiment. When we do the experiment, we find, as did Galileo, that equation (6) is obeyed to reasonable accuracy and equation (5) is not.

HOW MANY PRIME NUMBERS ARE THERE? AN EXAMPLE OF MATHEMATICAL REASONING

It takes some experience with mathematics to learn that seemingly simple ideas often have far-reaching and unexpected consequences. This is what we meant by referring to the *richness* of mathematics.

Some understandable and striking examples arise from the use of the ordinary whole numbers. We can all add, subtract, multiply, and divide. We perform the processes in a mechanical way, and most of us do not expect surprises, but the surprises are there. We know we can add and multiply any two whole numbers, but we cannot always subtract or divide, unless we allow negative numbers or fractions. Let us take a closer look at division. *Any* number can be divided by one with no remainder, and can also be divided by itself with no remainder.

$$143 \div 1 = 143$$
$$143 \div 143 = 1$$

Some numbers can be divided without remainder by other numbers as well: 12 can be divided by 2, 3, 4 and 6. A number like 12 is called *composite*. Others which cannot be so divided, such as 3, 7, or 17, are called *prime*.

How many prime numbers are there? Certainly as a number gets larger the chance that it will be divisible by some smaller number other than 1 increases. After all, the larger a number is, the more numbers smaller than it there are. One can determine whether a number is prime or not by trial and error, and make lists of prime numbers, but the larger the number we look at, the harder it is to tell if it is prime or not. In Table 5-1 we give a list of the prime numbers less than 100, and some information aout how many prime numbers there are that are smaller than 2000. When we look at this list, we do indeed find that they tend to be rarer the larger we go. We can conceive of two possibilities: either that the *number of primes* is infinite—in other words, no matter how large the numbers we look at,

Table 5-1
Prime Numbers

Prime numbers less than 100:

2 13 31 53 73
3 17 37 59 79
5 19 41 61 83
7 23 43 67 89
11 29 47 71 97

There are thus 25 primes less than 100.
Between 100 and 200 there are 21 primes, so the number between 0 and 200 is 46.
Between 200 and 400 there are 32.
" 400 and 600 " " 31.
" 600 and 800 " " 30.
" 800 and 1000 " " 29.
" 1000 and 1200 " " 28.
" 1200 and 1400 " " 26.
" 1400 and 1600 " " 29.
" 1600 and 1800 " " 27.
" 1800 and 2000 " " 25.

we will always find more primes—or else that the *number* is finite—that is, the list of primes comes to an end somewhere with the largest prime number of all.

Is the *number of primes* finite? While this question has a simple answer which we will give shortly, the reason for asking it is not for the intrinsic interest in the answer. From the point of view of scientific applications, it is not of much importance whether the number of primes is infinite or not.*[1] We ask it in order to show that such simple operations as multiplication and division of whole numbers can lead to complex and interesting questions, the answers to which are not at all intuitively easy to guess.

How in fact are we to find out whether the number of primes is finite? Do we have to continue to examine larger and larger numbers to see if we find no more primes after a given prime is reached? Clearly, this could not be a convincing proof, as we can never examine *all* numbers to see if they have divisors. We would run out of patience and time with some large number, never being sure that the next larger number, which we failed to examine, will not turn out to be prime.

In fact, the question can be answered by a simple logical argument that does not require going on forever. To answer it, we will ask the reader to accept on faith two theorems about composite numbers. Before we state the theorems, we note that among the different ways a composite number can be written as a product of smaller numbers multiplied together, there is at least one way that uses only prime numbers. Twelve can be divided by 2, 3, 4, and 6; it can be written as $2 \times 6, 3 \times 4,$ and $2 \times 2 \times 3$. The first two examples use the composite numbers 6 and 4; only the last one uses only prime numbers. The number 10 can be written only as 2×5, involving only prime numbers.

The theorems which we will ask the reader to take on faith are easy to understand and intuitively plausible. The first states, *There is only one way to factor a composite number into primes*. (We do not count the *order* of primes here; of course, 12 can be written as $2 \times 3 \times 2$ or $3 \times 2 \times 2$, but, whatever order we write the primes, there are only two 2's and one 3.)

The second theorem states, *If a number is divisible by a composite number, it is divisible by the prime numbers that when multiplied together make up the composite*. For example, since 96 is divisible by 12, it must also be divisible by 2 and by 3. This theorem in turn implies that *If a number is not divisible by any prime number (other than itself or 1) it is not divisible by any other number either; hence it must also be a prime number*. If these theorems are accepted as true, we can proceed.

Our answer to the question *Is the number of primes finite?* will be obtained as follows:

1. We will assume that the answer is "yes."
2. Then we will show that this assumption leads to a logical contradiction.
3. So we will conclude that the answer is "no."

---

* It has become important, however, in methods of constructing unbreakable codes for communications systems. See reference note 1 at the end of the chapter.

Let us assume that the number of primes is finite, so that there are only N prime numbers. Let us write the primes in order starting with 2, so that the sequence is 2, 3, 5, 7, . . . , P_N, where P_N, the Nth in line, is the largest prime number. (Remember, we do not know it yet; and what is more, we intend to prove that it does not even exist!) Now let us multiply all the primes together to produce a new number, q.

$$q = 2 \times 3 \times 5 \times 7 \times \ldots \times P_N$$

q is obviously divisible by 2, by 3, by 5 . . . , and by P_N, so it is not a prime.

But now consider the number $q + 1$. If we divide this by 2, the result is $q/2 + 1/2$. Since q is divisible by 2, therefore $q/2$ is a whole number, so the result of dividing $q + 1$ by 2 is a whole number + the fraction 1/2. Clearly $q + 1$ is not divisible by 2.

Now let us divide $q + 1$ by 3. The result of dividing $q + 1$ by 3 is $q/3 + 1/3$. But q is clearly divisible by 3 to give a whole number. So the division of $q + 1$ by 3 gives a whole number + the fraction 1/3. So $q + 1$ is not divisible by 3, either.

A little thought shows that the same reasoning applies in turn to 5, 7, and so on up to P_N. We will always obtain a whole number plus a fraction upon division.

We conclude that the number $q + 1$ is not divisible by any prime number. If $q + 1$ is not divisible by any prime number, it itself is prime. Now obviously $q + 1$ is larger than P_N. But this contradicts our assumption that P_N was the largest prime. Hence the assumption that a largest prime number exists cannot be true. *Therefore there must be an infinite number of primes.*

MATHEMATICS WITHOUT QUANTITIES—THE BRIDGES OF KOENIGSBERG

Another example of the simplicity and power of a mathematical argument is the following[2]: The German town of Koenigsberg is located on the river Pregl, at a point where two branches of the river join to form a single branch. The two branches upstream from the confluence are joined by a channel, creating an island. In the eighteenth century the various parts of the city were connected by seven bridges as shown in Figure 5-1. In those days the citizens of Koenigsberg, to pass the time on Sundays, used to go for walks over the bridges in an attempt to find a path that crossed each of them once and once only. It was not necessary to end at the starting point. By the time this problem attracted the attention of the mathematician Euler, not only had no one ever found such a path but also a belief was growing that it was impossible.

Euler addressed himself to this problem with the objective of either finding a path or proving that no such path existed. His approach began, as all mathematics begins, with the attempt to abstract from the real situation, with all its complexities and irrelevancies, a simpler but equivalent structure. It needed to be simple so that

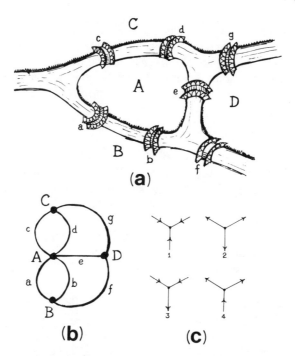

FIGURE 5-1. The bridges of Koenigsberg. (a) Schematic map of Koenigsberg. The bridges are denoted by small letters and the land areas of the city by capital letters. (b) An abstraction of the above map. The bridges are shown as lines, and the land areas are designated as points. (c) Walks leading in and out of three-fold vertices. Number 1 shows three walks entering a vertex and none leaving: it is impossible to do this while taking a walk. Number 2 is also impossible. Number 3 represents a vertex which has been entered twice and left once: it must represent the end of a walk. Similarly, number 4 must represent the starting point of a walk.

it could be solved, and equivalent in the sense that a solution to it was a solution also to the real problem.

Consider Figure 5.1a, in which the bridges are designated by small letters a, b, c, d, e, f, g and the land parts of the city by capital letters A, B, C, D. The areas of the land masses, the distances of the bridges from each other, and the widths of the rivers at various places would all seem to be details that are irrelevant to the problem we are trying to solve. Let us assume this, which permits redrawing the figure in such a way as to eliminate them. We replace land areas A, B, C, D by points and the bridges a, b, c, d, e, f, g between the land areas by lines connecting the points, to produce Figure 5-1b. The reader should compare Figures 5-1a and 5-1b carefully, to convince himself that except for distortion of the distances involved Figure 5-1a is the same as Figure 5-1b. Further—and this is the essence of applying mathematics to a physical problem—the reader has to decide for himself that the abstraction is close enough to the reality so that a solution of one is a solution of the other. Is the problem of finding in Figure 5-1b a path that runs over each line once and once only the same as the problem of taking a walk over the seven bridges of Koenigsberg that crosses each bridge once and once only?

Now let us consider any of the points A, B, C, and D, which we will call "vertices." Coming together at each vertex are a number of lines, each representing a bridge to be crossed. Suppose there exists a walk that traverses each bridge once. In such a walk each bridge is traversed in one and only one direction. This implies that each bridge may be marked by an arrow whose head points in the direction of the walk. At any one vertex, say C, the lines emanating from it therefore bear arrows pointing either into the vertex or out from it. C has three lines coming from it, so that the arrows showing the direction of the walk can have the following configurations (see Figure 5-1c):

Case 1: 3 lines can point inward.
Case 2: 3 lines can point outward.
Case 3: 2 lines can point in and 1 out.
Case 4: 2 lines can point out and 1 in.

We realize at once that all three lines cannot point in the same direction, because the walker cannot enter the land mass C three times without having left it once, nor can he leave it three times and never have come into it. If his walk started at C, there must be an outward arrow to show which bridge he left it by. Then he must return before he leaves it again, so there must be one inward arrow and two outward arrows. If his walk started elsewhere, he must arrive at C the first time (inward arrow), then leave to continue his walk (outward arrow). *Hence cases 3 and 4 are the only possible ones.* But this argument proves one more thing. At C, the number of outward arrows is either *one more* than the number of inward arrows or *one less*. In the first case the walker must have started his walk at C; in the second, he must have ended it there. Note that this is not true for a vertex with an even number of lines (bridges). Merely knowing that there were (say) two inward arrows (the walker arrived twice) and two outward ones (he left twice) would not allow us to conclude that he started his walk there. He may have, or he may not have.

So far we have looked only at vertex C, and concluded that because it is odd the walk must have either started or ended there. We note the same reasoning applies to the equivalent vertex B. But as we go on to A and D we note that they are odd also, with five and three bridges, respectively. This means that the walk must either start or stop in each of four places. *But since a walk can start in one place and one only, and end in one and one only, no walk at all is possible.* The problem is solved.

The above reasoning has shown that a walk is impossible if there are more than two odd vertices. This of course applies not only to Koenigsberg, but to any city built on rivers with bridges over them. Again, the mathematics has shown

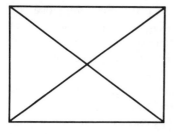

FIGURE 5-2. Can this figure be drawn without lifting the pencil from the paper or retracing any lines?

generality: we have solved the problem not only for Koenigsberg but for all other cities as well.

We have shown that our reasoning applies to any city in the world. It solves other problems also. Can Figure 5-2 be drawn without lifting the pencil from the paper and without retracing lines? A quick glance shows it has four odd vertices (each having three lines joined). The answer is no.

We have also solved the problem of finding a walk that ends at the same point it started: in this case there must be *no* odd vertices. We must end where we began: for each departure there must be a return, hence an even number of bridges at each vertex.

Notice that in the above analysis, the quantities we normally associate with mathematical reasoning—the area of the city of Koenigsberg, the distance from one point to another on the path, the angles between the entering and departing paths at a vertex—were all irrelevant. All that matters are the basic connections of the path. It would be possible to draw the map of the city and its bridges on a sheet of rubber, and then deform the sheet in all sorts of ways by pulling at its edges, and Euler's result would still be just as true. The branch of mathematics that deals with those properties of systems that remain the same in spite of distortions of distance and shape is called topology.

CALCULATION AND MATHEMATICS

The story of the bridges of Koenigsberg shows us two ways to solve mathematical problems. One is the direct, or "brute force" way: calculate what has to be calculated, divide the problem up into an exhaustive list of alternatives and explore each alternative in turn. The second is to look for some mathematical technique that bypasses all the tedious number-crunching that the first way requires. The citizens of Koenigsberg, on their Sunday afternoon outings, were following the first procedure—or would have been if they had kept careful records of all the different unsuccessful walks they took, and eventually showed that every *possible* attempt had failed. Obviously Euler's way—of standing back from the problem, abstracting the general features that matter from the messy details that do not, and then solving the problem simply and once and for all—is better. It is better because it takes less time, and it is better because it is beautiful, but it has one drawback. That such an approach will succeed can never be known in advance, and there are times when we need the answer right away and cannot wait for an ingenious mathematician to devise a shortcut.

The problem is a common one for scientists, particularly in the physical sciences, where important and useful information can often be obtained from tedious computational operations. Do we need the information now, or should we wait? For the scientist the choice may be a hard one, but for the mathematician there is no problem. The *raison d'être* of mathematics is to do just what Euler did. Most mathematicians are no less bored by adding up long columns of figures than the rest of us. They do not consider it their job, and are usually annoyed when non-mathematicians assume that it is.

The point may be illustrated by two episodes in the life of Karl Gauss (1777–1855), one of the greatest of mathematicians. Gauss was born a poor boy, the son of a bricklayer, in Braunschweig, Germany. The schoolmaster in the local school Gauss attended, a certain Herr Büttner, was a hard taskmaster who gave his classes practice in arithmetic by asking them to add up long sequences of large numbers. For his own convenience, so that he would not have to do the tedious arithmetic involved to check his pupils' almost invariably erroneous answers, the sequences of numbers he assigned his classes to add were chosen to form what is called an arithmetic series—the successive numbers in the long list differed by a constant amount. For example, the series 11, 14, 17, 20, 23, 26 is such a series, in which each term increases by 3. Büttner then made use of a well-known formula for the sum of such a series: the sum is equal to the number of terms times one-half the sum of the first and last terms. For the series given above, the sum is

$$6 \times \left(\frac{11 + 26}{2}\right) = 111.$$

In any event, Büttner wrote on the blackboard a list of large numbers forming such a series, and after finishing turned around to face the class, expecting as usual to have a free hour or so while his pupils sweated and struggled, to find little Gauss handing in his slate with the correct sum written out. Gauss had recognized the numbers as forming an arithmetic series, figured out on the spot the formula for the sum, and calculated it. Büttner, to his everlasting credit, though no mathematician himself, knew one when he saw one. With his own money he bought Gauss the best textbook on arithmetic then available and brought the boy's abilities to the attention of people who could help him in his career.

When Gauss was 24 years old, the minor planet Ceres (one of the asteroids) was discovered. Ceres was small, and, with the telescopes of those days, difficult to find. From the point of view of astronomy, and of the confirmation of Newton's laws of planetary motion, it was desirable to use data available on its position to calculate its future orbit. Gauss, with his amazing facility for arithmetic computations, took on the job. He was able to deduce Ceres's orbit successfully, receiving recognition from mathematicians and astronomers all over Europe. This led to his appointment in 1807 to the position of director of the Observatory in Gottingen. A major part of his professional life from his 24th year onward was spent on such calculations, which, on the one hand, no one in Europe could do as well as he, and, on the other, can be done with greater accuracy by a modern computer in an hour or so. The loss to mathematics by what we can in retrospect regard as an extraordinary waste of Gauss's valuable time can hardly be estimated.[3]

REFERENCE NOTES

1. Martin E. Hellman, "The Mathematics of Public Key Cryptography," *Scientific American*, August 1979, pp. 146–157.
2. Euler's article is reprinted in Scientific American, *Mathematics in the Modern World: Readings from Scientific American* (San Francisco: W. H. Freeman, 1968).
3. Eric T. Bell, *Men of Mathematics* (New York: Simon and Schuster, 1937).

SUGGESTED READING

Gardner, Martin. *Martin Gardner's New Mathematical Diversions from Scientific American*. New York: Simon and Schuster, 1966.
———. *Mathematical Magic Show: More Puzzles, Games, Diversions, Illusions and Other Mathematical Sleight-of-Mind from Scientific American*. New York: Vintage, 1977.
———. *Aha! Insight*. New York: Scientific American/W. H. Freeman, 1978.
Jacobs, Harold R. *Mathematics: A Human Endeavor*. 2nd ed. San Francisco: W. H. Freeman, 1982.
Kline, Morris. *Mathematics for Liberal Arts*. Reading, Mass.: Addison-Wesley, 1967.
Newman, James R. *The World of Mathematics*. New York: Simon and Schuster, 1956.
Scientific American. *Mathematics in the Modern World: Readings from Scientific American*. San Francisco: W. H. Freeman, 1968.

PROBLEMS

1. Invent another "mind-reading" trick like the one described on pp. 77–78.
2. In the following relations, is y proportional to x?
    (a) $y = 1/x$
    (b) $y = x + 1$
    (c) $y = -4x$
    (d) $y = \sqrt{x}$
    (e) $y = 0$

MATHEMATICS

3. Find the first two prime numbers larger than 100.
4. Is the number formed by multiplying any number of prime numbers together and adding 1 to the product a prime number? Test your conclusion using the list of prime numbers given on p. 83.
5. If you live in a city or town with rivers and bridges, make a schematic map, and find a "Koenigsberg walk" in it, or show that there is none.

6

Is Heat a Substance?

THE CONFLICT BETWEEN THE CALORIC AND KINETIC THEORIES OF HEAT

What Is Heat?

The scientific study of heat, like much else in science, has its roots in the most primitive experiences of daily life. One of the first things we learn is to recognize the difference between hot things and cold things, and for the rest of our lives that difference remains important. It is not surprising that the ancient Greeks, among other peoples, engaged in speculations about what it is that makes hot things hot. According to one of their theories, fire is one of four basic elements—the others being earth, air, and water—that make up all substances. We know today that this theory is wrong, that the elements are not earth, air, fire, and water, but oxygen, hydrogen, carbon, iron, and so on. The old theory sounds absurd to us now. But the reader, if asked how we know that oxygen is an element and fire is not, might be hard put to answer. The four-element theory seemed to its followers to give order and coherence to a confusing world, which is one of the conditions any scientific theory has to satisfy. One purpose of this chapter is to explain why we no longer believe that fire is an element.

There were by the eighteenth century—the period when the discoveries we will discuss were being made—some additional reasons besides natural curiosity for an interest in heat. First, there were important practical considerations. The industrial revolution was beginning, and engines which used heat to do work had been invented. Improvements in these engines were constantly being made, though without much help from any scientific theory of heat. A second reason was the gradual recognition of the crucial role played by heat in the various processes taking place in the earth's atmosphere. Winds are created when the heat of the sun warms the air near the ground; the air expands and rises, and air from colder areas rushes in to take its place. The sun's heat also evaporates water from oceans, rivers, and lakes; the water vapor rises, and later, when the air containing it is cooled, falls as rain or snow. Without the heat received from the sun, the earth would be a much quieter place, with no changes of the seasons or alternation of day and night, no rain or snow, no thunderstorms or hurricanes. The tides would still continue, and volcanoes would erupt from time to time, but not much else would happen on the time scale of human life.

The Caloric Theory

By the end of the eighteenth century, the French chemist Lavoisier had established the science of chemistry in a form we still recognize as basically correct. His discoveries followed from a basic concept he held about matter: *matter can neither be created nor destroyed.* Although chemical and physical changes—combustion, oxidation, dissolution, boiling, and freezing—may change the form and appearance of substances, the total amount of matter, as measured by weight, does not change.

Lavoisier was the first to describe combustion as a combination of oxygen in the air with various substances, and he identified a large number of the chemical elements: oxygen, hydrogen, nitrogen, sulfur, carbon, iron, and so forth. To this list he added the element *calorique,* the substance of fire and of heat. The view that heat is a substance was not a new one; it can be recognized as a restatement of the idea of fire as one of the elements. What was new, perhaps, at this time of the beginning of the atomic theory of matter, was the idea that heat, like other matter, may also be composed of atoms or particles. Not everyone who believed that heat was a substance believed it to be atomic; for some, heat was an indestructible fluid not necessarily composed of individual particles.

Since heat seemed to move around more freely than other kinds of matter, escaping, for example, through the walls of a container in which a hot body is placed, it was conjectured that its particles must be light and mobile, a hypothesis that explains not only why heat flows easily, but also why hot bodies do not seem to weigh much more or much less than cold bodies. Since matter can easily be warmed, it was concluded that the atoms, or fluid, of heat are attracted to the atoms of the ordinary chemical elements. The expansion of heated bodies was explained by either the hypothesis that this additional matter surrounds the ordinary atoms and increases the space between them, or by the hypothesis that the substance (or atoms) of heat is self-repelling—the latter hypothesis having been introduced to explain why heat flows so easily. (The quick fading of a red-hot piece of iron vividly demonstrated this ready flow.) Since the *substance* of heat was called *caloric,* the theory describing its properties is referred to as the *caloric theory.*

The Kinetic Theory

There was, however, another theory: heat was not a substance, but just a motion of the heated body. By some it was conceived as a kind of vibration of the body as a whole, much like the vibration of a tuning fork, but at a frequency too

IS HEAT A SUBSTANCE?

high for our ears to hear, and perceived only by the sensation of touch. By others, it was believed to be a motion of the atoms of which the body was composed, a much more chaotic motion than that of a tuning fork, one in which the individual atoms bounce back and forth in a continual series of collisions with their neighbors. The motion theory, or *kinetic theory,* was not quite so old as the theory that heat is a substance, but it was based on a very ancient observation: the production of heat by friction. Plato had stated, "For heat and fire . . . are themselves begotten by impact and friction: but this is motion. Are not these the origin of fire?"[1] In Elizabethan times Francis Bacon recognized the production of heat by frictional processes, such as hammering on an anvil, and concluded that "The very essence of heat . . . is motion and nothing else."[2] Galileo, Newton, and the chemist Robert Boyle (1627–1691) also held this view, as did the philosopher John Locke, who reasoned from the production of heat by friction: "The axle-trees of carts and coaches are often made hot, and sometimes to a degree, that it sets them on fire, by the rubbing of the naves [hub or axle] of the wheels upon them." He concluded that "heat is a very brisk agitation of the insensible parts of the object, which produces in use that sensation from whence we denominate the object hot. . . ."[3]

The Usefulness of the Wrong Theory

We will not pretend to maintain suspense as to the outcome of this controversy. The reader probably knows that the second view, that heat is atomic motion, has won out. The focus of our discussion is not on the conclusion itself, but on why and how we have reached this conclusion.

The caloric theory, *although wrong,* served a very important purpose in its time. A number of the significant properties of heat were discovered with its help. There are many other examples from the history of science of incorrect theories that led to useful results. We have described in a previous chapter how, before the establishment of the germ theory of disease, the spread of contagious diseases was attributed to "effluvia," emanations of toxic matter from the sick or the dead that can travel through the air. Because people concerned with public health believed this theory, they took sanitary measures to isolate the sick, create effective systems of garbage and waste disposal, and improve ventilation in factories and hospitals, all of which cut down the spread of disease.

Today, more than a century after the triumph of the kinetic theory of heat, we still use the phrase *flow of heat.* Liquids and gases flow: flowing is a behavior of substances. It is much harder to think of motion "flowing." The phrase is a legacy from the caloric theory, and it tells us that even though heat is the motion of atoms, many of its phenomena can be visualized in a very natural and simple way by thinking of it as a fluid substance.

At the end of the eighteenth century, the caloric theory was the prevailing one. There were good reasons for this: at that time it explained much of the then known properties of heat, while the kinetic theory seemed to explain very little. We will consider some of these properties.

Summary of What Is to Come

Although the material of this chapter has been written for the reader who has no prior training in the physical sciences, we do not expect that all such readers will necessarily find it easy going. For this reason we shall summarize here what is to come, so that some sense of the overall purpose can be maintained, and the reader will not be swamped by details.

In the next section, "Measuring Hotness," the concept of temperature is introduced. We learn how an instrument, the *thermometer,* was invented to substitute for and make quantitative our subjective impressions of what is hot and what is cold. We see further how an important principle of nature was discovered with the aid of this instrument: if bodies at different initial temperatures are brought together, they tend toward a state of uniform temperature; cold bodies warm up and hot bodies cool down until all are at the same temperature.

In the section "Heat and Heat Capacity," the necessity of a distinction between *temperature* and *heat* is pointed out. We recognize this distinction when we compare the heating effects of a small hot body and a large hot body. The two can be at the same starting temperature, yet the large body as it cools can warm its immediate neighborhood more than the small body. Temperature alone is insufficient to describe the effect of a hot body on its environment; its size matters also, and a new concept, heat, must be introduced. In the eighteenth century, as we noted, heat was regarded as a substance, and as such it could neither be created nor destroyed, but only moved from one place to another. It was said to be *conserved.*

The concept of *heat capacity* was introduced by Joseph Black, who noted that the heating effect of a hot body depended not only on its temperature and how much of it there was, but also on the chemical identity of the substance. The heat capacity of mercury is different from that of water, and that of copper differs from both. In order to define the heat capacity, Black assumed that heat was conserved. Since the concept of heat capacity proved useful, and enabled the results of various experiments to be correctly predicted, it supported the idea that heat is conserved, and hence the theory that heat is a substance, i.e., the caloric theory.

In the section "Latent Heat," we discuss a further application of the principle of the conservation of heat to such processes as the melting of solids and the boiling of liquids. Black discovered that melting and boiling take place with a large absorption of heat, but with no rise in temperature. These phenomena could be rationalized by the caloric theory, so they were regarded as giving it additional support.

In the next three sections, we turn to the work of Count Rumford, who was a believer in the kinetic theory of heat—that heat was a kind of internal motion of a hot body, and not a substance at all. In the section "Does Heat Have Weight?" we describe Rumford's attempt to answer this question. It would seem reasonable to expect that if heat is a substance, it should weigh something. Rumford performed a very careful experiment making use of the observations of Black on latent heat, and concluded that heat had no weight. This experiment failed to convince the

IS HEAT A SUBSTANCE?

believers in the caloric theory that heat was not after all a substance, and we discuss the reasons why.

In the section "Heat from Friction," we describe Rumford's most famous experiment, his observations on the production of heat by friction when boring cannons for the Bavarian army. Rumford's experiments seemed to show that an indefinite amount of heat could be produced by friction, which he interpreted as disproving the principle of conservation of heat—the foundation of the caloric theory.

In the section "Atomic Motion," we describe an experiment by Rumford which tends to show that in a liquid seemingly at rest, the atoms are actually in constant motion. The experiment is, in retrospect, one of the most convincing demonstrations of the correctness of the kinetic theory, but Rumford himself, because he was not very confident about what the experiment proved, did not press his point strongly. In this section, we attempt to give the reader a qualitative picture of just what kind of motion heat actually is, by considering the motion of atoms in gases, liquids, and solids.

In the last section, "Why Caloric Survived," we discuss why Rumford's experiments, which seem so convincing today, failed to convince his contemporaries, and why 50 years had to elapse before the kinetic theory was accepted. There were several reasons. Rumford's experiment on friction was qualitative rather than quantitative. He showed that friction produces heat, but was unable to describe quantitatively how much heat was produced from a given amount of motion. This was done 50 years later by Joule. Further, Rumford's explanation of how heat can be transmitted across empty space was not in accord with the theories then held about the nature of light and radiation. When a more correct theory of light was established, Rumford's ideas on heat radiation could be recognized as sound, and the kinetic theory of heat was accepted.

MEASURING HOTNESS

Making Things Quantitative

As we noted earlier, the problem of heat begins with our subjective perception of hot and cold. A lot of keen observation and deep understanding can be built on such simple and common experiences. But it is also true that sometimes a much deeper understanding can be obtained if we try to make our observations quantitative rather than qualitative, to be able to say not just "much hotter" but rather "four times hotter" or "100 degrees hotter."

The question of how to invent a quantitative measure for some intuitively-perceived concept is fundamental to all science. If we can talk about things numerically, we can think more precisely, and differences and nuances of behavior show up with greater clarity. The powerful logical processes of mathematics can illuminate the subject. Today, as a matter of course, we think of speeds in *miles per hour,* compare automobiles by their gasoline efficiencies measured in *miles per gallon,* and recognize that the *efficiency* in miles per gallon varies with the speed in miles per hour.

Let us turn to the problem of a quantitative measure for "hotness," and begin by looking at how hotness was perceived before the thermometer was invented. Francis Bacon, in his book *Novum Organum,* proposed a systematic procedure for studying nature scientifically.[2] As an example of his method, he analyzed the nature of heat. He made a list of situations in nature where heat was present, and a second list where heat was absent. His first list included the following:

1. The rays of the sun
2. All flame
3. Liquids boiling or heated
4. All bodies rubbed violently
5. Wool, skins of animals, and down of birds
6. Aromatic and hot herbs (which Bacon noted do not feel warm to the touch but "burn" the tongue)
7. Strong vinegar and all acids (which "burn" the eye, tongue, or other sensitive parts of the body)

We recognize today that some of these really are examples of the presence of heat, while others are not. Wool, for example, keeps a person warm, but not because it is in any sense "warm" itself. We still speak of "warm" clothing, but we know that such clothing is warm because it keeps the heat of the body in; it *insulates*. A wool blanket placed over a block of ice will keep it from melting rapidly on a warm day, but the wool doesn't "cool" the ice any more than it "warms" the person who wears a wool sweater. Bacon obviously relied on the purely sensory quality of hotness, and confused subjectively similar but quite different sensations. It is easy to criticize his mistakes, but we are in his debt for his view that nature should be studied by experiment and can be rationally understood by so doing. He himself was aware of the dangers of relying on sensory observation alone; he knew that whether something feels warm or cold depends not only on its hotness, but also on our skin's exposure to warm or cold bodies immediately before.*

To make a quantitative measure of anything, one needs to choose a standard. This had been done thousands of years earlier for such quantities as length and weight. For length, two marks an arbitrary distance apart were made on some object, and this distance was given a name such as a "foot" or a "meter." Then other distances could be measured by comparison with this standard distance. Having established a measure, we no longer had to rely on purely sensory concepts for distance. So the time it takes to walk a certain distance is a poor quantitative

---

* The reader can confirm this by a well-known experiment: Prepare three bowls of water, one cold, one lukewarm, and one hot. Place one hand in the cold water and one hand in the hot water. After a few minutes, place both hands in the lukewarm water. It feels both hot and cold at the same time.

IS HEAT A SUBSTANCE?

measure compared to the yardstick, though it is a useful piece of information for someone who wants to take a walk. How tired one gets or how long it takes to walk a certain distance may depend on the terrain, or on how tired one was to begin with.

Weight was done in much the same way; an arbitrary object was chosen and defined to be "1 pound," or "1 kilogram." Then it became possible to use a scale or balance to compare the weights of other objects to our standard.

Thermometers

Most things expand when heated. This is true for all gases and almost all liquids and solids. Galileo was the first to use the fact of this expansion as a way to measure hotness. In about 1592, he built a simple instrument using the expansion of air, which was employed by a physician friend of his to measure fever in his patients. During the next century, the use of liquids instead of air made instruments working on this principle easier to manufacture and use. Thermometers today do not differ much from those manufactured in Florence, Italy in the seventeenth

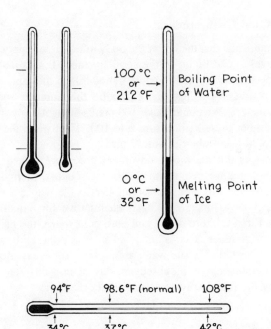

FIGURE 6-1. Thermometers. The thermometers with the larger bulb will show a greater change in height of the mercury column for a given change in temperature, as shown by the two thermometers on the upper left. Similarly, a narrower diameter of the bore will also give a large change in height for a small temperature change, as in the fever thermometer, bottom.

century. A bulb of glass containing the liquid (water, alcohol, or mercury) is attached to a long capillary (narrow-bore) tube of uniform cross-section (Figure 6-1). After most of the air has been pumped out, the other end of the capillary tube is sealed off to keep the liquid from evaporating and to keep dirt out of the instrument.

To use a thermometer, we bring it into contact with some body, for example, water in a kettle, and note the height of the liquid column in the capillary tube. If we heat the water in the kettle until we can feel by touch that it is hotter, and again bring the thermometer into contact with the water, we shall find that the liquid in the capillary tube rises higher than before. Our instrument thus passes a very important test: when "hotness" increases, it tells us so.

The change of height of the liquid in the tube depends on the size of the bulb attached to it at the bottom of the thermometer, as well as the narrowness of the tube itself. Fever thermometers, for example, which measure only a small range of temperature (95–110° F or 35–43° C—we will explain these degrees later), are made with very narrow capillaries. We quickly find that the thermometer indicates smaller differences in hotness than our senses can perceive. This is one advantage of having made an instrument to replace our sensory impressions.

Standardization and Calibration

We have established that the hotter the body to which we apply a thermometer, the higher the level of the liquid in the capillary tube. But to make our measure quantitative we need a standard. We could define this in the same way we defined the foot or the pound, by taking one particular thermometer and calling it the "standard thermometer." We could make 100 marks along its length, every quarter-inch, say, and number these marks from 0 to 100. Then when the height is at the 47th mark we would have a hotness of 47 units.

However, this is not the most convenient procedure. It requires comparing every new thermometer we manufacture with the standard thermometer. Instead, we can make use of an interesting observation possible with a thermometer even before we decide how to make reading it quantitative (or numerical). If a thermometer is placed in a mixture of ice and pure cold water, the height of liquid in the thermometer will reach a constant level and stay there. It will not change appreciably so long as both ice and water are present. If the experiment is repeated with the same thermometer on a subsequent day or in a different laboratory, the height reached by the liquid column is essentially the same.

A *Qualitative Measure*

This gives us a reference point: any other body can be described as hotter or colder than melting ice. But it is not yet a quantitative measure. It does not give us a numerical answer to the question, how much hotter is boiling water than ice?

There are other processes in nature that also maintain a constant temperature.

IS HEAT A SUBSTANCE?

The melting of solids in general shows this feature; for example, the melting of butter was once proposed as a reference point. However, the melting point of butter depends to some extent on how the butter was prepared or which cow the milk came from. The boiling point of water is more nearly constant, so it was chosen as a second reference point.*

A Quantitative Measure

We now have two reference points, the ice point and the boiling point of water, to which we give arbitrary numerical values. In the Celsius (Centigrade) system, the ice point is called 0 and the boiling point is called 100. The distance on the thermometer between 0 and 100 is divided into 100 equal parts called Celsius or Centigrade degrees. This is the system used for ordinary purposes in Europe, for example in weather reports and to measure fever; it is also used in scientific laboratories everywhere. The Fahrenheit scale was chosen with the ice point at 32 and the boiling point at 212. The distance between these numbers is divided into $212 - 32 = 180$ equal parts. Thus a Fahrenheit degree represents a smaller difference in temperature than a Celsius degree. Both choices are equally arbitrary. Marking these degrees on a thermometer is one example of what is called "calibration."

It is obvious that all thermometers, regardless of their dimensions and the liquid in them, will agree at 0° C and 100° C. They have to, because we defined things that way. What about in between? When a thermometer using mercury as the liquid reads 50° C, halfway between ice and boiling water, will an alcohol thermometer also read 50° C? Careful measurement shows that the two thermometers do not agree exactly, but the difference is small and requires careful measurement. In the eighteenth century this difference was not noticed.

Does the Thermometer Measure "Hotness"?

Does a thermometer always agree with our subjective feelings of hot and cold? We find that the correspondence is sometimes good and sometimes not. For example, how cold we feel on a winter day depends not only on the temperature as defined by the thermometer, but also on the force of the wind, and to some extent on the dampness. One can feel much colder on a windy 30° F day than on a still day at

---

* There is a difficulty: the boiling point varies with the atmospheric pressure. Water boils at a lower temperature on a mountain-top than at sea level. (This observation can be made with a thermometer even before we choose a numerical scale for temperature.) Since atmospheric pressure varies from day to day even at sea level, we have to correct for this variation. But the cheapness and ease of purification of water makes it worthwhile to put up with the inconvenience. We choose the boiling point of water when the atmospheric pressure is equal to the *average* value at sea level. This requires us to be able to measure atmospheric pressure in order to measure temperature; how we do that is another story.

10° F. As an indicator of how we should dress in winter, the thermometer, in spite of its numerical precision, does not always answer our purpose.

When we compare the hotness of different samples of a single substance such as water, the thermometer works. If *this* water feels hotter than *that* water, the thermometer rises higher in *this* water. The same is true if we examine two pieces of copper, or two rocks. But if we compare different materials it fails again. Anyone who has groped under the bed with bare feet for a pair of fur slippers on a cold winter morning knows that the slippers feel warm and the floor feels cold. But a thermometer brought in contact with the slippers and then with the floor would read the same. We are making the same mistake as Bacon when he said that wool, the skins of animals, and the down of birds are "warm."

It is clear then that the thermometer does not always agree with our sensations of hot and cold. We have several choices at this point. We can decide that the thermometer has failed, that it does not correctly measure the property it was designed to measure, and that we should try to develop some other instrument. Or we can conclude that the feelings of hotness and coldness have to be studied through our senses only, and that an attempt to quantify these sensations through the use of an instrument like the thermometer will necessarily miss their real essence. Or we can decide that the thermometer really is measuring something important, and that we will therefore stick with it, hoping ultimately to find some satisfying explanation of why some bodies feel warmer than others even though the thermometer tells us that they are at the same temperature.

The thermometer won acceptance for a number of reasons. One experimental observation that helped is the following: if equal weights of ice-cold water at 0° C and boiling-hot water at 100° C are mixed, the resulting warm water is found by the thermometer to have a temperature of about 50° C—"halfway" between the arbitrarily defined 0° C water and 100° C water. In general, two equal weights of water of different temperatures will when mixed reach a temperature halfway between the two starting temperatures. Other liquids behave in the same way.

These results are what we would have intuitively expected. We would have been surprised if our invention had given a different answer, and might have discarded it and looked for other ways to define hotness. Yet there was no logical basis for being certain in advance that the thermometer would give a reading halfway between the readings for the hot and cold water. The fact that it did led scientists to trust its usefulness.

The Equilibrium of Heat

An even more important discovery was made with the thermometer, one *not* expected on the basis of commonsense notions. If a number of objects, some hot and some cold (as determined by the thermometer) are placed in close proximity, they will all in a short time come to the same temperature. Those that were hot initially will have cooled; those that were cold will have warmed. The final tem-

IS HEAT A SUBSTANCE?

perature, uniform throughout the whole system, will be somewhere between the hottest and coldest initial temperatures of the separate objects. This tendency to reach a final uniform temperature is independent of the nature of the bodies: they can be made of the same substance or of different substances, and can be any combination of solids, liquids, or gases. After a uniform temperature is reached, nothing further happens; the temperature remains the same. Joseph Black (see p. 106) termed this state of affairs the *equilibrium of heat*.

Note that these objects need not all *feel* equally warm after reaching this state of equilibrium—remember the slippers and the cold floor. Generally, when metals are hotter than normal body temperature, they feel hotter than materials such as wood or wool that are at the same temperature as the metals, and when metals are cold, they feel colder. If we had trusted only our sensations, we would never have discovered the equilibrium of heat. Even without any theory of what heat is, it seems that the thermometer is telling us something very important about heat and its properties.

Science and Quantification

Making quantitative comparisons rather than qualitative ones is important in science, and has enabled us to make tremendous progress in some fields. There are some who feel that this is essential to science, and that no field of study deserves the name of science unless its concepts can be described by numerical measures. This is a view we consider too narrow, as will become obvious in other chapters. But there is no doubt that the physical sciences could hardly exist without such procedures. The thermometer has certainly helped make possible the successful study of heat.

Yet there are risks and uncertainties in quantification, as the reader may now understand. Arbitrary decisions were made in inventing and calibrating the thermometer. We might have done things in a hundred different ways. Some would have been equivalent or nearly the same, others would have been very different. We were lucky in the choices we did make, but even so we had to make compromises, and to accept the fact that the thermometer does not always agree with the subjective sensations that led us to invent it in the first place.

There are other examples of quantification in science that have not worked so well, or whose merits are matters of dispute. The intelligence quotient (IQ) is an example. The IQ test represents an attempt to provide a quantitative measure of an intuitive concept—innate intelligence. We all feel, on the basis of our experience with people, that some people are smarter than others, that they learn and think more quickly. It also seems reasonable that what people know, and how well they can solve problems, depend on some combination of innate ability and the results of training and experience.

The IQ test was designed to measure the innate component independently of training and experience. But does it really do this? Is it even possible to separate the two factors? Psychologists who study intelligence differ strongly among themselves about just what the IQ test actually measures, and some are very skeptical of those precise numbers, like 93 or 117, that are assigned to people on the basis of the test.[4]

Exact and Inexact Sciences

The sciences that have successfully made use of such quantitative instruments as thermometers—such as physics and chemistry—have been called the "exact" sciences, in contrast to those sciences—such as the social sciences—which have not so far been able to devise quantitative measures for most of their concepts.

There is no question that quantification is responsible for some of the success the exact sciences have had, but it is important to understand that the term *exact* can be misunderstood, if by it we mean mathematical exactness.

Any measurement we make with scientific instruments is of limited accuracy—the limits being set by the nature of the instrument, the property being measured, the patience of the investigator, and ultimately the laws of nature.

The geometrical theory of circles gives the length of the circumference as 2π times the length of the radius ($C = 2\pi r$). The number π can be calculated to as many decimal places as we have patience for—the latest computer results carry it to several million. But neither the length of the radius nor of the circumference can be measured with anything near such accuracy. Even when circles are drawn with sharp pencils and good compasses on high quality drawing paper, we can determine such lengths only to a few hundredths or thousandths of a centimeter. An experimental test of the theory of circles could corroborate it only to within such limited accuracy.

Now of course as methods of measurement improve, we can measure quantities to additional decimal places. When we do so, it is often discovered that theories that may have worked adequately to the fourth decimal place do not work to the sixth. This does not always mean that we must discard the theory. Usually we keep it unless there is a better theory to take its place. Instead of discarding it, we may content ourselves with the statement, "This theory is the best we have, but it works only to the fourth decimal place."

All scientific theories bear such a qualification. There are a few that work as precisely as we have been able to measure: the geometrical theory of circles works this well for the circles we draw on paper. With others, the best theories we have are not as exact as we can measure. The rain that falls during a day can be measured to 1/100 of an inch, but the meteorologist is happy if a heavy rain can be predicted a day in advance, without worrying about whether it will be a half inch or three-quarters of an inch. Einstein's theory of relativity has replaced Isaac Newton's earlier theory because it agrees better with experimental results. However, it is only under special circumstances—bodies traveling at very high speeds—that the ac-

IS HEAT A SUBSTANCE?

curacy with which we can measure such things as masses, times, and speeds permits us to detect the errors in Newton's laws. We will return to this question in Chapter 14.

The principle of the equilibrium of heat, that bodies placed in contact come to the same temperature, can be verified at best within the limited "exactness" of thermometers. The best mercury-in-glass thermometers can be read to 0.001° C. To the extent that certain instruments based on different principles are able to measure temperature a little more accurately than this, the principle of equilibrium can be further confirmed.

HEAT AND HEAT CAPACITY

The Invention of "Caloric"

Once we know that when different bodies at different temperatures are placed near enough to each other the hot one cools down and the cold one warms up, it is natural to conclude that "something" is being transferred between them.

Think of two blocks of iron, each with a thermometer attached, one at 100° C and the other at 0° C. When the blocks are placed in contact, one thermometer begins to fall and the other to rise. Is some substance (heat) leaving the hot block for the cold one? Is some substance (cold) entering the hot block? Could it be even more complicated, with transfers occurring in both directions? The mere observation that the thermometers, initially different, tend toward equality provides no clue.

In thinking about this problem people have tended to believe, almost surely for psychological reasons, that "something" goes out of hot bodies into cold ones. While some experiments seem to suggest a principle of coldness that can leave a cold body and enter a warm one, anyone who has boiled water over a fire, or warmed himself at one, is more likely to think of a principle of heat than of cold. This is one time that the intuitive guess has turned out to be scientifically correct, as we will see.

The eighteenth century was a period of increased scientific investigation of a number of phenomena—light, heat, electricity—all of which were thought of as substances. Because these substances moved around more freely than even the most fluid kinds of ordinary matter and because they could penetrate ordinary matter easily and rapidly (think of light traveling through a hard, solid piece of glass, or electricity traveling through a metal wire), they were described as "subtle fluids." The atomic theory of matter was being developed during this time, so it was natural for many scientists to think that these subtle fluids too were composed of atoms, but especially light ones. Others did not adopt such a specific picture of what a "subtle fluid" was.

Conservation of Heat

There is one important consequence of adopting a theory that heat is a substance.

The view at the time was that matter has one essential feature: it can neither be created nor destroyed. For example, wood is a complicated mixture of chemical compounds of the elements carbon, hydrogen, and oxygen, with small amounts of other elements. When wood is burned, its original compounds disappear, and water vapor and gaseous carbon dioxide are produced. But all the carbon originally in the wood is now in the carbon dioxide (CO_2), and all the hydrogen is now in the water (H_2O). If the wood and the air needed to burn it are weighed at the start, and if all the products (ashes, water, carbon dioxide, and unused air) are collected at the end of the combustion, it is found that the total weight of these products is the same as that of the air and wood. The individual elements are said to be "conserved."

According to the caloric theory, heat is also conserved. There is only so much of it in the universe, and if you want to accumulate it somewhere—say, to warm a kettle of water—you must get it from somewhere else, such as the fire you put the kettle over. The idea that heat is conserved proved very fruitful, in that it could account for a large number of experimental observations.

Joseph Black

The caloric theory was developed through the work of a large number of individuals, but it received considerable clarification and a definitive statement by the Scottish chemist and physician Joseph Black (1728–1799). Black, a professor at the Universities of Glasgow and Edinburgh, never published the results of his researches, though he described them in his lectures. Fortunately, his own lecture notes and notes of his students were prepared for publication by one of his colleagues at Edinburgh, John Robison.[5] Black himself was cautious as far as theory was concerned. He discussed both the caloric theory and the theory of heat as the motion of atoms, tending toward the caloric theory, but was relatively dispassionate in his appraisal of both. The reasons for his preference will be discussed as we go on.

Black was one of the first to recognize the importance of the tendency toward equalization of temperature in hot and cold bodies placed together: the "equilibrium of heat."

Heat versus Temperature

Another of Black's major contributions was in distinguishing between the concept of temperature and that of heat. The need for two separate concepts, heat and temperature, can be made clear by a simple argument. Imagine a furnace heated to red heat, containing two pieces of iron, one a small nail weighing a few grams

IS HEAT A SUBSTANCE?

and one a big chunk weighing several kilograms. Both are glowing at a red heat, and one might suspect that both are at the same temperature; direct measurement indeed shows that they are.

But the nail dropped in a pail of cold water will cool quickly without heating the water much. The large piece of iron, on the other hand, may heat the water enough to make it boil. Although the two pieces of iron are at the same temperature, *their respective abilities to heat a certain amount of water are not at all the same.* If one explains the changes in temperature when hot and cold bodies are brought together as a transfer of heat from the hot body to the cold body, the large piece of hot iron must have had more heat to lose.

One can distinguish therefore between the *intensity* of heat, as measured by the temperature of a body, and the total *amount* of heat it contains. This distinction is much like the one we use when referring to gases, between the pressure of a gas and the amount of the gas: an automobile tire and a bicycle tire may be at the same pressure and yet contain very different amounts of air.

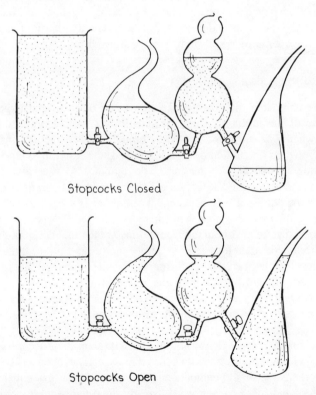

FIGURE 6-2. Water seeks its own level. The level of the water in the vessels acts like temperature if given a chance (by opening the stopcocks). Heat capacity is analogous to the capacity of the vessels.

A more cogent analogy to temperature is the principle "water seeks its own level." A large tank of water, containing water at a level of 10 feet, can be connected to a number of smaller vessels (Figure 6-2). When the faucet is opened, water rises in all of the smaller vessels to a height near 10 feet, and the water level in the tank falls slightly. The final height at equilibrium is the same in the tank and all the vessels it is connected to, but the *amount* of water in each vessel clearly is not: the bigger the diameter of the vessel, the more water it contains. The height of the water is like temperature; the volume of water in each vessel is like heat.

We have used this picture not only to make the distinction between temperature and heat easy for the reader to visualize, but also to show how readily the phenomenon of heat can be described by picturing heat as a substance. Today we know that heat is not a fluid but rather the motion of atoms of matter; however, we find it a lot harder to explain to a nonscientist what happens at temperature equilibrium in terms of the motion of atoms. In the eighteenth century, very little was known about atoms: how they moved, what they weighed, what their structure was. It was difficult then to draw any useful consequences from the idea that heat was atomic motion. Later, we will quote an argument Black gave against the kinetic theory, exactly on the grounds that it did not seem to explain as plausibly as the caloric theory how the atoms of which matter is composed take up or give off heat.

What Will the Final Temperature Be?

We have compared the heating effect on water of small and large pieces of hot iron. One of the problems Black dealt with was that of predicting the final temperature when *different* substances were brought into contact. We have already mentioned that when hot and cold water are mixed, the resulting temperature is an average: 1 kilogram of 100° C water and 1 kilogram of 0° C water give 2 kilograms of 50° C water. What about 1 kilogram of hot iron and 1 kilogram of cold water? Is the final temperature halfway between? The answer is a dramatic NO.

Early experiments were done with mercury and water, since mercury was a convenient substance to handle in the laboratories of the time; because it is a liquid, it can be stirred, and therefore reaches temperature equilibrium quickly. It was found that the final temperature on bringing together equal weights of the two liquids was very little different from the initial temperature of the water. For example, if the water was at 0° C and an equal weight of mercury was at 100° C, the final temperature after bringing them together was only about 3° C rather than 50° C.

Now mercury is a very dense liquid. One kilogram of it occupies only 1/13 the volume of a kilogram of water. It occurred to the early proponents of the caloric theory that perhaps we should compare equal volumes rather than equal weights of the two liquids, and indeed when we do so, the final temperature is more nearly halfway between the starting temperatures. But as Black pointed out, more accurate examination shows that even when equal volumes are compared, the final temperature is closer to that of the water.

IS HEAT A SUBSTANCE?

Here is Black's description of his reasoning:

> It was formerly a common supposition, that the quantities of heat required to increase the temperatures of different bodies by the same number of degrees, were directly in proportion to the quantity of matter in each [and thus to their weights]. . . . But very soon after I began to think on this subject (*anno* 1760), I perceived that this opinion was a mistake, and that the quantities of heat which different kinds of matter must receive, to raise their temperature by an equal number of degrees, are not in proportion to the quantity of matter in each, but in proportions widely different from this, and for which no general principle or reason can yet be assigned. This opinion was first suggested to me by an experiment described by Dr. Boerhaave in his *Elementa Chemiae* [1732]. . . . Boerhaave tells us, that Fahrenheit agitated together quicksilver [mercury] and water [of initially different temperatures.] From the Doctor's account, it is quite plain, that quicksilver, though it has more than 13 times the density of water, produced less effect in heating or cooling the water to which it was applied than an equal volume of water would have produced. He says expressly, that the quicksilver, whether it was applied hot to cold water, or cold to hot water, never produced more effect in heating or cooling an equal measure of the water than would have been produced by water equally hot or cold with the quicksilver, and only two-thirds of its bulk. He adds, that it was necessary to take three measures of quicksilver to two of water, in order to produce the same middle temperature that is produced by mixing equal measures of hot and cold water. . . . This shows that the same quantity of the matter of heat has more effect in heating quicksilver than in heating an equal volume of water, and therefore that a smaller *quantity* of it is sufficient for increasing the sensible heat of quicksilver by the same number of degrees. . . . Quicksilver, therefore, has less *capacity* for the matter of heat than water (if I may be allowed to use this expression) has; it requires a smaller quantity of it to raise its temperature by the same number of degrees. [Black's italics]

One should pay particular attention to Black's words: "the quantities of heat which different kinds of matter must receive, to raise their temperature by an equal number of degrees, are not in proportion to the quantity of matter in each, but in proportions widely different from this, and for which no general principle or reason can yet be assigned." He thus introduced the concept of *heat capacity*.

The Capacity for Heat

Science is among other things a search for simplicity and order; it would have seemed a simpler universe if, after being placed in contact, equal weights of any two substances originally at different temperatures always came to a temperature equilibrium halfway between their starting temperatures. But the experiments performed to test this expectation immediately revealed glaring contradictions. It would have seemed an equally simple universe if equal volumes of different substances behaved in this simple way. This turned out to be closer to the truth, at least when water and mercury were the substances compared. However, as Black realized, there is still a big enough discrepancy to cause a critical person to reject this alternative also. No two substances, whether compared by weight or by volume, have the same capacity for heat. The universe is less simple than we might have liked. Black's decision was to accept this lack of simplicity as a problem he could not yet solve: the capacity for heat of different substances was not predictable by any simple theory known to him. Let us therefore measure them, Black concluded, and hope that some day we will develop theories that will provide understanding of why they have the values they do.

Amount of Heat and Capacity for Heat

In Black's approach, the concept of the conservation of heat played an essential part. In this view, when a substance is placed in contact with colder bodies, the heat it loses must all go to those colder bodies. None can disappear. All the heat lost by the mercury on cooling from 100° C to 3° C is gained by the water, warming it from 0° C to 3° C. Further, in this view, if the water is then cooled again to 0° C, the same amount of heat that entered it from the warm mercury must be lost to some colder body.

In the above, we have by implication given a more precise meaning to the concept "amount of heat." Again we have found it necessary to choose a quantitative measure for an intuitive concept. We cannot weigh an amount of heat, or measure it out in a gallon jug. So we *define* it by the temperature rise it produces in a substance. If 1 kilogram of water rises 3° C in temperature, it has gained a certain amount of heat. If it rises 6° C, it has gained twice the amount. We may express this in an equation for the amount of heat, Q:

$$Q = C \times (t_2 - t_1)$$

where t_2 and t_1 are the final and initial temperatures, respectively, so that $t_2 - t_1$ is the *change* in temperature. C is a constant whose value we have not yet determined. However, this is a matter of our own choice; we can assign it any numerical value we want. It is as arbitrary as calling the boiling point of water 100 degrees Celsius. Unfortunately, the capital letter C will have two meanings throughout the remainder of this chapter: as a symbol for the Celsius degree of temperature and, in italics, as a symbol for specific heat. We hope the context will make clear which meaning is meant.

A convenient choice for C was, first, to choose water as the standard substance; second, to use 1 kilogram as the standard amount; and, third, to set C for water equal to 1. *This means that a unit amount of heat is defined to be the amount of heat that causes the temperature of 1 kilogram of water to rise 1 degree Celsius.* The name chosen for this unit was *Calorie*. It is familiar to us from its use to specify the energy value of foods.* In England the unit of heat was defined as that quantity

---

* Specifically, to say that a biscuit has 50 Calories means that if the biscuit (which chemically is composed primarily of carbon, hydrogen, and oxygen) is burned in enough additional oxygen to convert it all to carbon dioxide and water—this is just how the body metabolizes food—the heat produced is enough to raise the temperature of 1 kilogram of water by 50 degrees Celsius, or to raise the temperature of a 50-kg human being (110 pounds of mostly water) by 1 degree Celsius.

IS HEAT A SUBSTANCE?

that raises the temperature of 1 pound of water 1 degree Fahrenheit. This is called the *British thermal unit* (Btu) and is used for rating home heaters and air conditioners (1 Calorie = 3.9685 Btu).

Now let us return to the experiment where 1 kilogram of mercury at 100° C and 1 kilogram of water at 0° C are brought into contact and come to temperature equilibrium at 3° C. The mercury has cooled 97° C; the water has been warmed 3° C.

The quantity of heat lost by the mercury, Q_m, is by the conservation principle the same as that gained by the water, Q_w.

The constant C for water has been chosen to equal 1; we will use the symbol C_w for C to remind the reader that it is a constant associated with water. Mercury will have a different constant, which we denote by C_m:

$$Q_m = Q_w$$
$$C_m \times (100° \text{ C} - 3° \text{ C}) = C_w \times (3° \text{ C} - 0° \text{ C})$$

or:

$$C_m \times 97 = C_w \times 3$$

We see that the constant C_m for 1 kilogram of mercury must be only about 1/30 of the constant C_w for water. Since we chose C_w to be equal to 1, then $C_m = 0.03$ (3/97 is 0.031; for this discussion 0.03 is sufficiently accurate).

The constant C is what Black meant by the term "capacity for heat." As noted by him, it is different for each substance in nature, and the only way we can know it is to measure it by the procedure sketched above.

The modern term for the capacity for heat of a fixed amount of a substance—we use one kilogram as the standard amount—is *specific heat*. In modern usage the heat capacity of a body is its specific heat multiplied by its mass.

An Experimental Test

Now we can use our definition of specific heat, together with the principle of conservation of heat, to predict the final temperature when any substances at different initial temperatures are placed in contact and allowed to come to temperature equilibrium. The specific heat of copper, for example, is found to be 0.09 by an experiment comparing it to water; this is three times the specific heat of mercury. If 1 kilogram of mercury at 100° C and 1 kilogram of copper at 0° C are placed in contact, we would predict from the above approach that the final temperature would be 25° C.* We do the experiment, and the result agrees with the prediction.

---

* The final temperature, which is unknown, we call t. The change in temperature of the mercury is $100 - t$, and of the copper is just t, as it started at 0° C. The respective quantities of heat are as follows:

$$\text{Lost by mercury} = C \text{ (mercury)} \times (100 - t) = (0.03) \times (100 - t)$$
$$\text{Gained by copper} = C \text{ (copper)} \times t = (0.09) \times t$$

Since these are equal, by conservation of heat, the final temperature must be 25° C.

This justified to Black and the scientists of the time the concept of the conservation of heat and the caloric theory on which it was based.

Heat Capacity: Caloric or Kinetic Theory?

As mentioned, heavy substances like mercury and iron have much smaller specific heats than lighter substances like water. Black, assuming (correctly) that the individual atoms of the heavier substances were heavier than those of the lighter substances, found this hard to understand if heat is to be regarded as motion. He argued, plausibly, that the heavier an atom was, the more effort would be needed to get it moving a given amount, so that heavier substances should have higher specific heats.

Though the caloric theory did not provide any explanation of why heavier substances should have smaller rather than larger specific heats, at least it did not seem to predict the opposite of the experimental facts—so the heat-capacity concept gave support to the caloric theory in this way also. It took more than 50 years before the atomic theory of matter, which in Black's time was in its infancy, could be developed to the point where Black's criticism could be answered.

Practical Consequences of Heat Capacity

The concept of heat capacity as presented above may be difficult to understand at first, but it can be related to some common experiences, as can the concept of temperature. The specific heat of a substance is a measure of how much that substance resists a change in temperature. Substances of high specific heat require *more* heat to warm them (increase their temperature by a given number of degrees) and, conversely, give out more heat on cooling, than do substances of low specific heat. Water, the substance chosen as the standard for defining the unit of heat, the Calorie, happens to have one of the highest specific heats of any known substance. This fact has a number of important practical consequences.

One of the simplest is that water is the best substance to fill a hot water bottle with. When we use a hot water bottle we do not want its temperature too high, but we do want it to give out as much heat as possible while it cools. Water is the best substance to use because it has a high heat capacity. Mercury, aside from the fact that it is expensive and toxic, gives out only 0.03 times as much heat as water, kilogram for kilogram.

One of the problems in using the sun as a source of energy is its intermittent and irregular character. It shines by day and not by night, and less of its energy reaches the earth's surface on cloudy days. To make solar energy economically practicable, various methods of storing the energy overnight or for periods of several days have been proposed. One such proposal is to let the sun heat up large quantities of water, which is cheap, easily stored and circulated, and of course has a high specific heat.

IS HEAT A SUBSTANCE?

LATENT HEAT

Melting Ice

We have distinguished two concepts: temperature and heat. In the caloric theory, heat is a substance that flows from hot bodies (bodies at a high temperature) to cold ones, and it continues to do so until all the bodies are at the same temperature.

It is clear that to make the temperature of a body rise (in the absence of chemical reactions like combustion taking place within it) heat must be added to it from some other body. Does it follow that if heat is added to a body, the temperature *must* rise? The question sounds almost silly, but the following two statements are not logically equivalent:

1. To raise the temperature, heat must be added.
2. If heat is added, the temperature must rise.

It was one of Black's discoveries that the first statement does not logically imply the second and that the second is not always true.

We mentioned earlier the use of the melting point of ice as a fixed point for calibrating thermometers. This works as a good calibration point because the temperature of a mixture of ice and water remains constant regardless of the relative amounts of ice and water in the mixture. It does this in a warm room—the laboratory. The cold vessel containing the ice–water mixture is certainly receiving heat from adjacent objects in the laboratory, a fact we can show by applying a thermometer to the nearby objects; they will be cooled down below their original temperatures. Yet the *temperature* of the mixture does not rise while heat is being added to it! A well-stirred mixture of ice and pure water remains at 0° C until all the ice is melted. Only then does the temperature start to rise, eventually reaching that of the room. This is why ice cubes cool a drink and keep it cool, and it explains how an old-fashioned icebox works.

The fact that ice takes in heat from the environment as it melts without itself rising in temperature was discovered by Black. It had been previously thought that when ice was placed in a warm room its temperature would rise continuously until the melting point of 0° C was reached, at which point the ice would immediately melt, and the resulting water would continue to rise in temperature (see Figure 6-3).

Here is Black's description of his discovery:

> Fluidity [melting] was universally considered as produced by a small addition to the quantity of heat which a body contains when it is once heated up to its melting point; and the return of such body to a solid state as depending on a very small diminution of the quantity of its heat, after it is cooled to the same degree; that a solid body, when it is changed into a fluid, receives no greater addition to the heat within it than what is measured by the elevation of temperature indicated after fusion [melting] by the thermometer; and that, when the melted body is again made to congeal [solidify] by a diminution of its heat,

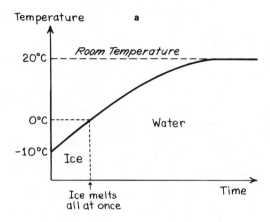

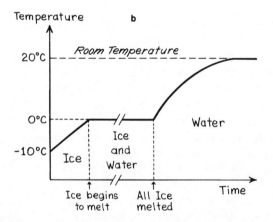

FIGURE 6-3. Latent heat. The upper figure (a) shows how the temperature of ice was believed to change as it melted. The lower figure (b) shows how it really behaves. Ice has a large latent heat, and takes a long time to melt. For this reason, the horizontal axis that represents time elapsed is shown with a break during the period when the ice is only partially melted.

it suffers no greater loss of heat than what is indicated also by the simple application to it of the same instrument.

This was the universal opinion on this subject, so far as I know, when I began to read my lectures in the University of Glasgow, in the year 1757. But I soon found reason to object to it, as inconsistent with many remarkable facts, when attentively considered; and I endeavored to show, that these facts are convincing proofs that fluidity is produced by heat in a very different manner.

The opinion I formed from attentive observation of the facts and phenomena is as follows. When ice, for example, or any other solid substance, is changing into a fluid by heat, I am of opinion that it receives a much greater quantity of heat than what is perceptible in it immediately after by the thermometer. A great quantity of heat enters into it, on this occasion, without making it ap-

IS HEAT A SUBSTANCE?

parently warmer, when tried by that instrument. This heat, however, must be thrown into it, in order to give it the form of a fluid; and I affirm that this great addition of heat is the principal, and most immediate cause of the fluidity induced.

And, on the other hand, when we deprive such a body of its fluidity again, by a diminution of heat, a very great quantity of heat comes out of it, while it is assuming a solid form, the loss of which heat is not to be perceived by the common manner of using the thermometer. The apparent heat of the body, as measured by that instrument, is not diminished, or not in proportion to the loss of heat which the body actually gives out on this occasion; and it appears from a number of facts, that the state of solidity cannot be induced without the abstraction of this great quantity of heat. And this confirms the opinion, that this quantity of heat, absorbed, and, as it were, concealed in the composition of fluids, is the most necessary and immediate cause of their liquidity. . . .

If we attend to the manner in which ice and snow melt, when exposed to the air of a warm room, or when a thaw succeeds to frost, we can easily perceive, that however cold they might be at the first, they are soon heated up to their melting point, or begin soon at their surface to be changed into water. And if the common opinion had been well founded, if the complete change of them into water required only the further addition of a very small quantity of heat, the mass, though of a considerable size, ought all to be melted within a very few minutes or seconds more, the heat continually to be communicated from the air around. Were this really the case, the consequences of it would be dreadful in many cases; for, even as things are at present, the melting of great quantities of snow and ice occasions violent torrents, and great inundations in the cold countries, or in the rivers that come from them. But, were the ice and snow to melt as suddenly, as they must necessarily do were the former opinion of the action of heat in melting them well founded, the torrents and inundations would be incomparably more irresistible and dreadful. They would tear up and sweep away every thing, and that so suddenly, that mankind should have great difficulty to escape from their ravages. This sudden liquefaction does not actually happen; the masses of ice or snow melt with very slow progress, and require a long time, especially if they be of a large size, such as are the collections of ice, and wreaths of snow, formed in some places during the winter. These, after they begin to melt, often require many weeks of warm weather, before they are totally dissolved into water. . . . In the same manner does snow continue on many mountains during the whole summer, in a melting state, but melting so slowly that the whole of that season is not a sufficient time for its complete liquefaction. . . .

If any person entertain doubts of the entrance and absorption of heat in the melting ice, he needs only to touch it; he will instantly feel that it rapidly draws heat from his warm hand. He may also examine the bodies that surround or are in contact with it, all of which he will find deprived by it of a large part of their heat; or if he suspend ice by a thread, in the air of a warm room, he may perceive with his hand, or by a thermometer, a stream of cold air descending constantly from the ice; for the air in contact is deprived of a part of its heat, and thereby condenses and made heavier than the warmer air of the rest of the room: it therefore falls downwards and its place round the ice is immediately supplied by some of the warmer air; but this, in its turn, is soon deprived of some heat, and prepared to descend in like manner; and thus there is a constant flow of warm air from around to the sides of the ice, and a descent of the same in a cold state from the lower part of the mass, during which operation the ice must necessarily receive a great quantity of heat.

It is, therefore, evident, that the melting ice receives heat very fast, but the only effect of this heat is to change it into water, which is not in the least sensibly warmer than the ice was before. A thermometer, applied to the drops or small streams of water, immediately as it comes from the melting ice will point to the same degree as when it is applied to the ice itself, or, if there is any difference, it is too small to deserve notice. A great quantity, therefore, of the heat, or of the matter of heat, which enters into the melting ice, produces no other effect but to give it fluidity, without augmenting its [temperature]; it appears to be absorbed and concealed within the water, so as not to be discoverable by the application of a thermometer.

Note that Black's recognition of the latent heat not only explains why the melting and boiling phenomena are good ways to hold temperature constant, and therefore good calibration points for thermometers: it also explains an important natural observation: during a thaw, snow may melt rapidly on reaching its melting point, but it will not melt instantaneously, because of the large quantity of heat required to melt it. If this were not so, we would be exposed to disastrous floods far more often than we actually are.

Why did others not see that the failure of snow and ice to melt at the moment the temperature of the surroundings rises above 0° C was a fact of great significance? Although obvious in any climate where appreciable snow falls in winter, nevertheless until Black it was generally neither thought to need any explanation, nor recognized as a fact that tells us something very important about the process of melting. This is one example of how a new scientific theory not only explains the experimental observations that led to its discovery, but can also explain other observations that either did not seem to be at all related, or were not thought to need an explanation.

Black noted also that a latent heat is required not just for melting ice, but to melt any solid. Further, he found that the boiling of liquids happens in the same way: when water is heated by a flame, the temperature rises until boiling begins; then the temperature remains constant at 212° F until all the liquid boils away. Changing 1 pound of water into 1 pound of steam requires a latent heat considerably greater (about 6 times) than the latent heat required to melt 1 pound of ice.

Latent Heat and Caloric

Still, the discovery of latent heat was a surprise. How does heat that does not cause the temperature to rise fit in with the caloric theory?

First, can we make a quantitative statement about the latent heat? Is there a definite quantity of heat needed to melt 1 kilogram of ice at 0° C to form 1 kilogram of water at 0° C? Black found the answer to be yes. He described an experiment in which warm water at 190° F was mixed with a nearly equal weight of ice. The final temperature was 53° F (note that this is much less than 111° F, which is halfway between 32° F, the starting temperature of the ice, and 190° F). Black's value for the quantity of heat absorbed by ice on melting is 139 Btu per pound, in very good agreement with the modern, more accurate value of 144 Btu.

We will demonstrate the principle of his procedure with a somewhat simpler

IS HEAT A SUBSTANCE?

calculation: If we add 1 pound of water at 170° F to 1 pound of 32° F ice, the ice does not all melt. We end up at 32° F with some ice left. If we use water at 180° F, we melt all the ice, but end up with the 2 pounds of water a few degrees warmer than 32° F. By trial and error we find that if the hot water is at about 176° F we will just melt the ice, but the 2 pounds of water will be at 32° F.

This tells us that the heat that left the pound of warm water—since the water was cooled 144° F (from 176° F to 32° F) the amount of heat was 144 Btu—was just sufficient to melt the pound of ice without raising its temperature. So 144 Btu represents the latent heat needed to melt 1 pound of ice—the ice absorbing this amount of heat melts without getting warmer.

Second, it was important to find out if heat is still conserved. Is all the heat we put into ice to melt it given out when we freeze the water? This was tried, and the answer was found to be yes.

Third, can the notion of latent heat be explained in terms of the caloric theory? A satisfying explanation was put forward by those who believed that caloric was composed of atoms: there must be a kind of chemical reaction between the atoms of caloric and the molecules of water. On melting, a molecule of ice was thought to react with a definite number of atoms of caloric to form a new molecule. The caloric atoms, being on the outside, would make the new molecules more fluid—they would slide past each other more easily.

It can thus be seen that the caloric theory, with its principle of the conservation of heat, and the concepts of heat capacity and latent heat, could predict quantitatively the results of a variety of experiments in the laboratory, and explain certain natural phenomena as well. As we stated earlier and will state again, a theory can be wrong and still successfully describe a wide variety of experiments and observations.

OTHER TRIUMPHS OF THE CALORIC THEORY: THE FLOW OF HEAT

The caloric theory, subsequently to Black's work, went on to even greater triumphs before its final downfall. Up to now we have discussed only the conditions reached when the temperature is the same throughout, after the flow of heat has been stopped. But we might also ask how *fast* heat flows. *How long does it take for the state of temperature equilibrium to be reached once hot and cold bodies are brought together?* In the period 1820–1830, J. Fourier developed a quantitative theory for the flow of heat that could be used to make very accurate predictions of the time it takes to reach a uniform temperature (the equilibrium of heat). His highly mathematical theory can be most easily visualized by thinking of heat as a fluid substance, capable, like water, of flowing "downhill" with a velocity proportional to the "steepness" of the hill.

Fourier's theory can be humorously but accurately summarized in the following way: To stay warm in winter, follow these rules:

1. Wear a wool sweater instead of a cotton sweater.
2. Wear two sweaters instead of one.
3. Stay indoors.

Rule 1 expresses the fact that the rate of heat flow depends on the material through which it flows. Substances through which heat flows slowly are said to be good insulators (or poor conductors). Metals are good conductors of heat, which is why cold metals feel colder and hot metals hotter than other substances. Rule 2 expresses the familiar fact that the thicker the layer of insulation, the slower the loss of heat, and Rule 3, the equally familiar fact that the greater the temperature difference the faster the heat loss from the warmer body.

Expressed as an equation, the three rules state:

$$\text{heat flow} = (\text{constant depending on the substance}) \times \frac{(\text{temperature difference})}{(\text{thickness of insulation})}$$

The constant, depending on the substance, is a small number for good insulators and a large one for good conductors. The ratio of the temperature difference to the thickness of the insulation is analogous to the "steepness" of the "hill" down which heat is flowing.

If the walls of a house are 6 inches thick, and the indoor temperature is being maintained at 68° F on a day when it is 32° F outside, the temperature difference is 36° F, and the "steepness" (or more scientifically the gradient) is 6° F per inch. If the outdoor temperature were 50° F, the gradient is decreased to 3° F per inch, and the heat loss is half as great.

THE EFFICIENCY OF STEAM ENGINES

About the same time as Fourier was doing his work on the flow of heat, an important discovery about steam engines and the amount of useful work that can be obtained from them was made by the French engineer Sadi Carnot (1796–1832). Carnot also used the flow of water as an analogy to the flow of heat. He noted that the amount of work obtainable from falling water is proportional to two things: the weight of water that falls, and the height through which it falls (in more modern terms, the potential energy change of the water).

$$\text{useful work} = \text{weight of water} \times \text{height of fall}$$

His analysis concluded that the useful work obtainable from a steam engine (or any other engine for obtaining useful work from burning fuel) must also depend on similar factors: the amount of heat absorbed from the burning fuel, and the fall in temperature during the operation of the engine. Steam and other heat engines had long been operated using two temperatures at different parts of the cycle: the temperature of the boiler (of a steam engine) and the outside ambient temperature at which the steam is condensed during the cycle. Carnot's point was that this lower temperature was not just an accidental feature of the way steam engines of the time were designed, but was essential to the process. For the engine to do useful work,

IS HEAT A SUBSTANCE?

there must not only be a high temperature but also a low one, just as to get power from water there must not only be water at a high level, but also a low level for it to fall to. Further, the greater the temperature difference used by a heat engine, the more work can be gotten from it.

The caloric theory has long since been discarded, but Carnot's analysis is still the foundation for the study of the efficiency of heat engines. More than that, it is the foundation of one of the most famous generalizations in science, the second law of thermodynamics.

RUMFORD: DOES HEAT HAVE WEIGHT?

Benjamin Thomson, Count Rumford[6]

We turn now to the experiments performed by Count Rumford in 1790–1800 to discredit the caloric theory. These experiments appear to us, with the advantage of hindsight, to offer fatal objections to that theory and provide overwhelming support to the theory that heat is motion. Yet it is a historical fact that the caloric theory continued to be followed by the great majority of scientists for another 50 years, and, as we have noted, significant discoveries were made with its help. Why the caloric theory was preferred for so long, in spite of Rumford's work, is an important question; and its answer will tell us a lot about how science is done.

Benjamin Thomson (1753–1814) was born in the colony of Massachusetts. Siding with the Loyalists during the American Revolution, Thomson acted as an undercover agent for the British governor in intelligence-gathering activities against the rebellious colonists in New Hampshire. When his activities were discovered, he first fled to Boston, where he joined the staff of the British General Gage, and continued surreptitious activities on behalf of the British in his home town of Woburn, Massachusetts. Letters in which he used invisible ink to transmit information to them still exist. When the British abandoned Boston in 1776, Thomson went to England where he became private secretary to the Secretary of State for the colonies. There is some reason to believe that while he held this position he was engaged in spying for the French. After the American Revolution ended, he was appointed by the Elector of Bavaria as colonel in his army, and settled in Munich, where he received the title of Count Rumford. Although he had made secret arrangements with the British Foreign Office to supply them with information as long as he should hold his commission in the Bavarian army, Thomson apparently reneged on his promise. He was a thoroughly unsavory person, but good moral character is not a prerequisite for scientists, or, for that matter, poets.

In any event Count Rumford did an effective job as aide-de-camp and later as Inspector General of Artillery to the Emperor. A lot of his research on heat came out of practical problems, among them the determination of the most efficient clothing for the soldiers, studies of nutrition, construction of field stoves and lamps, exterior ballistics (the behavior of bullets after they have left the gun barrel), and the manufacture of cannons.

His interest in the theory of heat had begun at the age of 17 when he read a treatise on fire by the famous Dutch chemist, Boerhaave, who suggested *that heat is a vibration of a heated body* at a frequency which is too high to be heard as sound but which is sensed by the skin as hotness. This idea remained with Rumford, and his initial experiments on heat many years later were performed in order to support it. Later, his ideas came to be more in accord with the modern view that heat is a chaotic motion of the individual atoms of matter.

Rumford's War against the Caloric Theory

As part of his campaign against the caloric theory, Rumford carried out a large number of experiments, three of which we will describe at length here:

1. He used the concept of latent heat in what was up to that time the best experiment ever performed to see if "caloric" had any weight, which he believed it should have if it was a material substance; within the limits of the accuracy of his experiment, no weight was detected.
2. He showed that mechanical work could, through friction, produce an apparently unlimited quantity of heat, making more definitive the observation of Bacon and Locke that heat can be produced by friction. If this experiment is accepted at face value, heat is not conserved: it can be produced at will, and thus cannot be composed of a substance that can be neither created nor destroyed.
3. He showed that if a solution of salt in water is overlaid with a less dense layer of pure water and left quiet, without stirring, the salt spreads upward into the pure water, thus suggesting that the atoms of matter are in constant motion, a fact consistent with the theory that heat is the motion of atoms.

Does Heat Weigh Anything?

One obvious way of testing the relative merits of the caloric and kinetic theories is to see if heat has weight. If it does, one cannot avoid accepting the view that heat is a substance. Unfortunately, if it does not, one need not necessarily reject the caloric theory. There are a number of reasons why this is so. First, there is a purely practical issue: no experiment for measuring the weight of heat can ever give us an unambiguous answer that heat has no weight. Any device for weighing anything has a certain limit of accuracy, so that the best result the experiment could give is that "Heat weighs less than this device can detect." It could be like trying to weigh a mosquito on a butcher's scale; such a scale will read zero if anything much less than a few grams is placed on it. As we will see, Rumford informed us exactly how sensitive his weighing balance was, and we can only conclude that if the heat he tried to weigh had any weight, it was less than this.

However, there is another reason why a zero result for the weight of heat need not have convinced a believer in the caloric theory to give it up. During this time,

IS HEAT A SUBSTANCE?

the concept that anything deserving the name of "substance" had to have weight was not fully accepted, nor do we accept it fully today. There were a number of "substances" under study then in addition to heat, such as light and electricity, which are difficult or impossible to weigh as the term "weighing" is ordinarily understood. We know today that light has some of the characteristics of a substance, including the properties of being deflected by a gravitational field and exerting a pressure as a gas does, but it could not have been weighed in Rumford's experiment. Thus a direct attempt to weigh heat would have been decisive only if it gave a positive result. The negative result found by Rumford left the question open. Still, Rumford's work is worth reading about as an example of an experiment in which the value of the result is determined by the care and attention to detail with which it was done.

Weighing heat is not easy. The obvious method is to weigh a body first when it is cold and then when it is hot, to see if there is a difference in the two weights. Naturally, one wishes to use as sensitive a weighing device as possible. The most sensitive such instrument available to Rumford was the balance. This is an instrument with a rigid beam, supported in the center on a knife edge, and with two attachments out toward the ends of the beam at equal distances from the center. The two bodies whose weights we are comparing are to be hung from the attachments (see Figure 6-4). If the weights of the bodies are equal and if their distances from the center of the beam are equal, the beam will balance. If the weights are not equal, weights are added to the lighter side until a state of balance is achieved. Such instruments, capable of high accuracy, were being made in Rumford's time.

Let us imagine that we start with two bodies that have equal weights at room temperature, so that they just balance each other. We want to heat up the body on one side of the instrument. Immediately we run into problems. The hot body will

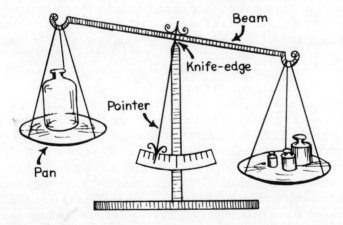

FIGURE 6-4. Analytical balance. The figure is schematic, and in the interest of clarity many important features are omitted.

heat up the half of the beam it is hanging from, causing that half to expand. This will give the hot body greater leverage, and it will seem to weigh more even though it hasn't really gained in weight. Further, hot bodies warm the air next to them. The heated air rises and, if the balance is a sensitive one, the resulting air currents can cause it to move erratically. Such currents are called convection currents, and Rumford was one of the first to study carefully their role in cooling hot bodies and in tranferring heat from one place to another.

A further problem arises from our desire to raise the temperature of the body as high as possible to get the largest possible amount of heat into it. Since we expect heat to weigh little if anything, the more we have, the more easily we can detect any change in weight. But a hot body can undergo changes in weight for extraneous reasons: iron rusts more rapidly when hot; since rusting is a chemical combination of the iron with oxygen from the air, the iron will gain weight as it rusts; materials containing moisture will lose it when they are heated, and hence lose weight. In the weighing of very cold bodies, these problems are replaced by others: not only will there still be convection currents and unequal expansion of the beam of the balance, but also condensation of moisture on the body.

Now Black had found that when ice melts, the amount of heat it absorbs *without any change in temperature taking place,* is 139 Btu per pound (the modern more accurate value is 144 Btu per pound). Thus we can compare the weight of water when liquid at 32° F with that of the ice it has formed at 32° F, after it has lost 139 Btu for each pound. In Rumford's experiment, a large amount of water in a glass vessel was placed on one arm of a balance, and a counterweight of some material that does not freeze at 32° F was placed on the other arm. Rumford used a mixture of water and alcohol, the alcohol serving, as it still does today, as an antifreeze. The pure water was then frozen to ice, losing a large amount of latent heat in the process; the counterweight had not lost any heat. Then his intention was to observe whether the weight of the water that had frozen had changed enough to upset the balance. Because the whole experiment was done in a room at 32° F, there should have been no problem with convection currents in the air, unequal heating of the beam, and so forth.

A similar experiment had been carried out earlier by a British physician, George Fordyce, in 1785, who found that water on being frozen *increased* in weight by a small amount. Rumford's first experiment, to his surprise, confirmed Fordyce's result. This was a jolt to Rumford. He measured the change in weight carefully: it was about 0.003% of the initial weight of the water. Since the result was contrary to his expectation, he immediately set out to look for possible sources of error in the experiment. The first possibility he considered was that the right and left arms of the balance, which happened to be treated differently during manufacture, might have contracted differently on cooling. This meant that two sides which were in perfect balance at one temperature might have gotten out of balance at a different temperature.

IS HEAT A SUBSTANCE?

Rumford tested this by filling two brass globes with nearly identical weights of mercury at 61° F and bringing them to a state of balance by adding small weights, as needed, to the lighter side. The balanced globes were then brought into a cold room (26° F) and left overnight. Since the substances on both arms of the balance were the same, the same amount of heat would be lost from both sides on cooling from 61° F to 26° F. Any apparent difference in weight that might have developed could be attributed to some extraneous cause, such as unequal changes in length of the balance arms. No change was found; Rumford's confidence in the balance was restored.

He then returned to the purpose of the experiment itself. His first results, similar to those of Dr. Fordyce, had seemed to show that bodies losing heat gained weight. *Heat seemed thus to have negative weight.*

To test this in a different way, he did an experiment not involving latent heat. He noted that the specific heat of water is about 30 times that of mercury, so that if we place a weight of water on one arm of the balance and an equal weight of mercury on the other, and then cool them, the water loses 30 times as much heat as the mercury. This time Rumford found no change in relative weight—the two weights remained in balance. His prejudice in favor of the weightlessness of heat having been confirmed, he then returned to the experiment in which water was frozen and latent heat lost.

Rumford conjectured that the flaw in his previous attempt to study this was his uncertainty as to whether the two glass globes had had time to reach the temperature of the cold room; there might have been temperature inequalities that caused convection currents in the air and upset the balance. To avoid this source of error, he inserted small thermometers in both the water and the alcoholic solution used to balance it. Then the final balancing of the two, using small pieces of silver to obtain exact equality of both sides, was not carried out until some hours after both liquids had reached the temperature of the cold room.

> This being done, the bottles were all removed into a room in which the air was at 30° [Fahrenheit—hence below the freezing point of water], where they were suffered to remain, perfectly at rest and undisturbed, forty-eight hours; the bottles A and B being suspended to the arms of the balance. . . .
>
> At the end of forty-eight hours, during which time the apparatus was left in this situation, I entered the room, opening the door very gently for fear of disturbing the balance; when I had the pleasure to find the three thermometers, viz. that in the bottle A,—which was now inclosed in a solid cake of ice,—that in the bottle B, and that suspended in the open air of the room, all standing at the same point, 29° F., and the bottles A and B remaining in the most perfect equilibrium.
>
> To assure myself that the play of the balance was free, I now approached it very gently, and caused it to vibrate; and I had the satisfaction to find, not only that it moved with the utmost freedom, but also, when its vibration ceased, that it rested precisely at the point from which it had set out. . . .
>
> Having determined that water does not acquire or lose any weight upon being changed from a state of fluidity to that of ice, and vice versa, I shall now

take my final leave of a subject which has long occupied me, and which has cost me much pains and trouble; being fully convinced, from the results of the above-mentioned experiments, that if heat be in fact a substance, or matter,—a fluid *sui generis,* as has been supposed,—which, passing from one body to another and being accumulated, is the immediate cause of the phenomena we observe in heated bodies,—of which, however, I cannot help entertaining doubts,—it must be something so infinitely rare, even in its most condensed state, as to baffle all our attempts to discover its gravity. *And if the opinion which has been adopted by many of our ablest philosophers, that heat is nothing more than an intestine vibratory motion of the constituent parts of heated bodies, should be well founded, it is clear that the weights of bodies can in no wise be affected by such motion.* [Our italics]

It is, no doubt, upon the supposition that heat is a substance distinct from the heated body, and which is accumulated in it, that all the experiments which have been undertaken with a view to determine the weight which bodies have been supposed to gain or to lose upon being heated or cooled, have been made; and upon this supposition,—but without, however, adopting it entirely, as I do not conceive it to be sufficiently proved,—all my researches have been directed.

The experiments with water and with ice were made in a manner which I take to be perfectly unexceptionable, in which no foreign cause whatever could affect the results of them; and the quantity of heat which water is known to part with, upon being frozen, is so considerable, that if this loss has no effect upon its apparent weight, it may be presumed that we shall never be able to contrive an experiment by which we can render the weight of heat sensible. . . .

I think we may very safely conclude, that ALL ATTEMPTS TO DISCOVER ANY EFFECT OF HEAT UPON THE APPARENT WEIGHTS OF BODIES WILL BE FRUITLESS.[7]

Rumford's conclusion was a rash one. All he really proved was that whatever weight heat might or might not have was less than his sensitive balance could detect. But although rash, he was basically right, and his conclusion that heat had no weight was accepted even by the supporters of the caloric theory. Nobody tried to weigh heat again after that.

HEAT FROM FRICTION

The Boring of Cannons

Rumford's experiments on the production of heat by friction in the boring of cannons at the Munich arsenal are his most famous (see Figure 6-5). That heat can be produced by friction was, as noted earlier, a fact that had been known for a long time, and those who believed in the caloric theory were well aware of it. However, their explanation differed from Rumford's. What he accomplished in these experiments was to make the point more incisively than had previously been done, and to keep the issue alive at a time when the caloric theory was predominant. He also included some experiments that tended to refute the explanations through which believers in the caloric theory tried, without sacrificing the principle that heat is conserved, to account for the production of heat by friction.

IS HEAT A SUBSTANCE?

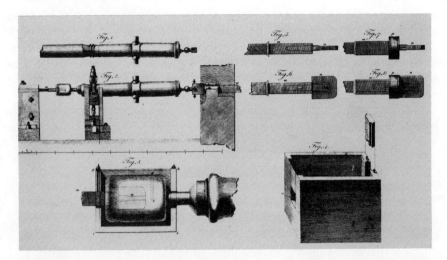

FIGURE 6-5. An illustration from Rumford's paper, "An Inquiry Concerning the Source of the Heat which is Excited by Friction," showing the apparatus used by him in the cannon boring experiment. Figure 1, upper left, shows the cannon as received from the foundry, and Figure 2, below, shows it mounted in the machine used for boring. (Reproduced with the permission of Harvard University Press.)

It frequently happens that in the ordinary affairs and occupations of life, opportunities present themselves of contemplating some of the most curious operations of Nature; and very interesting philosophical experiments might often be made, almost without trouble or expense, by means of machinery contrived for the mere mechanical purposes of the arts and manufactures. . . .

Being engaged lately in superintending the boring of cannon in the workshops of the military arsenal at Munich, I was struck with the very considerable degree of Heat which a brass gun acquires in a short time in being bored, and with the still more intense Heat (much greater than that of boiling water, as I found by experiment) of the metallic chips separated from it by the borer.

The more I meditated on these phenomena, the more they appeared to me to be curious and interesting. A thorough investigation of them seemed even to bid fair to give a farther insight into the hidden nature of Heat; and to enable us to form some reasonable conjectures respecting the existence, or non-existence, of an igneous fluid,—a subject on which the opinions of philosophers have in all ages been much divided.[8]

One possible explanation that supporters of the caloric theory could offer for the large amount of heat evolved on boring the cannons is that the boring process converts the bulk metal into little chips and shavings, and this process could explain the heat given off. It has been assumed that a particular substance—brass, for example—will have a heat capacity dependent only on the amount of substance and not on its state of subdivision. However, what if this should prove not to apply to metal when finely divided, i.e., what if the heat capacity of a pound of cannon metal in the form of chips or little shavings were less than that of a pound of the bulk metal? Then it would follow that grinding up a pound of metal would cause some heat to be released, much as breaking a glass bottle full of water spills the water. The released heat would raise the temperature of the metal and of neighboring bodies.

To test whether this explanation could be correct, Rumford compared the specific heats of chips and bulk metal:

> From whence comes the Heat actually produced in the mechanical operation above mentioned?
>
> Is it furnished by the metallic chips which are separated by the borer from this solid mass of metal?
>
> If this were the case, then according to the modern doctrines of latent Heat, and of caloric, the capacity for Heat of the parts of the metal, so reduced to chips, ought not only to be changed, but the change undergone by them should be sufficiently great to account for all the Heat produced.
>
> But no such change had taken place; for I found, upon taking equal quantities, by weight, of these chips, and of thin slips of the same block of metal separated by means of a fine saw, and putting them at the same temperature (that of boiling water) into equal quantities of cold water (that is to say, at the temperature of $59\frac{1}{2}°$ F), the portion of water into which the chips were put was not, to all appearance, heated either less or more than the other portion in which the slips of metal were put.
>
> This experiment being repeated several times, the results were always so nearly the same that I could not determine whether any, or what change had been produced in the metal, in regard to its capacity for Heat, by being reduced to chips by the borer.[8]

Rumford performed an even more decisive test: he used a blunt borer on an already hollowed cannon so that only a small amount of metal chips was formed. The heat continued to pour forth at the same rate.

Could the heat have been produced by a chemical reaction of the metal with air during the boring? Rumford next immersed the whole system in water. The result was spectacular, in the literal sense of providing a spectacle. The heat produced was enough to boil the water.

> The result of this beautiful experiment was very striking, and the pleasure it afforded me amply repaid me for all the trouble I had had in contriving and arranging the complicated machinery used in making it.
>
> The cylinder, revolving at the rate of about 32 times in a minute, had been in motion but a short time, when I perceived, by putting my hand into the water and touching the outside of the cylinder, that Heat was generated; and it was not long before the water which surrounded the cylinder began to be sensibly warm.
>
> At the end of 1 hour I found, by plunging a thermometer into the water in the box . . . , that its temperature had been raised no less than 47 degrees; being now 107° of Fahrenheit's scale.
>
> When 30 minutes more had elapsed, or 1 hour and 30 minutes after the machinery had been put in motion, the Heat of the water in the box was 142°.

IS HEAT A SUBSTANCE?

> At the end of 2 hours, reckoning from the beginning of the experiment, the temperature of the water was found to be raised to 178°.
> At 2 hours and 20 minutes it was at 200°; and at 2 hours 30 minutes it ACTUALLY BOILED!
> It would be difficult to describe the surprise and astonishment expressed in the countenances of the bystanders, on seeing so large a quantity of cold water heated, and actually made to boil, without any fire.
> Though there was, in fact, nothing that could justly be considered as surprising in this event, yet I acknowledge fairly that it afforded me a degree of childish pleasure, which, were I ambitious of the reputation of a grave philosopher, I ought most certainly rather to hide than to discover.[8]

Rumford described an attempt at a quantitative calculation of the amount of heat produced in his experiment: he concluded that nine wax candles, burning continuously, would produce the same heat per hour as his two horses turning the borer inside the cannon.

Rumford's conclusion was that the supply of heat that can be produced by friction is inexhaustible. Therefore, the caloric theory, requiring that the universe contain a fixed amount of heat, undestroyable and uncreatable, had to be wrong.

> By meditating on the results of all these experiments, we are naturally brought to that great question which has so often been the subject of speculation among philosophers; namely,—
> What is Heat? Is there any such thing as an igneous fluid? Is there anything that can with propriety be called caloric?
> We have seen that a very considerable quantity of Heat may be excited in the friction of two metallic surfaces, and given off in a constant stream or flux in all directions without interruption or intermission, and without any signs of diminution or exhaustion.
> From whence came the Heat which was continually given off in this manner in the foregoing experiments? Was it furnished by the small particles of metal, detached from the larger solid masses, on their being rubbed together? This, as we have already seen, could not possibly have been the case.
> Was it furnished by the air? This could not have been the case; for, in three of the experiments, the machinery being kept immersed in water, the access of the air of the atmosphere was completely prevented.
> Was it furnished by the water which surrounded the machinery? That this could not have been the case is evident: first, because this water was continually receiving Heat from the machinery, and could not at the same time be giving to, and receiving Heat from, the same body; and, secondly, because there was no chemical decomposition taken place. . . .
> And, in reasoning on this subject, we must not forget to consider that most remarkable circumstance, that the course of the Heat generated by friction, in these experiments, appeared evidently to be inexhaustible.
> It is hardly necessary to add, that anything which any insulated body, or system of bodies, can continue to furnish without limitation, cannot possibly be a material substance; and it appears to me to be extremely difficult, if not quite impossible, to form any distinct idea of anything capable of being excited and communicated in the manner the Heat was excited and communicated in these experiments, except it be MOTION.[8]

This last paragraph is Rumford's dramatic conclusion. It is a denial of the principle of the conservation of heat, the cornerstone of the caloric theory.

In the next paragraph, Rumford acknowledges the vagueness of his own view that heat is motion, and the difficulty of drawing inferences from it.

> I am very far from pretending to know how, or by what means or mechanical contrivance, that particular kind of motion in bodies which has been supposed to constitute Heat is excited, continued, and propagated; and I shall not presume to trouble the Society with mere conjectures, particularly on a subject which, during so many thousands years, the most enlightened philosophers have endeavoured, but in vain, to comprehend.
>
> But, although the mechanism of Heat should, in fact, be one of those mysteries of nature which are beyond the reach of human intelligence, this ought by no means to discourage us or even lessen our ardour, in our attempts to investigate the laws of its operations. How far can we advance in any of the paths which science has opened to us before we find ourselves enveloped in those thick mists which on every side bound the horizon of the human intellect? But how ample and how interesting is the field that is given us to explore!
>
> Nobody, surely, in his sober senses, has even pretended to understand the mechanism of gravitation; and yet what sublime discoveries was our immortal Newton enabled to make, merely by the investigation of the laws of its action!
>
> The effects produced in the world by the agency of Heat are probably *just as extensive,* and quite as important, as those which are owing to the tendency of the particles of matter towards each other; and there is no doubt but its operations are, in all cases, determined by laws equally immutable.[8]

Why Rumford Did Not Win

Rumford's results did not close the question. He had argued with his characteristic enthusiasm from an experiment lasting a few hours that the supply of heat was inexhaustible. While he had proved that the breaking down of the metal into chips or shavings could not have been the source of the heat, he had not ruled out to the satisfaction of the supporters of the caloric theory the possibility that the surfaces of the metal were being somehow altered by the boring process, and that this alteration was the origin of the heat.

Rumford was right in his conclusion, but the supporters of the caloric theory were not being foolish or obstinate in raising objections to his work. There were too many phenomena that the caloric theory could explain that the kinetic theory, at that time still only a vaguely formulated notion, could not.

The conflict between rival scientific theories is often of this kind. It is not usually a simple matter of which theory fits the "facts" better. Each theory will fit one set of facts well and another set badly, and it is left to the intuitive judgment of the scientists of the time to choose which set of facts is the important one to explain. It took 50 years from the time of Rumford's experiments for the kinetic theory to develop to the point where it could account for those phenomena that the caloric theory explained so well.

IS HEAT A SUBSTANCE?

Rumford's quantitative experiment had compared the heat produced by the work of the horses to the heat which would be provided by a certain number of wax candles. He did not establish a quantitative relation between the *work done* and the *heat produced*. This latter experiment was done by James Joule 50 years later. Joule was able to do this because in the intervening years, the concept of mechanical energy had been refined, and second, electrical energy and its production by mechanical means—as in a dynamo—had been discovered. Joule was able to show that a fixed amount of mechanical energy always gives a fixed amount of heat. Even more strikingly, Joule showed that electricity can produce heat—a fact now known by anyone who has ever used an electric toaster or a flatiron. He then proved that if a fixed amount of mechanical energy is used to produce an electric current, and the current is used in turn to produce heat, the amount of heat is the same as that which the mechanical energy would have produced directly. Although his experiments, because of their quantitative character, were more convincing than Rumford's, the conclusions Joule drew from them met with both indifference and opposition before they were finally accepted.

MOLECULAR MOTION

Do Atoms and Molecules Move?

At the time Rumford did his experiment on friction, what he meant by motion was a vibration of the entire heated body. As he continued his work, he moved closer to our modern notion of heat as chaotic atomic motion. His ideas were not clearly formulated in his own mind, but a way of supporting the claim that heat is a motion of the individual particles of matter seems to have occurred to him, namely, to show that even in a substance at rest, in which no motion is obvious to the eye, the atoms are moving about all the time. Although nothing was then known about the size or speed of motion of atoms and molecules, Rumford, to demonstrate this motion, devised an experiment that was simple in concept, but—again from the advantage of hindsight—decisive in results. Rumford himself did not have the confidence in this experiment that he did in the other experiments we described; he was aware of the vagueness of his own ideas and did not press his point.

Salt water and fresh water can be easily mixed to form a more dilute, uniformly salty water by shaking or stirring the two liquids together. (To some extent, the mixing process can be followed by eye, because pure water and salt water affect light differently. Optical effects seen as these liquids are mixed are similar to those seen when boiling water is poured into a glass, a teaspoon of sugar is added, and the mixture is stirred. As the solid sugar dissolves, it forms a concentrated sugar solution in its vicinity; stirring mixes this with the less dense water above.) What would happen, though, if we did not stir the solutions?

Molecules and Atoms

It is necessary at this point to explain the distinction between atoms and molecules.

All matter in the universe is made up of combinations of some 100 different kinds of elements. These elements are composed of particles called atoms; all the atoms of any one element have identical *chemical* properties.*

The units of which matter is composed are not usually the individual atoms, but rather tightly bound combinations of more than one kind of atom. These combined units are called molecules. Some substances have molecules with two or three atoms, but larger numbers are more common. Water has three atoms in the molecule (H_2O), common sugar (sucrose) has 45 ($C_{12}H_{22}O_{11}$), and proteins have thousands, or tens of thousands.† When sodium chloride (NaCl) dissolves in water, it breaks up into separate, electrically charged *atomic* units of Na and Cl. For simplicity in what follows, we describe salt in water as though it is composed of salt "molecules" with the chemical formula NaCl, which is how Rumford would have visualized a salt solution.

Demonstration of the Constant Motion of Molecules

Rumford placed a layer of salt solution, which is more dense than pure water, at the bottom of a glass vessel. Over it he carefully placed a layer of distilled water, so that little mixing occurred. If the molecules were at rest, the salt molecules, being heavier than the water, would stay on the bottom indefinitely. If, however, they were constantly in violent motion, the salt molecules would tend to spread gradually through the overlying layer of water to produce a uniform solution, just as though the vessel had been stirred.

To observe what was happening, Rumford added a droplet of oil of cloves, which is only slightly denser than pure water, but less dense than either the initial salt solution or the more dilute salt solution produced by stirring. The droplet of oil therefore sank to the bottom of the water layer but floated on top of the salt solution (see Figure 6-6).

> When we mix together two liquids which we wish to have unite, we take care to shake them violently, in order to facilitate their union; it might, however, be very interesting to know what would happen, if, instead of mixing them, they were simply brought into contact by placing one upon the other in the same vessel, taking care to cause the lighter to rest upon the heavier.
>
> Will the mixture take place under such circumstances? and with what degree of rapidity? These are questions interesting alike to the chemist and to the natural-philosopher.

---

* In most elements, the atoms are not all identical in *mass,* although this circumstance has very little effect on chemical properties. Atoms of the same element which differ in mass are called isotopes.

† The difference between an atom and a molecule is portrayed in Figure 10-1.

IS HEAT A SUBSTANCE?

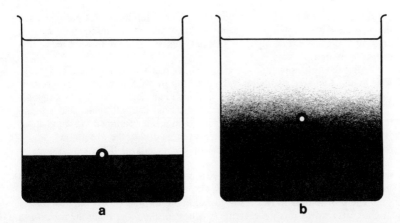

FIGURE 6-6. Although sodium chloride—ordinary table salt—is colorless, we portray it here by shading. The darker the shade the more concentrated the salt. Initially, the boundary between the dense salt solution and the less dense water is sharp, and the oil droplet, of intermediate density, floats at the boundary between the two. As time goes on the boundary ceases to be sharp. Salt spreads upward, increasing the density of the liquid above the original boundary, and the oil droplet floats upward accordingly.

The result would depend, without doubt, on several circumstances which we might be able to anticipate, and the effects of which we might perhaps estimate à priori. But since the results of experiments, when they are well made, are incomparably more satisfactory than conclusions drawn from any course of reasoning, especially in the case of the mysterious operations of Nature, I propose to speak before this illustrious Assembly simply of experiments that I have performed.

Having procured a cylindrical vessel of clear white glass . . . I put it on a firm table in the middle of a cellar, where the temperature, which seemed to be tolerably constant, was 64 degrees of Fahrenheit's scale.

I then poured into this vessel, with due precautions, a layer of a saturated aqueous solution of muriate of soda [sodium chloride—ordinary table salt], 3 inches in thickness, and on to this a layer of the same thickness of distilled water. This opertion was performed in such a way that the two liquids lay one upon the other without being mixed, and when everything was at rest I let a large drop of the essential oil of cloves fall into the vessel. This oil being specifically heavier than water, and lighter than the solution of muriate of soda on which the water rested, the drop descended through the layer of water; when, however, it reached the neighbourhood of the surface of the saline solution it remained there, forming a little spherical ball which maintained its position at rest, as though it were suspended, near the axis of the vessel. . . .

[H]aving observed, by means of the scale attached to the vessel, and noted down in a register, the height at which the little ball was suspended, I withdrew, and locking the door, I left the apparatus to itself for twenty-four hours. . . .

As the little ball of oil, designed to serve me as an index, was suspended a very little above the upper surface of the layer of the saturated solution, this showed me that the precautions which I have taken were sufficient to prevent the mixing of the distilled water and the saline solution when I put one upon the other, and I knew that this mixture could not take place subsequently without causing at the same time my little sentinel, which was there to warn me of this event, to ascend. . . .

After the little apparatus mentioned above had been left to itself for twenty-four hours, I entered the cellar, taking a light in order to note the progress of the experiment, and I found that the little ball had risen [about $1/4$ of an inch].

The next day, at the same hour, I observed the ball again, and I found that it had risen about [$\frac{1}{4}$ of an inch] more; and this it continued to ascend about [$\frac{1}{4}$ of an inch] a day for six days, when I put an end to the experiment. . . .

But without spending more time on the details of these experiments, I hasten to return to their results. They showed that the mixture went on continually, but very slowly, between the various aqueous solutions employed and the distilled water resting upon them.

There is nothing in this result to excite the surprise of any one, especially of chemists, unless it is the extreme slowness of the progress of the mixture in question. The fact, however, gives occasion for an inquiry of the greatest importance, which is far from being easy to solve.

Does this mixture depend upon a peculiar force of attraction different from the attraction of universal gravitation, a force which has been designated by the name of chemical affinity? Or is it simply a result of motions in the liquids in contact, caused by changes in their temperatures? Or is it, perhaps, the result of a peculiar and continual motion common to all liquids, caused by the instability of the equilibrium existing among their molecules?

I am very far from assuming to be able to solve this great problem, but it has often been the subject of my thoughts, and I have made at different times a considerable number of experiments with a view of throwing light into the profound darkness with which the subject is shrouded on every side.[9]

In the next-to-last paragraph above, Rumford considers three alternative explanations of the phenomenon. The first is that there is some attractive force in the pure water that draws the salt upward. The second is that fortuitous fluctuations in temperature in the liquid, arising, for example, from drafts in the room or changes in temperature from day to night, might have produced convection currents in the liquid that caused mixing of the salt solution and the water. Both of these possibilities were plausible in terms of the scientific knowledge of the time, and would surely have been offered as counterarguments by supporters of the caloric theory, if Rumford had been dogmatic in his conclusion that he had really provided additional disproof of that theory. The third hypothesis, although vaguely worded, clearly implies the possibility that a state of continual molecular motion was the cause of the mixing. Rumford's final paragraph is revealing of his own lack of conviction about just what his experiment really proved.

That what was observed in this experiment, a phenomenon called *diffusion,* is exactly the consequence of the random chaotic motion of molecules was only established in a thoroughly convincing way by Albert Einstein in 1905, although Rumford's intuitive explanation was accepted when the kinetic theory finally won out over the caloric theory. It is unfortunate that Rumford did not follow this up with another obvious experiment—to show that if the experiment were repeated at

IS HEAT A SUBSTANCE?

a higher temperature the salt would spread through the water faster. Had he performed this experiment, he would have found a spectacular increase in the rate of motion of the salt, far greater than he would have dared to expect; surely this result would have delighted him, and perhaps would have influenced the acceptability of the kinetic theory as well. The reasons for this rapid increase of rate with temperature are too technical to be discussed here.

Heat as Molecular Motion

It will help the reader to understand the significance of Rumford's experiment, and to have a clearer idea of the kinetic theory of heat, if we take some time to describe molecular motion.

We said earlier that heat is the chaotic motion of molecules constantly colliding with each other. In Figure 6-7a we show a gas, in which the molecules are on the average fairly far apart. A single molecule will travel in a straight line for some distance until it runs into another molecule. The reader may visualize what happens then by imagining that the molecules act just like billiard balls. After the collision, the molecule will be traveling in a different direction, probably with a different

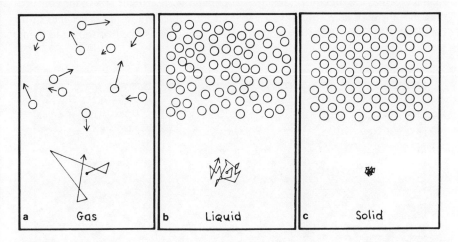

FIGURE 6-7. Molecular motion in the three states of matter. In the gas (a), molecules are very far apart on the average. The arrows are meant to indicate the speed and direction of each molecule. The path of each molecule consists of long, straight sections (between collisions) separated by abrupt changes in direction at each collision, as shown by the diagram at the bottom. In (b) the liquid is shown. Arrows representing the speeds are omitted. The molecules are much closer together than in the gas, but still irregularly arranged. The path is much more tortuous. The molecules suffer collisions more frequently and take longer to move an appreciable distance (bottom). In the solid (c), the molecules are packed in a regular pattern, and although they are continually colliding with their neighbors, they do not escape from their starting position, as shown in the bottom diagram.

speed. The arrows are drawn to show at one moment of time direction in which each molecule is moving, and how fast (the length of the arrow).

In the bottom part of the figure, we show the path of one of the molecules over a period of time: at each collision there is a change in direction. One can see that if we wait sufficiently long, any given molecule can travel a long distance from its starting point.

In Figure 6-7b we show a liquid. The molecules are on the average much closer to each other than in the gas. The path of any given molecule is much more tortuous than in the gas, but there is enough empty space for the molecules to get around each other, and, again, a molecule can travel a long distance from its starting point, although it takes longer.

In Figure 6-7c we show the situation in a solid. The typical solid differs from the typical liquid in two important ways: the molecules are arranged in a periodic pattern, like the tiles on a bathroom floor, and there is very little empty space. The result is that the molecules are trapped by their neighbors, against which they are constantly bumping, but from which they rarely escape. The path of one such trapped molecule can be seen at the bottom of the figure.

Figure 6-8a shows the situation at the start of Rumford's experiment. The water molecules are shown as open circles and the salt molecules as filled circles. Real molecules are, of course, much smaller in proportion to the size of the container than those shown here. Initially, all the salt molecules are in the lower half of the container. Through time, they tend to wander along paths like that shown in Figure

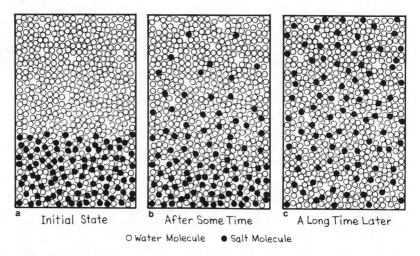

FIGURE 6-8. Salt diffusion in water. The salt molecules are portrayed as black circles and the water molecules as white ones. Initially (left), the salt molecules are only in the bottom half of the container, but as time goes on they spread upward (center), and eventually are uniformly distributed throughout the container (right).

IS HEAT A SUBSTANCE?

6-7b, so that some of them reach the upper half of the container (Figure 6-8b). After enough time, they will be uniformly distributed in the container, as in Figure 6-8c. They continue wandering after this happens, but, on the average, at any subsequent time, half will be in the upper half of the container and half will be in the lower half.

We know today that the temperature of a substance is related to the average speed of the molecules: the higher the temperature, the greater the average speed.

THE CONVERSION OF MOTION, THROUGH FRICTION, INTO INVISIBLE MOLECULAR MOTION

The conversion of ordinary motion into heat, which according to the kinetic theory is the chaotic motion of molecules, is a little harder to visualize than the kinetic theory explanation of diffusion.

We will attempt to do this, in a rather dramatic and fanciful way, by asking the reader to visualize one of those horrifying multiple-car collisions on a superhighway. Such collisions are unfortunately common when cars are travelling too fast in proportion to the mean distance between them. When such an accident occurs it is all too obvious where the energy of motion of the cars has gone—into crushed fenders, smashed-in auto bodies, broken glass and bent metal. For our visualization we will imagine a different kind of car, composed of a hard, elastic, undentable metal, so that when one of these cars hits a concrete wall at 60 miles an hour it simply bounces back at 60 miles an hour in the opposite direction. In short the car is a kind of idealized billiard ball.

Let us further imagine a perfectly straight section of superhighway, 10 lanes wide and with concrete walls at the sides which prevent cars from leaving the road, along which a stream of cars 10 cars wide and 100 cars long is travelling at a 60-mile-an-hour speed.

Now let us assume an observer in a space satellite many miles above the earth is using radar to look for large metallic objects on the earth's surface. The radar is not sensitive enough to distinguish the individual cars in the procession, but instead records it as a blur, 10 times longer than it is wide, travelling at 60 m.p.h. in a certain direction.

At this point a major earthquake occurs, which happens to dump one huge pile of large boulders from a mountainside directly into the path of the procession, and a second pile directly behind it. The first row of cars hits the pile in front, and each car recoils backwards. Because of the irregularity in shape and placement of the boulders the cars do not bounce back in exactly the opposite direction from their previous one, but at various angles. Almost immediately collisions take place between the first row cars and those of the second row, throwing them backwards or to the side, though because of the irregularity some of the second row cars may escape such a collision only to run into the boulders directly instead. In a short time all 1000 of the cars will have been involved in collisions. They are, on the average, still going at 60 miles per hour, but in all directions instead of straight ahead. Further, they will continue to collide with each other, with the two piles of boulders blocking the road, and with the concrete walls of the highway. All the forward motion of the original procession has been lost. But the motion is still there, though expressed as chaotic motion in all directions.

Meanwhile, the observer on the satellite, watching his radar screen, no longer sees motion. The forward motion of the metal object has stopped suddenly, and he cannot see the erratic motion of the 1000 cars.

The slowing down of moving objects by friction, though not always as dramatic as in our collision–earthquake combination, takes place in much the same way. To show that the original easily seen motion has been transferred to invisible chaotic atomic and molecular motion, we use a thermometer. To produce an appreciable rise in temperature, we must stop a lot of motion rather rapidly—use a high speed drill in tough metal, or fire a bullet into a thick wooden plank. Under these conditions the conversion of motion to heat is easily observed.

When a ball rolls along the ground, only the atoms at the surface of the ball are engaged in little local "collisions" with the atoms of the ground. The conversion of the ball's motion to the motion of the atoms of both the ball and the ground is a more gradual process, allowing a ball to roll quite a distance, at least on smooth ground, before it comes to rest.

WHY CALORIC SURVIVED

How Does Heat Get through a Vacuum?

We have shown that at least two of Rumford's experiments raised very serious questions about the caloric theory, and that his experiment on the weight of heat at least added something to the plausibility of a kinetic theory. Why then was the kinetic theory not immediately accepted? Is it a question of the rigid conservatism of a scientific establishment resisting the innovations of a brilliant and creative individual?

The fact is that Rumford was not ignored. He was well recognized and influential in his day. He played a major role in strengthening scientific research in England in his founding of the Royal Institution in about 1800, and succeeded in having the young chemist Humphrey Davy, an early supporter of his ideas on heat, appointed as director. Rumford's work was not disregarded by his opponents. They simply felt that the caloric theory had more to offer. We have already stressed how certain phenomena could be explained better by caloric theory than by the primitive kinetic theory, such as the conservation of heat in experiments not involving friction.

Ideally, conflicts between opposing scientific theories can be resolved if the

IS HEAT A SUBSTANCE?

opponents can agree on a single experiment whose outcome will be accepted as decisive. But far more often, the conflict is not so clear-cut: each of several opposing theories will have strengths in some areas and weaknesses in others. Which theory particular scientists will adopt will depend on which strengths they think important and which weaknesses they are willing to overlook—obviously a highly subjective judgment.

There was one aspect of heat that played a major role in the ability of the caloric theory to withstand Rumford's criticisms: *the radiation of heat*. It was not until ideas about the radiation of heat changed that the kinetic theory finally displaced its rival.[10]

It was well known that heat could be transmitted by radiation through a vacuum as well as by direct contact between hot and cold bodies: Rumford himself had been one of the discoverers of this. Now if heat is the motion of atoms, how can it be transmitted through space where there are no atoms? Rumford argued, with no solid scientific basis (but with prophetic insight), that the vibratory motion that he believed heat to consist of creates what he referred to as "undulations in the ether"—wave motions that could travel through empty space and transmit heat without actual contact between the molecules of hot and cold bodies. But this hypothesis requires that heat exist in two forms, not one: in matter it is the motion of the matter, and in space empty of matter it is "undulations in the ether." This was less plausible than the caloric theory view, which was simpler—radiant heat is just the atoms of caloric leaving hot bodies and traveling through empty space to land on cold bodies. In particular, the "undulations in the ether" seemed very fanciful. There was no reason at that time to believe that such phenomena would occur, except to save Rumford's theory of heat as a vibratory motion.

As we have indicated earlier and will indicate again, when scientists are faced with facts which contradict a theory they favor, they are not obligated at once to reject the theory. It is always possible to offer an additional hypothesis to explain why these facts really fit in with the theory in spite of appearances. But the additional hypothesis should itself be plausible, and should be supported by independent evidence. Otherwise there is no limit to the number of additional hypotheses that can be invented to save a bad theory. This principle of parsimony in hypotheses—the fewer the better—was first formulated by the fourteenth-century philosopher, William of Ockham, and is known as Occam's Razor.[11]

What Is Light?

During the first half of the nineteenth century, two important discoveries about the nature of light were made. First, light, which had been suggested by Isaac Newton to be, like caloric, a kind of substance composed of small particles, was found to be better described as a wave motion—an "undulation in the ether." Second, it was discovered that radiant heat is a form of light, but a form invisible, as it happens, to the human eye. The reader may be familiar with what is called infrared light—light that cannot be seen but can be detected by photography. The radiant heat observed by Rumford is in fact a form of infrared light.

Thus, as Rumford had guessed, there is such a thing as radiant heat, a wave motion capable of being emitted by heated matter and traveling through empty space. It is therefore different from heat as molecular motion, and is capable of warming up cold matter that it falls on. Once these new ideas about light and heat were accepted, the way was clear for the acceptance of heat as molecular motion. In this new climate, Joule's quantitative repetition of Rumford's experiment could be recognized as destroying the caloric theory once and for all.

The reader may be justifiably annoyed at this point. The experiments of Black and Rumford, which we have described in detail, and which appeared to play central roles in the conflict between the caloric and kinetic theories, had nothing to do with the nature of light, yet this conflict was not resolved until the nature of light was better understood. Science often progresses in this way. Scientific theories are not independent structures, like unattached buildings, that can be characterized as sound or unsound independently of other buildings in the vicinity. Rather, they are interconnected and mutually supporting, even though the interdependence is often hidden until one theory collapses and the weakness of others is revealed for the first time.

Thus new discoveries in related fields, or other changes in intellectual climate arising from less tangible sources, can lead to a reevaluation and rejection of theories once thought solidly supported.

The Objectivity of Scientists

We have stressed the positive side of Rumford's scientific work: his incisiveness, his choice of significant experiments, his prophetic guesses which could not be rationally justified by the scientific knowledge of the time. But he was opinionated in his views, and often wrong. His commitment to the motion theory of heat was so strong that he often jumped to incorrect conclusions where a more cautious person might have held back. He believed, for example, that liquids and gases could not conduct heat directly, but transferred it only if they themselves were in motion produced by uneven heating. In this he was completely wrong.

He believed that cold could be radiated as well as heat. The experiment he performed to detect cold radiation seemed to show its existence, but we interpret it differently today (see Chapter 13). Apparently he believed this because of his early view that our sense perception of the temperature of a body is caused by a vibration of the body: cold bodies vibrate at a low frequency and should emit low-frequency "cold waves" through space.

There is a common belief that a good scientist is objective, not emotionally involved in his theories, and quite willing to discard them if experimental facts require it. It should be obvious that Rumford was no such ideal scientist, and was

IS HEAT A SUBSTANCE?

effective precisely because he was not. If he had not cared deeply, he would have done much less. Self-criticism is a desirable virtue, but there is no automatic rule that every scientist must possess it. It is the job of the consensus of scientists to decide which theory works and which one does not, and in the long run this tends to correct the mistakes of the enthusiast and to retain what is worthwhile in his work.

It should also be noted that Black was much more nearly the stereotypical ideal scientist—cautious, careful, dispassionate. He too was a great scientist. There is room in science for both.

THE LAWS OF ENERGY

We would like to close this chapter with a brief discussion of the concept of energy. The triumph of the kinetic theory of heat as a result of the work of Joule and Kelvin in England and of the German scientists Mayer and Helmholtz involved a recognition that heat, rather than being a unique phenomenon unrelated to mechanical considerations, was in fact a form of energy, itself a concept derived from Newton's laws of mechanics.

Energy is not an easy term to define; we will do so in a somewhat indirect way, first stating two of its essential properties, and then giving examples of its various manifestations.

1. Energy comes in many different forms, and can be converted from one form to another.
2. Throughout the various interconversions, the total amount of energy cannot change; energy can be neither created or destroyed: it is *conserved*.

The Forms of Energy

Let us first list some of the forms of energy we have already mentioned in this chapter:

1. The energy bodies possess by virtue of being in motion: kinetic energy. Both moving automobiles and moving atoms have kinetic energy. The motion of automobiles we can see; the motion of atoms is invisible but the evidence for it, although indirect, is overwhelming.
2. Electrical energy. This form of energy is usually measured (and paid for) in kilowatt-hours (1 kilowatt-hour costs about $0.10 in the Eastern United States as this book is going to press).
3. Radiant energy, the radiation emitted by hot bodies and by other sources such as fluorescent lights and radio stations, is composed of electro-mag-

netic waves which carry energy. Electromagnetic radiation includes light, infra-red and ultra-violet radiation, X-rays, γ-rays, and radio waves.

In addition to the above forms of energy we can recognize that energy is stored—is present in a *potential* form—in many bodies or substances that are not actually in motion. The energy stored in fuels, for example, becomes manifest when we burn the fuels to heat our houses or propel our automobiles. Among examples of energy in a latent or potential form are the following:

4. Fuels like coal, oil, and wood, and for that matter foods, which are the fuels that supply the energy for the activities of living organisms. Because fuels and foods release their energies by undergoing chemical reactions (usually but not exclusively with the oxygen of the air), they are said to contain *chemical energy*.
5. Another form of chemical energy is stored in electric batteries. When the battery is in use, a chemical reaction taking place inside it produces an electric current.
6. Elastic energy is stored in the mainspring of an "old-fashioned" watch (as opposed to digital and quartz watches which run on electricity), in the spring of a set mouse-trap, and in a drawn bow.
7. Gravitational potential energy is stored in any body that can fall down toward the earth (or the sun, or another planet). An example is the water stored behind a dam for use in a hydroelectric power station.
8. Nuclear energy is present in the atoms of uranium, radium, plutonium, hydrogen, and in many other atoms in a potential form, and is released when the atoms undergo a spontaneous radioactive decay (as happens with elements such as radium) or are induced to do so by bombardment with high-energy particles (as happens to hydrogen or plutonium in nuclear explosions or nuclear power plants).

Finally, and most relevant to this chapter, what we have called "heat" is the energy gained by the atoms and molecules of matter when the temperature is raised. This form of energy is called thermal energy. We have already mentioned that atoms and molecules in motion have kinetic energy. They possess potential energy also, arising from the forces they exert on one another. When the temperature of a body is raised, its atoms gain kinetic energy, and usually potential energy as well. The two together make up the thermal energy of the body. Temperature is proportional to the average kinetic energy per atom.

The Transformations of Energy

Having listed different forms of energy, we can now describe some of the transformations that occur. To do so, we will begin with Rumford's experiment on

IS HEAT A SUBSTANCE? 141

the boring of cannons. The kinetic energy of the boring tool was converted to thermal energy of both the tool and the cannon. This thermal energy, in one of his experiments, was allowed to flow into water surrounding the cannon. Enough thermal energy was absorbed by the water to cause it to boil, converting it to steam. The H_2O molecules in steam have a higher potential energy than in liquid water (the latent heat of boiling is a measure of this potential energy). The borer acquired its kinetic energy from the kinetic energy of two horses, who in walking in a circle were converting the chemical energy stored in their bodies, through the contraction of their muscles, into motion. The chemical energy stored in their bodies was obtained from the chemical energy of the fodder they had eaten—hay, grass, or oats—which in the living plant had been obtained from a chemical reaction using chlorophyll as a catalyst to convert radiant energy received from the sun into stored chemical energy. The sun radiated this energy, just as any heated body radiates energy, losing an equivalent amount of its thermal energy. The thermal energy of the sun was in turn produced by nuclear reactions taking place in the center of the sun (the conversion of hydrogen to helium). Through all these transformations, the total amount of energy does not change—none is lost, all can be accounted for.

Measuring Energy

We cannot claim that energy cannot be either created or destroyed unless we know how to measure it. For reasons partly of history and habit, and partly of convenience, different forms of energy may be measured in different units. We have mentioned Calories and British thermal units as measures of thermal energy, and kilowatt-hours as a measure of electrical energy. When electrical energy is converted to thermal energy by an electric heater, the number of Calories of thermal energy produced is always in strict proportion to the kilowatt-hours. The modern, accurate result of repeating Joule's original experiment is that 1 kilowatt-hour always gives exactly 860.42 Calories. Energy is measured in scientific laboratories in units called joules, which were named after Joule, but these were not the units he used in his researches.

The Laws of Thermodynamics

The laws governing the transformations of energy form a branch of physics called thermodynamics. The principle that energy cannot be created or destroyed is the first law of thermodynamics, and it is one of the most remarkable generalizations in all of science, providing a single framework relating nuclear reactions with the growing of oats, the charge and discharge of batteries, the motion of planets in their orbits, and the motion of the atoms composing the planets.

The importance of energy for human needs was widely recognized long before the law of its conservation was accepted. For hundreds (and perhaps thousands) of years, ingenious inventors have unwittingly tried to violate this law by designing

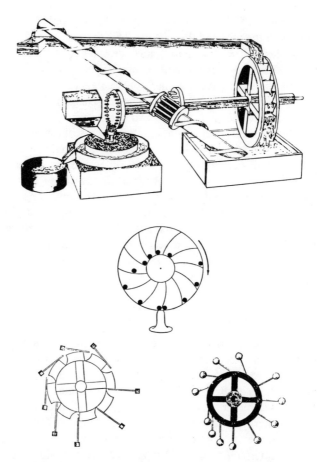

FIGURE 6-9. Perpetual motion machines. The top device runs by water power; some of the power is used to pump the water back uphill, and the rest for other purposes. The other three devices are variants of the same basic idea: a wheel which is kept in a state of permanent unbalance by balls or weighted arms which always exert more leverage on one side of the wheel than the other.

machines that would create unlimited amounts of energy. Such a device is called a perpetual motion machine. A Sanskrit manuscript of the 5th Century A.D. gives the first recorded description of such a device, and by the time of the Renaissance many were being proposed.[12] Some of them used the energy generated by falling water to pump the water back uphill, leaving some extra energy for other purposes. Other designs were based on wheels kept somehow in a state of unbalance in the earth's gravitational field, so they would keep turning indefinitely. Some of these devices are shown in Figures 6-9 and 6-10. None has ever worked.

IS HEAT A SUBSTANCE?

FIGURE 6-10. (a) An elaborate American invention, by a Mr. Horace Wickham, Jr. of Chicago, designed in 1870. (b) The simplest of all devices: a gravity-neutralizing substance is placed under one half of a wheel. The other half always feels the force of gravity, and the wheel keeps turning.

REFERENCE NOTES

1. Plato, *The Dialogues of Plato,* trans. B. Jowett, 2 vols. (New York: Random House, 1937).
2. Francis Bacon, *The New Organon and Related Writings,* ed. Fulton H. Anderson (Indianapolis: Bobbs-Merrill, 1960).
3. John Locke, "Elements of Natural Philosophy." In *The Collected Works of John Locke,* ed. C. Baldwin, 12th ed. (London: Printer, 1824).

4. Arthur Jensen, *Bias in Mental Testing* (New York: Free Press, 1980); Leon J. Kamin, *The Science and Politics of I.Q.* (Hillsdale, N.J.: L. Erlbaum, 1974).
5. Joseph Black, *Lectures on the Elements of Chemistry*. 3 vols. (1806; facsimile reprint, New York: AMS Press, 1984).
6. The biographical material on Rumford is from the book by S. C. Brown, *Count Rumford, Physicist Extraordinary*. (Westport, Conn.: Greenwood Press, 1979).
7. Count Rumford, "An Inquiry into the Weight Ascribed to Heat," in *The Collected Works of Count Rumford*, ed. S. C. Brown, vol. I (Cambridge, MA: Harvard University Press, 1968), pp. 27–48.
8. Ibid., "An Experimental Inquiry Concerning the Source of the Heat Which is Excited by Friction," pp. 3–26.
9. Ibid., vol. 2 "Of the Slow Progress of the Spontaneous Mixture of Liquids Disposed to Unite Chemically with each Other." pp. 74–79.
10. S. G. Brush, "Should the History of Science be Rated X?" *Science* 183 (1974): 1164.
11. William of Ockham, 1300–1348. In his original wording: "A plurality must not be asserted without necessity." Paraphrased more felicitously by John Ponce of Cork in the seventeenth century: "Entities should not be multiplied without necessity." In *Familiar Quotations*, John Bartlett, 14th ed. (Boston: Little, Brown, 1968).
12. Arthur W. J. G. Ord-Hume, *Perpetual Motion: The History of an Obsession* (New York: St. Martin's Press, 1977).

SUGGESTED READING

Brown, Sanborn Conner. *Count Rumford, Physicist Extraordinary*. Westport, Conn.: Greenwood Press, 1979.

―――. *Benjamin Thomson, Count Rumford*. Cambridge: MIT Press, 1979.

Brown, Sanborn Conner, ed. *The Collected Works of Count Rumford*. vol. 1; *The Nature of Heat* (1968); vol. 2; *Practical Applications of Heat* (1969). Cambridge: Harvard University Press, 1968–1969.

Cardwell, D. S. L. *From Watt to Clausius: The Rise of Thermodynamics in the Early Industrial Age*. Ithaca: Cornell University Press, 1971.

Layzer, David. "Heritability Analysis of I.Q. Scores: Science or Numerology?." *Science* 183 (1974): 1259; Herrnstein, Richard. "I.Q." *The Atlantic* (September 1971): 43; Sowell, Thomas. "The Great I.Q. Controversy." *Change* (May 1973): 33.

Roller, Duane. "The Early Development of the Concepts of Temperature and Heat. The Rise and Decline of the Caloric Theory." In *Harvard Case Histories in Experimental Science*. 2 vols. Edited by J. B. Conant and L. K. Nash. Cambridge: Harvard University Press, 1957.

Romer, Robert H. *Energy: An Introduction to Physics*. San Francisco: W. H. Freeman, 1976.

PROBLEMS

1. (a) State how temperature is defined and measured.
   (b) Explain the difference between the Fahrenheit and Celsius (Centigrade) scales.
   (c) This country is in the process of switching from Fahrenheit to Celsius. Why is the Celsius scale better?
2. Which of the following experimental facts were used, at the time of the controversy between the two theories, to support the kinetic theory, which were used to support the caloric theory, and which supported neither?
   (a) The Specific heat of mercury is 0.03 Cal/kgm ° C.
   (b) Ice melts at 0° C.

IS HEAT A SUBSTANCE?

(c) Heat flows faster through metals than through wood. — K
(d) Heat can be radiated through a vacuum. — C
(e) The work that can be obtained from a waterfall is proportional to the product of the weight of the water and the height of the waterfall.
(f) Light is a wave motion (an undulation). — K
(g) Water seeks its own level. — C
(i) Heat flows readily. —
(j) The work that can be obtained from a heat engine depends on the temperature *difference* in the engine.

3. Why was Rumford's Theory of the radiation of heat less plausible at the time he proposed it than the Caloric theory of the radiation of heat? Why did it become more plausible later?
4. 4 kgm of water at 50° C and 10 kgm of Iron (Specific heat = 0.1 cal/kgm-°C) at $-10°$ C are brought into contact. What is the final temperature?
5. 0.5 kgm of Aluminum at 100° C is dropped into 1 kgm of water at 20° C. The final temperature is 27.3° C. What is the Specific heat of Aluminum?
6. 0.1 kgm of ice at 0° C is dropped into 1 kgm water at 100° C. The latent heat of fusion of ice = 80 Cal/kgm. What is the final temperature? Hint: solve this problem in two steps. First, imagine the ice melts to form water at 0° C, cooling the hot water to a temperature t_1. Then solve the problem of finding the final temperature when 2 kgm of water at t_1 is mixed with 0.1 kgm of water at 0° C.
7. The Specific heat of Iron is 0.1 Cal/kgm-°C. Suppose the unit of heat had been defined as the amount of heat needed to raise the temperature of 1 kgm of Iron 1° C. Call this unit the I.T.U. (Iron Thermal Unit). What would the Specific heats of Iron and water be, using this unit?
8. A student measuring the latent heat of melting of ice is careless in weighing, so he adds 50 grams of ice to the water in the calorimeter when he thinks he only added 40. Will his calculated value for the latent heat be larger, smaller, or the same as the correct value? Explain your reasoning.
9. Distinguish transfer of heat by (a) conduction (b) convection (c) radiation, and give examples of each.

EXPERIMENTS

Many significant experiments on heat and energy are simple to do, and require equipment almost any college physics department will have on hand.

Ones we have used include measurements of Specific heats of metals, the latent heat of fusion of ice, and the mechanical and electrical equivalents of heat (equivalent to Joule's experiments). An experiment testing Newton's law of cooling can be used to introduce the concept of exponential decay and the use of a logarithmic graph (see Chapter 15).

7

Probability

HOW TO DEAL WITH UNCERTAINTY

In previous chapters the role of logic and mathematics in science was discussed, but no attempt was made to teach any specific knowledge of either. In this and a subsequent chapter we discuss two closely related branches of mathematics—probability and statistics—which play such a central role throughout science that we felt it worthwhile to try to explain some of their basic concepts.

By now the reader will have recognized the provisional and uncertain character of scientific truth—the fact that no scientific theory can ever be proven true beyond any possibility of doubt. There is nothing certain in science but uncertainty. This may lead one to believe that mathematics, representing the best kind of language with which to discuss exact things, may not be of much use when talking about the limits to exactness.

However, the opposite is true. Probability and statistics have been developed for just this purpose of talking precisely about impreciseness.

HOW TO GAMBLE AND WIN

The subject developed first was probability, and it began historically as a means of deciding how to bet in games of chance. The essence of any game of chance is ignorance—the inability of the player to predict precisely what faces of the dice will come up on the next throw or what cards the next hand will contain. This does not rule out the possibility that a sufficiently skilled and unscrupulous person will be able to control the outcome of the next throw of the dice, or cheat at cards. But if this happens the outcome is determined—there is no more ignorance, at least not for the crooked player. Nor is the possibility ruled out that a clever physicist, who studies exactly how the dice are held by the player and how fast they are thrown and in what direction, might be able to predict what will happen.

But most times that dice are thrown, the player does not make the effort to control the outcome. It is this state of ignorance that probability deals with.

HEADS OR TAILS?

Let us choose for simplicity the toss of a coin. In our ignorance, we are forced to content ourselves with the statement, "It will come out either heads or tails, and there is no reason to expect one result more than the other."

This is a statement of the greatest possible ignorance. We have predicted nothing, and neither result we get on tossing the coin will surprise us. It might seem that there is no mathematical way to talk about our helplessness. But there is—*if we change the question we ask*. Instead of asking what the next toss of the coin will show, we ask a different question: what will happen *on the average* if we toss a coin a large number of times? If the statement we made above about heads and tails being equally likely is really true, we can make very precise statements indeed about what will happen on the average if we toss the coin a thousand or a hundred thousand times.

The reader already knows the answer to the question—in a large number of tosses, heads will come up about half the time. The proof that this is a reasonable expectation—even though the reader knows it already—is provided by the theory of probability, which is also able to provide us with the answers to much more complicated questions, which our intuition and experience do not.

For example, if we toss a coin four times, what are the chances of getting two heads and two tails? The answer is that 3/8 of the time we will get this result, but this is not something we would have guessed offhand. We will prove it a few pages hence. More complicated problems can be posed and solved: what are the chances of throwing a 10 in dice? What are the chances of getting a 7 before a 10 in a sequence of throws of the dice? What are the chances of being dealt a royal flush in poker, or 13 spades in bridge?

Numerical Magnitudes

Probability theory deals with such phenomena by assigning a numerical magnitude to the chance of a particular outcome in a "trial"—the toss of a coin, the throw of a die, or the dealing of a card from a deck. Since we believe that heads or tails are equally likely, we say that heads has a 50% probability. So, of course, does tails. An impossible event—drawing an ace from the deck if you hold 4 aces already in your hand—can be said to have 0% probability. And an event that *must* happen, such as getting *either* heads *or* tails, can be assigned 100% probability.

Actually, it has been found more convenient to use fractions and decimals, rather than percentages. We can say equally well that heads comes up 50% of the time, or that heads comes up 1/2 of the time. So our scale runs as follows:

Impossible event	0
Heads on toss of a coin	0.5
Absolutely certain event	1.0

Are Tosses Independent?

Now we can go on to discuss the probabilities of more complex events, made up of sequences of simpler events, such as one toss of a coin. But to do this we must first decide whether the simple events are *independent* or not. For example, we plan to toss a coin 4 times and try to predict how often on the average we will get 2 heads. We begin with our assumption that the chance of heads on the first toss is 0.5. Suppose we do get heads on the first toss. What is the chance of getting heads on the second toss? Is it less than before, and the chance of tails greater?

The "Law of Averages"

There is a popular belief that the theory of probability provides us with a "law of averages" which ensures that if we get heads on the first toss of a coin, this must be compensated for by a better than even chance of getting tails on the second toss, so that in the long run the heads and tails will even out.

There is indeed a result in probability theory known as the "law of averages" (more accurately known as the law of large numbers). We will discuss it later in this chapter, but for the present it is sufficient to state that a coin does not remember whether it came out heads or tails the last time it was tossed, nor do our fingers possess some capacity we are not conscious of to bias the toss. All our experience with coins shows that the probability of getting heads with a coin is 0.5 regardless of the *results of any preceding tosses*. Each individual toss of a coin is an *independent* event, independent of previous and future tosses. Fortunes have been lost in gambling because this basic concept was not understood.

Suppose we take a deck of 52 cards, with the usual 13 cards of each suit, and ask what the probability is that the first card drawn from it is a spade. Since there are 13 spades among the 52, the probability is 13/52 or 1/4 or 0.250. Now we draw the card and examine it, and let us say that it is indeed a spade. Now we ask the probability that the second card drawn is also a spade. There are only 51 cards left, of which only 12 are spades. The probability that the second card is a spade is less than 1/4: it is 12/51 = 0.235, not 0.250. If the first card drawn had been a heart, the probability of the second card's being a spade could be calculated from the fact that there are 13 spades in the 51 remaining cards: it would be 13/51 = 0.255. In this example, we can see that the probability of drawing a spade as the second card is *dependent* on the outcome of the first drawing.

If, after drawing the first card, we had returned it to the deck and shuffled the deck before the second drawing, then it would not have mattered whether it had been a heart or a spade: the probabilities in this case would be *independent*.

Another example of dependent probabilities is provided by the weather: suppose, in January, we wish to predict whether it will rain on May 15. We might reasonably estimate the probability by consulting past weather records to determine what fraction of days in May has been rainy in the past; let us say the answer is 0.25. But if it is May 14, and it is raining, the probability of rain on May 15 is higher than 0.25. Rainy days tend to bunch together, and so do sunny days. The probability of rain on May 15 is not independent of what the weather was on May 14.

Sequences of Tosses

The tosses of coins really are independent, as shown by extensive experience, so we can predict the probability of the outcome of several tosses of a coin by the following simple rule, which we ask the reader to take on faith:

1. *The probability of two independent events both occurring is the product of their separate probabilities.*

This rule can be illustrated best by examples:

Suppose we toss a coin twice and ask, "What is the probability of both tosses giving heads?" Since the 2 tosses are independent, and the probability of heads in each is 1/2, the probability of both being heads is $1/2 \times 1/2 = 1/4$.

Obviously, the probability of 2 tails will also be 1/4.

The probability of getting 1 head and 1 tail is a little more complicated. The probability of getting heads on the first toss and tails on the second is given by our rule above: $1/2 \times 1/2 = 1/4$. But there is another way we can get 1 head and 1 tail: get tails first and then heads. The probability of this result is also 1/4.

We notice that the two outcomes, heads followed by tails (HT) and tails followed by heads (TH), are *mutually exclusive*—we may get one or the other in two tosses of a coin, but not both.

We give a second rule of probability, which again we ask the reader to take on faith.

2. *The probability that either one or the other of two mutually exclusive events will occur is the sum of their separate probabilities.*

There are two ways to get the result 1 head and 1 tail, and they are mutually exclusive: each has a probability of 1/4, and the probability that one or the other will occur is the sum, or 1/2.

This second rule is very useful because there are many situations in which we are not interested in the exact outcome of a sequence of events, but only in less detailed questions: how many heads on the average will occur in 1000 tosses without regard to the order in which the heads occur? What are our chances of drawing *any* 4 hearts in a poker hand of 5 cards? Either of these outcomes can occur in

PROBABILITY

many different but mutually exclusive ways, and our rule (2) permits their probabilities to be calculated.

As an example we consider the outcome of tossing a coin 4 times. The possible results are listed in Table 7-1. To make the distinction between a detailed outcome, such as HHHT, and a less detailed outcome, such as "3 heads and 1 tail," we use the terms *simple* and *compound*.

There are 16 possible simple outcomes. They are mutually exclusive, and each has a probability, according to rule (1), of $1/2 \times 1/2 \times 1/2 \times 1/2 = 1/16$. If we are interested only in the *numbers* of heads and tails, we can see that there are 6 different outcomes with 2 heads and 2 tails, and only one with 4 heads. The probabilities are, respectively, $6 \times 1/16$, or $6/16$ ($=3/8$), and $1 \times 1/16$, or $1/16$. Three heads or 3 tails can occur in 4 ways; thus each has a probability of $4/16$. We note that if we add all the probabilities up we have $1/16 + 4/16 + 6/16 + 4/16 + 1/16 = 1$, corresponding to certainty. One of the listed outcomes above *must*

Table 7-1
Tossing a Coin Four Times

Simple outcomes	Compound outcomes	Number of simple outcomes corresponding to each compound outcome	Probability of each compound outcome
H H H H	4 heads	1	1/16 = 0.0625
H H H T H H T H H T H H T H H H	3 heads 1 tail	4	4/16 = 1/4 = 0.25
H H T T H T H T T H H T T H T H T T H H H T T H	2 heads 2 tails	6	6/16 = 3/8 = 0.375
T T T H T T H T T H T T H T T T	1 head 3 tails	4	4/16 = 1/4 = 0.25
T T T T	4 tails	1	1/16 = 0.0625
		16[a]	1.00[b]

[a] Total number of mutually exclusive simple outcomes.
[b] Sum of the probabilities of the mutually exclusive compound outcomes.

occur in 4 tosses of a coin. Note that we have now shown why the outcome of 2 heads and 2 tails will occur 3/8 of the time.

A Proof of the Obvious

We are now prepared to show how to prove a statement already obvious to the reader: in a large number of tosses, heads will tend to occur half the time.

Let us begin this discussion with what seems like a paradox, which has already been raised in our discussion in the preceding paragraph. Suppose someone tosses a coin and gets 10 heads in a row. We would be quite startled, so much so that we would wonder if we are the victims of a trick. But what if he had gotten the following sequence: HHTHTTHTTH, which has 5 heads and 5 tails. We would not have been in the least surprised. But perhaps we ought to have been. The second sequence is not one bit easier to get than the first (10-heads) sequence. Our calculations above tell us that each has a probability of 1/2 multiplied by itself 10 times (1/1024). Let us try to make this plausible by considering what chances we have to get this specific sequence.

If we toss tails the first time, we have failed. If we get heads first, then we must get heads on the second toss, also—we have another chance to fail. If we get through the first 2 tosses successfully with 2 heads, we now toss the coin a third time. This time we must get tails. And so on for the rest of the tosses. It is obviously no easier than getting 10 heads in a row.

The reason the second sequence does not surprise us is a psychological one—we do not really notice the detailed order of the heads and tails, but we estimate roughly that there are about equal numbers of each. We look at the outcome crudely, not paying attention to the exact order of heads and tails. We are unconsciously classifying all compound outcomes of 10 tosses that have about half heads as though they were a single result. Now there are many simple outcomes that correspond to this result and there is only one simple outcome for 10 heads. So it is much more probable that one of the *many* 5-heads outcomes should be observed than the *single* 10-heads outcome. Therefore, by experience, we have learned to expect a 5-heads result more often than a 10-heads one; we find a 5-heads result unsurprising, and a 10-heads one surprising.

It can be proved easily, though we will not attempt it here, that a fifty-fifty division into heads and tails corresponds to a greater number of simple outcomes than any other and hence is more probable. This can be understood intuitively by referring to Table 7-1, where the outcomes of 4 tosses are listed.

The Bell-shaped Curve

In Figure 7-1 we show these results graphically. The heights of the rectangles represent the probabilities of the various results for 4 tosses. As we have pointed

PROBABILITY

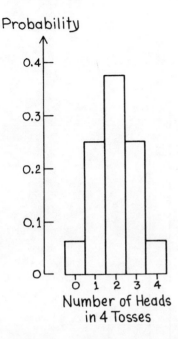

FIGURE 7-1. The probability of getting a certain number of heads in 4 tosses of a coin. Heads can come up any number of times from 0 to 4.

out, 2 heads and 2 tails is more probable than any other result because it can occur in more different (and equally probable) ways than any other result.

In Figure 7-2 and Table 7-2 we show the same result for 10 tosses, and in Figure 7-3 for 100 tosses. In this last figure, besides drawing individual rectangles for each outcome, we have also drawn a smooth curve through the centers of the tops of the rectangles.

The bell-shaped curve of Figure 7-3 occurs commonly in problems of probability when the number of different possible outcomes is large, as in 100 or more tosses of a coin. It is often called the Gaussian curve after the mathematician K. F. Gauss (1777–1855), who first studied its properties. It appears in nature under many other conditions as well. Whenever we have a large population classified according to some characteristic that can be measured numerically, and members of the population are divided up into groups according to this characteristic, the bell-shaped curve often (although not invariably) appears. For example, Figure 7-4 shows the result of measuring the heights of 1000 college students, and listing them according to height as measured to the nearest inch.

Using the figures, or tables like Table 7-2, we can calculate other probabilities. For example, in 10 tosses of a coin, what is the probability of getting *at least* 5 heads? To find this, use rule (2): add the individual probabilities of getting 5, 6, 7, 8, 9, or 10 heads; the result is 0.623.

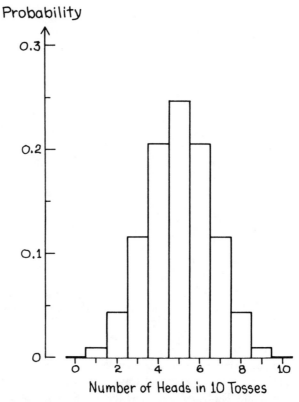

FIGURE 7-2. The probability of getting a certain number of heads in 10 tosses of a coin.

Table 7-2
Ten Tosses of a Coin

Outcome	Number of different simple outcomes corresponding	Probability (as fraction)	Probability (as decimal)
10H, 0T	1	1/1024	0.0009766
9H, 1T	10	10/1024	0.009766
8H, 2T	45	45/1024	0.04395
7H, 3T	120	120/1024	0.1172
6H, 4T	210	210/1024	0.2051
5H, 5T	252	252/1024	0.2461
4H, 6T	210	210/1024	0.2051
3H, 7T	120	120/1024	0.1172
2H, 8T	45	45/1024	0.04395
1H, 9T	10	10/1024	0.009766
0H, 10T	1	1/1024	0.0009766

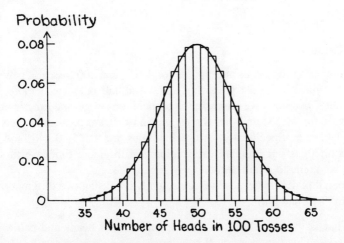

FIGURE 7-3. The probability of getting a certain number of heads in 100 tosses. The smooth bell shaped curve drawn through the tops of the rectangles. is called a Gaussian curve (after the mathematician Gauss). The curve is symmetrical about its highest point and it falls off equally rapidly on both sides.

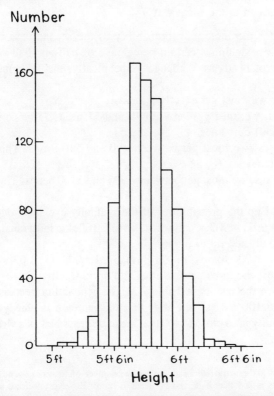

FIGURE 7-4. The number of individuals of a given height in a given group of 1000 college students. Note the general resemblance to the coin-tossing probabilities (see Figure 7-3).

Another Paradox

On comparing the tables and figures for 4, 10, and 100 tosses, a new paradox emerges. We have made the point that a half-and-half outcome is more probable than any other, but when we compare the probabilities of getting *exactly* half heads and half tails for 4, 10, and 100 tosses we find that the probability gets smaller rather than greater. Specifically, the probabilities are 0.375, 0.246, and 0.0796. Yet this contradicts our intuitive expectation that the more times we toss a coin, the closer we come to a half-and-half division.

The paradox occurs because we are actually confusing two different meanings of the term "half-and-half":

1. What is the probability of getting *exactly* half heads and half tails in N tosses? This decreases as N gets large, as we stated above.*
2. What is the probability that the number of heads is within, say, 1% of exactly half of the total number of tosses as N gets large? This second probability is the one that increases as N increases. (We will use the figure 1% as an arbitrary example. If we had chosen 5% or 0.1% the argument would proceed in the same way.)

The reason for the increasing probability in the second case is that the requirement that we be within a certain percentage of half-and-half corresponds to different *numbers* of heads as N changes. Specifically, the 1% condition implies the following:

In 10 tosses, we must get exactly 5 heads.

In 100 tosses, we must get between 49 and 51 heads. This corresponds to 3 possible (compound) outcomes: 49, 50, and 51.

In 1,000 tosses, we must get between 490 and 510 heads. This corresponds to 21 possible outcomes.

In 10,000 tosses, we must get between 4900 and 5100 heads. This corresponds to 201 outcomes.

When we add up the probabilities of the mutually exclusive outcomes in the last case, 4,900, 4,901, 4,902, . . . heads, up to 5,100, the individual probabilities, though each is small, add up to 0.956.

In the case of 1,000 tosses, the 21 probabilities add up to 0.493.

For 100 tosses, the three probabilities add up to 0.236.

We summarize the results in Table 7-3. The probability decreases slightly as we go from 10 to 100 tosses, but after that the increase is steady. By 100,000 tosses, the probability is so close to 1.0 that it can be described as a virtual certainty.

---

* Note that we cannot get exactly half heads unless the number of tosses is even. In 99 tosses, an outcome of 49 heads is less than half, and 50 heads is more. However 49 heads and 50 heads are equally probable in this case, and each is more probable than any other outcome.

Table 7-3
Probability of Getting Exactly Half Heads, and of Getting within 1% of Half Heads

Number of tosses	Probability of exactly half heads	Number of outcomes within 1% of half heads	Probability of getting within 1% of half heads
10	0.246	1	0.246
100	0.0796	3	0.236
1,000	0.0252	21	0.493
10,000	0.0080	201	0.955
100,000	0.0025	2,001	1.000[a]

[a] Actually a little less than one: 10 decimal places would be needed to show the difference.

The Law of Averages Justified

Suppose we start to toss a coin 100,000 times, and by chance the first 10 tosses come out all heads. We have some intuitive feeling that as we keep on tossing we should approach 50% heads. But the sense in which we approach 50% heads is the second one above: the probability of approaching to within 1% of exactly half should get greater and greater. Those 10 extra heads are about half of 21, about one-twentieth of 201, about one two-hundredth of 2,001. Even though the continued tossing will not lead to 10 extra tails to compensate for the 10 heads, on a *percentage* basis those 10 heads make less and less of a difference. Eventually, they are swallowed up in the 1% figure.

As Tippet puts it, the law of averages works by a *swamping* effect rather than a *compensating* effect.[1]

Uncertainty Remains

Our discussion of probability started with a statement of ignorance—we could predict nothing about the outcome of one toss of a coin. But from this simple starting point we were able to predict a lot about sequences of many tosses. Notice, however, that something of our original uncertainty remains. We can say that 10 heads in a row is an improbable event (1/1024), but we do not say that it will never occur. In fact, saying the probability is 1/1024 means that, on the average, 10 heads in a row will turn up once every 1,024 times we toss a coin 10 times. Of course, we cannot predict which sequence of 10 tosses will yield 10 heads.

We cannot even state that in 1,024 attempts 10 heads *must* come up once. They need not, or alternatively there is a small probability that they might occur 5 times in the 1,024 trials. We can only talk about the *chances* of various outcomes, and never reach certainty about what has not yet happened. Anything *can* happen.

Black Balls and White Balls

The tossing of a coin has two possible outcomes and extensive experience has taught us that they are equally probable, so each outcome is assigned a probability of 1/2. When we draw a card from a bridge deck and ask what is the probability of getting a particular suit—hearts, spades, diamonds, or clubs—again each is equally probable, and each has a probability of 1/4.

But we can also deal with events where the different outcomes are not equally probable. A simple example is the following: Imagine an enormously large container holding a large number of balls, 20% of which are white and the remainder black. If we reach in blindfolded and draw out a ball, there are two outcomes possible—white ball or black ball—but they are not equally probable. We expect to get white balls 20% of the time, so the probability of getting a white ball is 1/5 and that of getting a black ball is 4/5.

The individual outcomes for drawing 4 balls in succession are given in Table 7-4. They are the same as the outcomes in 4 tosses of a coin, and, as before, there are more simple outcomes for 2 white and 2 black than any other. However, since white balls are so much less likely than black balls, the probabilities are not the same. Rule (1) tells us the probability of 4 whites is $(1/5)^4 = 0.0016$, while the probability of 4 blacks is $(4/5)^4 = 0.4096$. The probability of any one simple

Table 7-4
Drawing Four Balls

Simple outcome	Probability of each	Compound outcome	Probability of each
B B B B	0.4096	4B	0.4096
B B B W	0.1024		
B B W B	0.1024	3B, 1W	0.4096
B W B B	0.1024		
W B B B	0.1024		
B B W W	0.0256		
B W B W	0.0256		
W B B W	0.0256		
B W W B	0.0256	2B, 2W	0.1536
W B W B	0.0256		
W W B B	0.0256		
W W W B	0.0064		
W W B W	0.0064	1B, 3W	0.0256
W B W W	0.0064		
B W W W	0.0064		
W W W W	0.0016	4W	0.0016

PROBABILITY

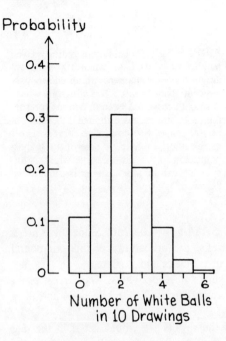

FIGURE 7-5. Probabilities of outcomes in drawing 10 balls from a container with 20% white balls.

Table 7-5
Probability that out of 10 Balls Drawn from a Bag Containing a Given Percentage of White Balls, N Are White

N	Probability of N for given percentage of white balls	
	20%	10%
0	0.107	0.349
1	0.264	0.387
2	0.302	0.194
3	0.201	0.057
4	0.088	0.011
5	0.026	0.0014
6	0.0055	0.0001
7	0.0008	*
8	0.0001	*
9	*	*
10	*	*

* Smaller than 0.0001.

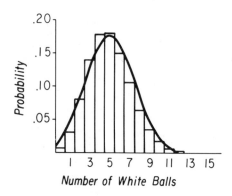

FIGURE 7-6. Probability of getting a given number of white balls when 100 balls are drawn from a container with an enormously large number of balls, 5% of which are white. A smooth curve has been drawn to represent the probabilities, even though $5\frac{1}{2}$ balls out of 100 cannot be white. The curve falls off more rapidly to the right side of the highest point than to the left. A curve which is unsymmetrical in this fashion is said to be "skewed."

outcome with 2 whites and 2 blacks is $(1/5)^4 (4/5)^2 = 0.0256$. Since there are 6 simple outcomes with 2 whites and 2 blacks, the probability of this compound result is 0.1536, less than the probability of 3 blacks and 1 white.

Figure 7-5 and Table 7-5 give the results for 10 drawings. The largest probability, as we would guess intuitively, corresponds to 20% white balls and 80% black.

We have also included in Table 7-5 the results we would get if in the bag there were 10% white balls, rather than 20%. This time the result of largest probability is to get 1 white ball out of 10 drawn. Figure 7-6 shows the results for 100 drawings from a bag with 5% white balls. The curve is clearly different in shape from the curve of Figure 7-3. That curve was symmetrical about the highest point: it fell off equally rapidly on both sides. The curve in Figure 7-6 shows that the probability of getting white balls falls off more rapidly on the right side than on the left. Drawing 5 white balls is most probable, but 6 is less probable than 4.

ANOTHER MEANING OF PROBABILITY

In the above, we have described the concept of probability in terms of processes that can be repeated an indefinitely large number of times, with our degree of ignorance and consequent inability to predict the outcome repeated each time.

But the term *probability* is used in another seemingly quite different sense, both by scientists and by laymen, to refer to events that take place only once and can never be repeated. What is the probability that the Democratic candidate will win the United States presidential election in 1992? What are the chances of the Soviet Union's swimming team winning all the gold medals in the next Olympic games? No one would estimate the Democratic chances in 1992 by the fraction of Democratic victories in earlier elections. Instead, one looks at the unique features of the situation in the year 1992. Who is the Democratic candidate? Who is his opponent? What are the issues and how do the American voters feel about them?

PROBABILITY

There will be only one 1992 election, and the circumstances prevailing in that year have not previously occurred and will not be repeated.

Yet we talk about the probability of such events, and are willing to bet on the outcomes provided we get the odds we consider reasonable, just as we are willing to bet on the toss of a coin or the turn of a card. Probability in this case represents a highly subjective judgment by an individual, and no two individuals are likely to agree exactly on the proper odds for such a bet, nor is there a way to prove one right and the other wrong. A person who bets even money on the Democratic candidate loses his bet if the Democrat loses, but his estimate that the odds were even before the election has not been proven wrong by this outcome.

Although this second meaning of the term *probability* does not lend itself to much in the way of mathematical analysis, it is a common usage nevertheless.

We used *probability* in this second sense when we said that the result of an experimental test of a scientific theory will make that theory either less probable or more probable. The change in a theory's probability which results from an experiment cannot easily be assigned a numerical magnitude. It is entirely a subjective judgment of the scientific community, and scientists often differ about it. Probability in the first sense has to do with trials that could be repeated an indefinitely large number of times. As far as we can tell, each toss of a coin is like any other. But when it comes to testing scientific theories, there are only unique events, like the United States election of 1992. There are not millions of universes, in some of which heat has been found to be a substance and in others the motion of atoms. So when we talk about the relative probabilities of these two theories we cannot mean what we mean when we talk about coins, cards, or black balls.

APPENDIX 7-1: COMPUTING AVERAGES

Introduction

In considering the outcome of a large number of tosses of a coin we stated three conclusions from the theory of probability that sounded contradictory, but were not.

1. The probability of getting *exactly* half heads* is greater than the probability of any other outcome.
2. The probability of getting *exactly* half heads gets less and less the more times you toss the coin.
3. The probability that the number of heads will be within a *specified percentage* (say 1%) of exactly half heads gets greater the more times you toss the coin (the "law of averages").

---

* In an *even* number of tosses; see the footnote on p. 156.

We want now to explore some points related to the second statement, which necessarily implies that the more you toss the coin the more likely it becomes that you get an outcome *other than* exactly half heads.

To do so we will calculate some averages. Let us imagine a number of repetitions—trials—of the experiment of tossing a coin N times. In any one trial we may or may not get $N/2$ heads; the outcomes of the various trials need not be the same, of course, but if we perform a large number of trials we can calculate an *average* outcome. Naturally we expect that the average number of heads will be $N/2$, but can we prove this?

First let us consider the simplest case: 1 toss, repeated many times. There are only two equally probable outcomes:

Outcome	Number of heads
H	1
T	0

In a large enough number of trials, each of the above outcomes will occur, on the average, equally often. If, for example, we make 1000 trials of the toss, the number of heads will be close to 500. The average number of heads per trial is, like any other average, this result divided by the number of trials (1000), giving the result 0.5. Note that we could have obtained the same result by taking the average of the two equally likely outcomes in the table above:

$$\text{average} = \frac{1 + 0}{2} = 0.5$$

In a 2-coin toss the equally probable outcomes are:

Outcome	Number of heads
HH	2
HT	1
TH	1
TT	0

The average over the outcomes is

$$\frac{2 + 1 + 1 + 0}{4} = 1$$

and if we had performed a large number of trials of a 2-coin toss, each of the outcomes would have been obtained about 1/4 of the time, giving the same average.

We can easily continue this line of reasoning to 5 or 6 tosses, but the table gets longer, and the arithmetic more tedious. Ten tosses has 1024 equally likely

PROBABILITY

outcomes, and by this time our patience would be exhausted. How are we to prove the result for 500 tosses or 1,000,000?

Actually we can, if we stop doing computations and use mathematics instead.

Tricks with Mirrors

To do so, let us first change our approach a little. Instead of calculating the number of heads in N tosses, let us calculate the *difference* between the number of heads and the number of tails. For example, in a 10-coin trial, one possible outcome is 7 heads and 3 tails. The difference between the number of heads and the number of tails is $7 - 3 = 4$. Another way to put this is to imagine that our coins, instead of being marked with heads on one face and tails on the reverse, are marked with the symbols $+1$ and -1. Then when we toss 10 coins we just add the numbers. The result is of course the same as if we had subtracted the number of tails from the number of heads. Let us use the symbol D for the outcome. The two-coin toss gives the following result.

Old outcome	New outcome	Number of heads	D
HH	$+1\ +1$	2	2
HT	$+1\ -1$	1	0
TH	$-1\ +1$	1	0
TT	$-1\ -1$	0	-2

We can now average the Ds rather than the number of heads. If the average D is zero, this is equivalent to an equal number of heads and tails. Our average D is calculated the same way as the average number of heads, by a simple average over the equally likely outcomes.

Let us consider the 10-coin case. Our list, as we mentioned, has 1024 individual outcomes. One outcome on that list is 10 heads, corresponding to a D of $+10$. Another equally likely outcome is 10 tails, with $D = -10$. The average D is equal to a fraction, the numerator of which contains the sum of all 1024 D values, and the denominator of which is the number 1024 itself. The two outcomes we have considered so far add up to zero.

Now consider all the outcomes for which there are 9 heads and 1 tail. The number of such outcomes is equal to the number with 9 tails and a head. To show this, consider one of each of them:

(a)	H	H	T	H	H	H	H	H	H	H		
(b)	T	T	H	T	T	T	T	T	T	T		
or (a)	$+1$	$+1$	-1	$+1$	$+1$	$+1$	$+1$	$+1$	$+1$	$+1$	($D = +8$)
(b)	-1	-1	$+1$	-1	-1	-1	-1	-1	-1	-1	($D = -8$)

(a) and (b) are equally likely. Thus for every D on the list with a value of $+8$, there is another with a value of -8. It is as though each outcome with a positive D has a mirror image with a negative D. All of these add up to 0. Notice that we have not bothered to say how many 9-head outcomes there are: we have had no need to.

So it should be clear that we can extend the argument to every D that is not zero. For every positive value of D on the list there is a negative D that just cancels it. Thus the average D of the 10 tosses is zero.

It should also be clear that our argument is as applicable to a million tosses as to 10.

We have thus proved that the average $D = 0$ for *any number* of tosses, which is of course equivalent to proving that the average number of heads in N tosses is $N/2$. The kind of argument we have used, which appeals to some symmetry within the problem, is called a "symmetry argument." One can see that, when applicable, a symmetry argument is very powerful and very economical. Such arguments are used frequently in modern physics and modern mathematics; in fact there is a branch of mathematics called group theory that deals exclusively with the mathematical consequences of symmetry.

To summarize, we can see that though *exactly* $N/2$ heads in N tosses gets less and less likely the larger N is, the average result is always $N/2$—there is a cancellation of "positive" and "negative" outcomes.

The Average of the Unaverage

We now want to find a different average, to describe the likelihood of outcomes other than $N/2$ (or $D = 0$). Looking at the table for 2 tosses we note the equal likelihood of HH and TT. Both of these differ from an outcome with an equal number of heads and tails by 1 toss.

The Ds are respectively $+2$ and -2, and differ by 2 units from $D = 0$. There is a sense in which a D of $+2$ and a D of -2 are equally far from a D of 0, and this suggests that we can express the tendency to get a result different from $N/2$ (or $D = 0$) by averaging over the Ds but disregarding the $+$ and $-$ signs. In other words when D is positive, we leave it alone, and when D is negative we change it to positive. We use the symbol $|D|$ for the D calculated in this fashion: $|D|$ is called the "absolute value" of D. The average over the 2-coin toss is now given by

$$\frac{2 + 0 + 0 + 2}{4} = 1.$$

As a further example we perform the averages over the 4-coin toss in Table 7-6.

PROBABILITY

Table 7-6
Various Averages for a Four-Coin Toss

Outcome	Number of heads	D	|D|	D^2
H H H H	4	4	4	16
H H H T	3	2	2	4
H H T H	3	2	2	4
H T H H	3	2	2	4
T H H H	3	2	2	4
H H T T	2	0	0	0
H T H T	2	0	0	0
H T T H	2	0	0	0
T H H T	2	0	0	0
T H T H	2	0	0	0
T T H H	2	0	0	0
T T T H	1	−2	2	4
T T H T	1	−2	2	4
T H T T	1	−2	2	4
H T T T	1	−2	2	4
T T T T	0	−4	4	16
16 outcomes	Sums	0	24	64

Averages: $\text{Av } D = \dfrac{0}{16} = 0$

$\text{Av } |D| = \dfrac{24}{16} = 1.5$

$\text{Av } (D^2) = \dfrac{64}{16} = 4$

$\sqrt{\text{Av } (D^2)} = 2$

The average of $|D|$ is thus a measure of the tendency to get a different result from $N/2$ for the number of heads. We have found Av $|D|$ for $N = 2$ is 1, and Av $|D|$ for $N = 4$ is 1.5. It is not hard to show that as N increases Av $|D|$ does also, but we will not do so. The reason for the D^2 in Table 7-6 will be explained shortly.

A New Headache: The Root-Mean-Square

We hope the idea of using $|D|$ as a measure of the deviation from the expected average made sense, and we wish this were all the mathematics we had to impose on the reader.

Unfortunately things are not so simple. For a variety of reasons which we cannot give here, mathematicians have concluded that the average of $|D|$ is not the best quantity to use for this purpose.

We used $|D|$ rather than D to prevent cancellations because of opposite signs. There is another way in mathematics to eliminate negative signs, namely to square the number in question:

$$2^2 = 2 \times 2 = 4 \qquad (-2)^2 = -2 \times -2 = 4$$

So in our procedure we could have also avoided cancellations by squaring the D values. Now squaring a number not only ensures that it has a positive sign, but also changes its magnitude:

$$3^2 = 9 \qquad (-4)^2 = 16 \qquad \left(-\frac{1}{10}\right)^2 = \frac{1}{100}$$

So after we have squared the Ds, and obtained the average of the squares by dividing by the number of Ds on our list, we must undo the effect of this squaring on the magnitude, by taking its square root. The various procedures for a 2-coin toss are displayed in the following table:

Outcome	Number of heads	D	$|D|$	D^2
HH	2	2	2	4
HT	1	0	0	0
TH	1	0	0	0
TT	0	−2	2	4
Sums		0	4	8
Averages:		$\frac{0}{4} = 0$	$\frac{4}{4} = 1$	$\frac{8}{4} = 2$

$$\sqrt{2} = 1.414$$

Note that the square root of the average of the squares is not equal to the average $|D|$, but a little larger. This is most usually the case. The awkward expression, "the square root of the average of the squares," is usually referred to as the "root-mean-square" and abbreviated to "r.m.s."

We have said that the reasons for using the r.m.s. of the Ds rather than the average $|D|$ are beyond the scope of this book. We will however ask the reader to accept that this is the preferable and standard procedure for describing how often deviations from the average occur.

More Tricks with Mirrors

We have found that the average for D^2 (before taking the square root) was 2 for $N = 2$, and the earlier table shows that it is 4 for $N = 4$. It is tempting to conjecture that this is a general result: that the average of D^2 for N tosses is just N. It is, and we will prove it, again by a symmetry argument.

Let us consider many trials of a 2-coin toss. The coin, as we suggested before, is to be thought of as having a $+1$ on one face, and a -1 on the other. Before we toss the two coins we do not know which of the two possible outcomes we will get on either. Let us represent the outcome of coin 1 by the symbol Y_1, which may be either $+1$ or -1, and the outcome of coin 2 by Y_2.

Thus we have:

$$D = Y_1 + Y_2$$
$$D^2 = (Y_1 + Y_2)^2$$

High school algebra tells us that the result of squaring $(Y_1 + Y_2)$ is $Y_1^2 + 2Y_1Y_2 + Y_2^2$. While the result is an elementary one, it is worth working out in full detail.

Algebra	Arithmetic
$(Y_1 + Y_2)$	$(7 - 2)$
$\times (Y_1 + Y_2)$	$\times (7 - 2)$
$Y_1Y_2 + Y_2^2$	$-14 + 4$
$Y_1^2 + Y_2Y_1$	$49 - 14$
$Y_1^2 + 2Y_1Y_2 + Y_2^2$	$49 - 28 + 4$
	$= 25$

(Since $Y_1Y_2 = Y_2Y_1$)

We could also represent the result in a table, in which the boxes contain the *individual* products of the multiplication, and the total result is the sum of the entries in each of the boxes:

	Y_1	Y_2			7	-2
Y_1	Y_1^2	Y_1Y_2		7	49	-14
Y_2	Y_1Y_2	Y_2^2		-2	-14	4

The advantage of such a tabular display is not obvious for the two-coin toss, but if the reader attempts to work out in full the algebraic result for D^2 for a 6-coin toss, $(Y_1 + Y_2 + Y_3 + Y_4 + Y_5 + Y_6)^2$ he will see that the table makes it easier to keep track of all the individual products and avoid arithmetic errors.

Now we can fill in at least *some* of the numerical values. Although we do not know if Y_1 is $+1$ or -1, Y_1^2 must be $+1$. Similarly $Y_2^2 = 1$. Thus,

	Y_1	Y_2
Y_1	1	Y_1Y_2
Y_2	Y_1Y_2	1

and $D^2 = 2 + 2Y_1Y_2$.

The next trial of the 2 coins gives a new D^2, for which Y_1Y_2 may or may not be the same as for the first trial, depending on what Y_1 and Y_2 actually are (whether the coins turned up $+1$ or -1).

The results are as follows:

	Outcome
First trial	$(2 + 2Y_1Y_2)$
Second trial	$(2 + 2Y_1Y_2)$
Third trial	$(2 + 2Y_1Y_2)$
Mth trial	$(2 + 2Y_1Y_2)$

where the terms $2Y_1Y_2$ in each trial need not have the same value. They are either $+2$ or -2, with equal probabilities.

The average of D^2 is obtained by adding up all the outcomes of the M trials and dividing by M. There are clearly M 2's in the sum, and M terms of the type $2Y_1Y_2$. The terms $2Y_1Y_2$ can be $+2$ or -2, and have an equal probability of being either:

Y_1	Y_2	$2Y_1Y_2$
$+1$	$+1$	$+2$
$+1$	-1	-2
-1	$+1$	-2
-1	-1	$+2$

Clearly, by symmetry, we tend to have complete cancellation, and these terms give a net contribution of *zero* to the average of D^2.

PROBABILITY

Thus we have proved again, and by a more tedious argument than before, that for $N = 2$ the average of D^2 is 2.

The advantage of doing the problem this way, however, is that unlike the direct approach by tabulation of the outcomes, which becomes more and more tedious the larger N becomes, this approach is easily extended to any number of tosses.

Let us display a 6-coin toss:

	Y_1	Y_2	Y_3	Y_4	Y_5	Y_6
Y_1	1	Y_1Y_2				
Y_2	Y_1Y_2	1			Y_2Y_5	
Y_3			1			
Y_4				1		
Y_5		Y_2Y_5			1	
Y_6						1

Note that we have not bothered to fill in every box—the reader can do so if he wishes.

We see first that the boxes on the "principal" diagonal of the table always contain the number 1. If there are N coins, there are N such boxes in the table. All other boxes contain products of the Y values for two separate independent tosses. The average of D^2 is thus obtained by preparing one such table for each trial of the 6-coin experiment, and adding up all the boxes for all M such tables, then dividing by M:

$$\text{Av}(D^2) = \frac{M \times 6 + \text{terms } Y_1Y_2, Y_1Y_3, Y_2Y_5, \ldots}{M}$$

By exactly the same symmetry argument, the net value of the sum of the terms Y_1Y_2, etc., tends toward zero after division by M, and we conclude $\text{Av}(D^2) = 6$ for the 6-coin case. And we see the same argument applies to *any* N value, giving us finally

$$\text{Av}(D^2) = N \text{ for } N \text{ tosses}$$

$$\text{r.m.s. of } D = \sqrt{\text{Av}(D^2)} = \sqrt{N}$$

Here we can add, with a sigh of relief, the letters *Q.E.D.*, the abbreviation of the Latin phrase *quod erat demonstrandum* (which was to be proven).

APPENDIX 7-2: APPLICATIONS OF PROBABILITY THEORY TO MOLECULAR DIFFUSION

Introduction

Probability theory has a very great number of applications in science. The theory plays a major role in genetics, for example, and is indispensable in one of the most important branches of physics, statistical mechanics, which is concerned with deducing the properties of matter from the properties of the individual atoms it is composed of. "Statistical mechanics" is a misnomer: the field should have been called "probabilistic mechanics." It is the present-day descendant of the kinetic theory of heat.

We will give two simple but important examples of how probability is applied to these subjects. In this Appendix we apply it to molecular motion; in an Appendix to Chapter 8 we apply it to genetics.

Molecules in Motion

In Chapter 4, on the kinetic theory of heat, we described one of Rumford's experiments that suggested that the atoms of matter are in constant motion. In this experiment Rumford placed a layer of fresh water over a layer of salt water, and showed that even though the system was left undisturbed, the salt eventually spread upward into the fresh water until the concentration of salt in the water was uniform throughout the vessel. The result, in time, was the same as if the two liquids—salt water and fresh water—had been shaken up together, or stirred briskly.

Rumford considered three possible explanations for this phenomenon. The first was that the system was not really undisturbed: the small changes in temperature of the room, or vibrations from noise, produced some mixing of the two solutions. The second was that there was an attractive force in fresh water that drew the salt molecules upward against the force of gravity. The third, stated only vaguely by him, was that the water and salt molecules were in constant chaotic motion, colliding with each other frequently and rebounding, and that this erratic motion led to the gradual spreading of the salt molecules through the water.

There are a number of strategies we might pursue in order to decide which of these alternatives to believe. We could, for example, try to do the experiment more carefully, taking great pains to avoid any small changes in the temperature of the laboratory, and shielding it from noises or other sources of vibration that might cause mixing of the two solutions. We might, as suggested earlier, do an experiment to see whether a higher temperature causes the salt to spread through the solution more rapidly. Such experiments are among the fairly obvious possibilities suggested by the several hypotheses we are considering. Both of them would be important and informative.

However, there is a more indirect and very powerful alternative. We asserted

PROBABILITY

in the preceding chapter that theories often have unsuspected and far-reaching logical consequences, that these consequences can be discovered by the use of logic and mathematics, and that they provide additional opportunities for testing the theory experimentally.

We will describe here how certain consequences of the central assumption of the kinetic theory—that molecules are in constant motion—can be deduced with the aid of the theory of probability. Predictions can then be made about the behavior of the salt molecules, predictions unsuspected by those who originally proposed the kinetic theory. Next we can compare the predicted behavior with experimental observation; when we find agreement, our belief in the kinetic theory will be greatly strengthened.

The Tortuous Path: A Random Walk

In Figure 7-7 we represent the path that an individual salt molecule might follow in a liquid. We can see that the path is a tortuous one, with frequent changes of direction and speed as collisions with other molecules occur. It would be very difficult to predict the path of one molecule in detail—we would have to know a lot more than we are likely to about the speed and direction of all the molecules with which the one we are interested in will collide. And further, since no two molecules will follow the same path, we may legitimately wonder why we should

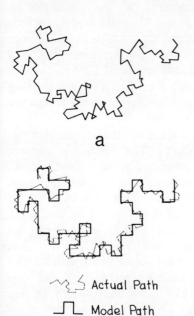

FIGURE 7-7. (a) Actual path of a molecule in a liquid (in two dimensions). (b) An approximation to the actual path composed of steps of equal length, and taking place only in one of four directions: up, down, left, or right.

want to predict the path of a single molecule in detail even if we could. The information hardly seems worth the effort it would cost.

The situation is similar to the one we encountered when trying to predict whether a coin would fall heads or tails if held a certain way and tossed in a certain direction with a certain force: maybe we could predict it, but why bother? Yet from our very ignorance in advance of whether the coin would come up heads or tails, we were able to predict from probability theory a lot about the *average* behavior in many tosses of a coin. We might try to do the same with the path of the molecule. The tortuous path is shaped by the collisions with other molecules. We might imagine that the path is composed of a series of short, relatively straight portions, traced out by the molecule *between* collisions, separated by abrupt unpredictable changes in direction that occur *during* the collisions (see Figure 7-7). Such a path, composed of independent steps, is called a "random walk." We can regard each segment of path as an "outcome" like the toss of a coin, but somewhat more complicated: instead of two outcomes, the segments of path can have various lengths and can be in any direction in space. It sounds too hopelessly difficult to apply probability: it isn't so hopeless as it looks, but it takes more mathematics than we can employ here.

Making a Model

We can try to make some simplifying assumptions. We can pretend that the motion of the molecule is simpler than it really is, simple enough to apply the little bit of probability theory that we have learned, and we can try to predict the average behavior of the molecules. Once having done this, we can ask the questions: How far off are we? How much in error are our conclusions because of the simplifications we have made? This process of simplifying a problem to make it solvable and then asking how closely the simplified but imaginary system approximates the real one is a common procedure in science.

It is common procedure because the real world is so infinitely complex that we do not try to answer all possible questions about it. The ones we try to answer must meet two conditions. First, they must be interesting questions, and, second, an answer must be possible. Now very often an answer is not possible if we ask for too much accuracy or detail. Our simplified picture of reality is often called a "model": it leaves out many of the complex details of reality, but we hope it can describe the actual behavior closely enough to be useful, and to make us feel that we understand the real phenomenon better. The *American Heritage Dictionary* gives as one definition of the word *model*, "a small object, usually built to scale, that represents some existing object." Toy trains are models of real trains. Some, quite primitive, are merely a few crudely painted blocks with wheels, pulled by a string. Others are detailed scale models, running on tracks, stopping at stations, and making appropriate noises. Though the resemblance can be amazing, no one would confuse them with the real thing.

Some models in science are literally that, as when an engineer makes a scale model of a planned bridge in order to study the stresses on the various cables and stanchions under different conditions of loading. The term is used more metaphorically when a psychologist, wanting to understand the influence of crowded living conditions on human beings, uses a colony of rats crowded together in a cage as a simpler and more controllable system to study. Sometimes the model is a mental image rather than a real object or group of objects, as when we speculate what will happen a hundred years from now if population continues to grow at a rate proportional to its size.

Some scientists have expressed the view that all scientific theories are models. This is a metaphorical use of the word that serves the purpose of reminding us that our theories are at best approximate descriptions of reality and that any of them may some day turn out to be wrong. The best we can claim for them is that they have worked so far in a practical sense. We prefer to distinguish a model from a theory on the basis that the element of approximation in a model is conscious and deliberate: we know what we are leaving out, but hope we can get along reasonably well without it.

Molecules, Dice, and Coins

In Figure 7-7 we represented schematically the path a single salt molecule might traverse as it collides with the other molecules of the system. This figure is, of course, a two-dimensional representation of a three-dimensional situation: the path traced out by the molecule can be better visualized as a crumpled, hopelessly tangled ball of yarn. We show in the figure the facts that the lengths of the steps (which represent motion between successive collisions) vary and that the molecule can go in any direction after a collision.

Now we will try to simplify this extremely complicated motion by pretending, first of all, that it is composed of steps or jumps of equal length taking place in equal time intervals. Second, we will assume that instead of taking place in any direction at all, each step will consist of motion in only one of six directions: up or down, left or right, forward (toward the observer) or backward. Each of these six directions, which the molecule can take on any step, will be equally probable, and will occur at random.

This is a drastic simplification of the real motion. But we do note that an imaginary molecule moving in this restricted way can get from any starting point in the container to any other part of the container, just as the real molecule can. In Figure 7-7 we compare the complicated path of a real molecule and a hypothetical path of a molecule moving according to our model, to show that the path of the molecule in the model can approximate the real one to some extent.

Now we must discuss the probability of a molecule moving in a particular path. We can "produce" a path in several ways. One is to toss a die for each step of the path and move the molecule according to the outcome as follows:

Result of tossing a die	Molecular step
1	up
2	down
3	right
4	left
5	forward
6	back

An even simpler way to produce a path is to toss a coin three times. The first toss is used to decide whether the first step will be up or down, the second toss chooses between right or left for the second step, and the third toss chooses between forward and back for the third step. This triple toss is repeated a large number of times. The two methods—die and coin—do not produce identical paths. Using a die, it is possible to have, say, three up steps in a row. Using the coin, a step up or down must be followed by a step either to the right or the left.

We now have a means to produce an indefinitely large number of paths of the model molecules. Every time we toss a die or a coin 1,000 times, we produce a possible path of 1,000 steps. If we repeat our sequences of 1,000 tosses 500 times, we have described 500 possible 1,000-step paths that molecules may traverse. We are not interested in the details of any one path, but rather in the *average* behavior of a large number of molecules.

Now we have made a model of the diffusion of the molecules; it is simpler than reality, but because it is simple we can use the mathematics we have described for coin tossing in order to predict what will happen to the salt molecules.

How Fast Does It Happen?

The qualitative result predicted by our model, that the salt spreads upward in the water, is too unsurprising to be really convincing. It is a result we were expecting the model to give almost before we worked out any of the mathematical details. The model and the theory behind it become convincing only to the extent that they can successfully predict more things than we knew at the start.

We turn therefore to a quantitative question: does our simple model correctly describe how fast the diffusion of the salt occurs? We have, after all, represented a complicated path with jumps of different lengths taking place at equal intervals. Obviously, our description of the individual path of any one molecule is incorrectly given by our approximation. But it is not hard to imagine that in spite of the unequal length of the jumps in the real path, there must still be an average jump length, and similarly that in spite of the varying time intervals between collisions, there must be an average time interval. In the spirit of our approximation, we can assume the uniform jump length of our simplified model to be the same as the average jump length of the real path, and the equal time intervals of the model to be equal to the average time interval between collisions on the real path.

PROBABILITY

Now we ask, what can we predict with this model that we did not know already?

The Random Walk in One Dimension

Since we have dealt at such length with coin-tossing, to the exclusion of dice, cards, and other chance processes, let us use for our model of the random walk a coin-tossing analogy.

For simplicity we will consider only the steps made by the molecule parallel to one particular direction; that is, steps to the left or to the right only. Our coins, as described in the Appendix 7-1, will bear the symbols $+1$ and -1: a $+1$ will mean that the molecule will take one step to the right, and -1 that it will take one step to the left.

Our laboratory experiment, rather than duplicating the one performed by Rumford, will have to be a little different. We can imagine a long narrow glass tube

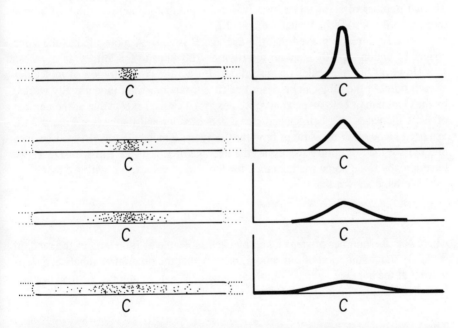

FIGURE 7-8. Diffusion of salt in a "one-dimensional" container (a long narrow tube). The container is, of course, three-dimensional, but one dimension, the length, is much greater than the other two, and we pay attention only to the motion of the molecules in this dimension. Initially the molecules are localized in the center, but as time goes on they spread out. In a long enough time (not shown in the figure) they will be uniformly distributed throughout the tube. In the early stages of diffusion, their distribution follows a Gaussian curve. On the left, the concentration is schematically indicated by the number of dots shown in the tube; on the right the concentration at each point in the tube is shown by the height of the curve at that point.

containing water. At the start of our experiment we introduce some salt solution right at the center of the tube, so that all the salt molecules are concentrated in a very narrow region of the tube (see Figure 7-8). We will refer to this starting point in the tube as the "origin." The molecules begin their random walks. Some move to the left, some to the right. As time goes on they spread out in the tube, getting on the average further and further from the origin.

How fast do they spread out? Since the individual molecules follow different paths, the most we can hope to predict is the *probability* of finding a particular molecule at a given distance from the origin. Since we are imagining the experiment being done on a very large number of molecules, the *number* of molecules found at a given distance is proportional to the *probability* that any one molecule will end up at that distance. How does the probability depend on distance?

Applying Our Mathematics

In the previous appendix we already have all the basic mathematics we need. The *net* number of steps taken by a molecule (number of steps to the right minus number to the left) is just what we called D.

We have already proved that the average D is zero. A molecule has the same chance of ending up 100 steps to the right as 100 steps to the left.

However the molecules do spread out. In fact the concentration of molecules at each point in the tube can be described (if our random-walk model really works) by the Gaussian or bell-shaped curve we described earlier, except that as the number of steps increases, the bell-shaped curve gets broader and flatter (see Figure 7-8). We need some other measure of how far the average molecule travels, as time goes on, and we may of course use either the average of $|D|$, or the r.m.s. of D. As we have already done some mathematics for the r.m.s. values, we will use that.

We have shown that

$$Av\,(D^2) = N$$

where N is the number of steps taken by each molecule as it undergoes the random walk. By the assumption of our model, namely that the time interval between steps is fixed at the average value, all molecules take the same number, N, of steps. The r.m.s. is just the square root of the average of D^2: $\sqrt{Av\,(D^2)}$.

The Experimental Test

How can we test this result by an experiment? What in fact can we measure? We have introduced our diffusing substance, in this case salt, into the center of the glass tube. Initially the salt is concentrated at the origin. As time goes on, its concentration at the origin decreases. We can by chemical analysis determine at any given time the number of salt molecules at each point of the tube. Some are close to the origin, some are far away.

PROBABILITY

Consider the following table, purporting to represent the outcome of an experiment in which 1,000 molecules were placed at the origin, and some time later, say one hour, the number of molecules at various distances from the origin were determined chemically. The distances were measured to the nearest millimeter, with the following results:

Final position	Number of molecules	D^2	Number × D^2
10mm	1	100	100
9	1	81	81
8	4	64	256
7	6	49	294
6	18	36	648
5	38	25	950
4	65	16	1040
3	91	9	819
2	110	4	440
1	115	1	115
0	123	0	0
−1	105	1	105
−2	94	4	376
−3	89	9	801
−4	53	16	848
−5	42	25	1050
−6	26	36	936
−7	10	49	490
−8	6	64	384
−9	3	81	243
−10	0	100	0
	1000 Total		

Sum of squared distances = 9976

$$\text{Av}(D^2) = \frac{9976}{1000} = 9.976$$

$$\text{r.m.s.} = \sqrt{\text{Av}(D^2)} = 3.158$$

As a practical matter we do not observe individual molecules, but we can perform the experiment by using molecules of a substance which has a distinctive color, for example a dye. Since we can measure the *intensity* of the color at different points in our long, narrow tube, we can "count" the molecules that way. Our calculation of r.m.s. distance $\sqrt{\text{Av}(D^2)}$ is done just as for the example of the 1,000 molecules above.

There is one important prediction we can make from our model already: the shape of the histogram representing the number of molecules at each distance should resemble the shape of the curve giving the outcomes of a large number of coin-tosses, the so-called bell-shaped or Gaussian curve, as shown in Figure 7-3. Is this prediction fulfilled?

The answer, from many experiments on diffusion of molecules under the conditions in which we would expect our model to work, is emphatically *yes*. It is pretty much what we find in nature (see Figure 7-7).

What We Do Not Know

But unfortunately we lack important information needed to predict precisely the outcome of the experiment.

What we *do not* know is this:

1. We said that the step length in the coin-tossing procedure must be related to some average step length for the real path of the molecules. But we have no idea of what that length should be.
2. Our formula for $Av (D^2)$ is in terms of the number of steps N, but we can't measure N; all we can measure is the time elapsed. We have no idea how many steps the molecules make in that period of time. We would need some average "step" duration and it is not observable.

We can cope with our ignorance by the usual mathematician's trick. We give symbols to the unknown quantities and plunge ahead, hoping somehow to get that knowledge eventually.

So we call the unknown average step length L, and the unknown average time it takes to make a step T.

Our formula from coin-tossing probabilities was:

$$Av (D^2) = N$$

This formula was obtained by assuming step-lengths of 1 unit: a head represented a step of one unit of length to the right and a tail one unit to the left, that is, 1 inch or 1 centimeter. Let us do a simple calculation if the step is L units, rather than 1 unit. We will do the calculation for a two-step walk:

Coin outcome	Distance	Distance squared
HH	$2L$	$4L^2$
HT	0	0
TH	0	0
TT	$-2L$	$4L^2$
		$\overline{8L^2}$

$$\text{Average} = \frac{8L^2}{4} = 2L^2$$

With a step of unit length (when $L = 1$) the average of D^2 was 2.
We can see that for steps of length L the formula becomes:

$$Av\,(D^2) = NL^2$$

We now need a relation between N and the time t of the experiment. If T is the time of one step, then the number of steps $N = t/T$. As an example, if the time of one step is 0.01 seconds, and 1,000 seconds have elapsed, the number of steps taken is $1{,}000/0.01 = 100{,}000$ steps.

So, combining these results with our original formula we have

$$Av\,(D^2) = L^2 \frac{t}{T}.$$

Neither L nor T are known: we do not lose any knowledge we have by combining them into a new, equally unknown quantity K: $K = \dfrac{L^2}{T}$ and

$$Av\,(D^2) = Kt.$$

What Is Testable

We do not know K, but still there is something testable. Does $Av\,(D^2)$ increase in proportion to time t? Even if we do not know K we can see that doubling t will double $Av\,(D^2)$, tripling t will triple $Av\,(D^2)$, and so on.

That is an experiment we can do. We can measure $Av\,(D^2)$ by the procedures described earlier, and can measure the time elapsed t. We can repeat this for a time period twice as long, or three times as long, and see if $Av\,(D^2)$ doubles or triples.

When we do the experiment described, the prediction is confirmed. $Av\,(D^2)$ does increase in direction proportion to t. But something else is gained. Although we did not know K before the experiment, we know it once the experiment is done. As an example, K for the diffusion of salt in water (Rumford's experiment) is found in the laboratory to be about .001 when distance is measured in millimeters and time in seconds. This does not seem very informative. It would tell us what the average step length is only if we already knew the average step time, but we do not know that either. Or conversely, if there were some reason to believe that the average step length were, say, 10^{-6} millimeters, we could deduce the average step time as 10^{-9} seconds. This is not very satisfactory.

The kinetic theory prediction that $Av\,(D^2)$ increases in proportion to t has been confirmed, and this adds considerable plausibility to the kinetic theory. But if only the kinetic theory could tell us what K should be, then if this turned out to be close to the observed K, our confidence in the theory would be so much greater!

Can the kinetic theory be used to predict the value of K? Yes. Does it come out close to the observed K? Yes. To understand the calculation requires much more mathematics and physics than we dare to introduce here, but it was performed for gases in a straightforward way in the nineteenth century, and modern computer experiments have made it possible to perform it for liquids also. The calculation uses the laws of physics to determine the average features of those very complicated paths illustrated in Figure 7-7.

We owe the treatment of molecular motion in these terms, and the recognition that it is directly related to the kinetic theory's picture of molecules in constant and chaotic motion, to Albert Einstein, who published his findings in 1905. It was an achievement that would have established a reputation for him as one of the outstanding scientists of his time, even if he had done nothing else.[2]

REFERENCE NOTES

1. L. H. C. Tippet, *Statistics* (Oxford: Oxford University Press, 1968).
2. Albert Einstein, *Investigations on the Theory of the Brownian Movement* ed. R. Furth, trans. A. D. Cowper (New York: Dover, 1956).

SUGGESTED READING

Articles by A. J. Ayer, "Chance," Warren Weaver, "Probability," and Mark Kac, "Probability," in Scientific American, *Mathematics in the Modern World*. San Francisco: W. H. Freeman, 1968.
Chapters on probability in the books by Harold R. Jacobs, *Mathematics: A Human Endeavor*. 2nd ed. San Francisco: W. H. Freeman, 1982, and by Morris Kline, *Mathematics for Liberal Arts*. Reading, Mass.: Addison-Wesley, 1967.
Weaver, Warren. *Lady Luck: The Theory of Probability*. New York: Dover, 1982. (Originally published, 1963.)

PROBLEMS

1. A die is a wooden cube the six faces of which are numbered by dots from 1 to 6. When the die is tossed on to a flat surface, any one of the faces has an equal chance of being uppermost. The numerical value of the uppermost face is the "outcome." Make a table of the possible outcomes of tossing two dice. What is the probability that the sum of the "outcomes" of the two dice is 7?
2. Joe forgets his girlfriend's phone number in Manhattan, so he dials a 7-digit number at random. What is the probability that he gets the right number on one attempt?
3. Max tosses two dice and Kevin tosses two coins. What is the probability that simultaneously both dice come up with the same number and both coins come up with the same face?
4. Which is more probable: to get 5 heads in 10 tosses of a coin, or 5,000 heads in 10,000 tosses? Explain.

5. A computer program is written that picks one of the letters A, E, or T at random (i.e., with equal probability). What is the probability that the word AT is produced? Give your reasoning.
6. A four-step, two-dimensional random walk is constructed as follows: A coin is tossed 4 times. The first and third steps are taken either to the right (heads) or to the left (tails). The second and fourth steps are taken either up (heads) or down (tails). What is the probability that the walk will end at the starting point?
7. There are 4 aces in a deck of 52 cards used in the game of blackjack. What is the probability that the first two cards dealt to the first player are both aces?
8. In a certain large city in the United States, the Jewish population is 10%. On a certain street, 3 people are stopped at random and asked if they are Jewish. All 3 answer "yes." Is the probability that the next person that walks by on this street is Jewish greater than 1/10, less than 1/10, or equal to 1/10? Explain your answer.
9. A man has a quarter, 3 dimes, a nickel, and a penny in his pocket. He reaches in and selects two coins at random. What is the probability that their net value *exceeds* 15 cents?
10. What is the probability of tossing a coin 8 times, and getting 7 heads in a row, followed by a tail?
11. An urn contains a very large number of balls, of which 2/3 are black and 1/3 are white. Three balls are drawn. What are the probabilities of each possible outcome?
12. A random walk is constructed using the above urn. A black ball gives rise to a step to the right $(+)$, a white ball a step to the left $(-)$. What is the average distance from the start after 3 steps?

EXPERIMENTS

Any number of experiments in probability may be done by tossing coins or dice, or drawing cards from a deck. Unfortunately such experiments quickly become tedious, a problem only partly alleviated by shaking a large number of coins in a container and then throwing them out on a table, or by similar expedients. One experiment we have used at Yeshiva College is to have each student toss 20 coins at a time, count the heads, repeat 50 times, and prepare a frequency table for comparison with both theoretical prediction and the combined outcome of all students in the class.

A more efficient procedure, if the students have access to a computer, is to simulate coin or dice tosses using the capacity of the computer to generate random numbers.

8
What Is Madness?
The Scientific Study of Mental Disorders

SECTION I: WHO IS MAD?

The words *psychology* and *psychiatry* are derived from the Greek word for "soul." The fields described by these words are those in which we attempt to study scientifically the most intimate and characteristic aspects of being human—how we feel, how we think, and how what we feel and think are expressed in what we do.

Studying these problems scientifically is neither the first nor the only way that they have been approached. They have been the concern of philosophy, literature, and religion for thousands of years. There are many who are skeptical of the value of a scientific approach to such deep and personal questions, and who feel that the attempt to be objective and precise will either miss the important things entirely, or distort them in the process of studying them. But the suffering and anguish of mental disorders are real enough, and for some the attempt to be scientific is justified by the hope of alleviating that suffering and anguish. For others the intellectual challenge is enough justification—can we gain a deep understanding of such complex problems through science?

We recognize in our daily experiences that there are people who behave strangely, who have a distorted view of the world, feel persecuted without any objective reason, have hallucinations, become hopelessly depressed and withdraw from friends and family, or commit suicide, when little, if any, objective basis for their feelings is apparent. We have learned to label such people "insane." But as we study such people more closely, and as we study ourselves more closely, we come to wonder how sharp are the boundaries between what is sanity and what is insanity.

Where do we draw the line—how do we decide who is sane and who is not? This question may not even seem to be a scientific one, but in this chapter we will see how central a role it plays in science. In less colloquial language, it is the problem of *classification*.

CHAPTER 8

A DEPRESSED GENIUS

The following quotation from the autobiographical writings of Tolstoi illustrates the problem:[1]

> The truth lay in this, that life had no meaning for me. Every day of life, every step in it, brought me nearer the edge of a precipice, whence I saw clearly the final ruin before me. To stop, to go back, were alike impossible; nor could I shut my eyes so as not to see the suffering that alone awaited me, the death of all in me, even to annihilation. Thus I, a healthy and a happy man, was brought to feel that I could live no longer, that an irresistible force was dragging me down into the grave. I do not mean that I had an intention of committing suicide. The force that drew me away from life was stronger, fuller, and concerned with far wider consequences than any mere wish; it was a force like that of my previous attachment to life, only in a contrary direction. The idea of suicide came as naturally to me as formerly that of bettering my life. It had so much attraction for me that I was compelled to practice a species of self-deception, in order to avoid carrying it out too hastily. I was unwilling to act hastily, only because I had determined first to clear away the confusion of my thoughts, and, that once done, I could always kill myself. I was happy, yet I hid away a cord, to avoid being tempted to hang myself by it to one of the pegs between the cupboards of my study, where I undressed alone every evening, and ceased carrying a gun because it offered too easy a way of getting rid of life. I knew not what I wanted; I was afraid of life; I shrank from it, and yet there was something I hoped for from it. . . . (pp. 28–30)
>
> I could not attribute a reasonable motive to any single act, much less to my whole life. I was only astonished that this had not occurred to me before, from premises which had so long been known. Illness and death would come (indeed they had come), if not to-day, then tomorrow, to those whom I loved, to myself, and nothing would remain but stench and worms. All my acts, whatever I did, would sooner or later be forgotten, and I myself be nowhere. Why then busy one's self with anything? How could men see this, and live? It is possible to live only as long as life intoxicates us, as soon as we are sober again we see that it is all a delusion, and a stupid one! In this, indeed, there is nothing either ludicrous or amusing; it is only cruel and absurd. . . .
>
> I was like a man lost in a wood, and who, terrified by the thought, rushes about trying to find a way out, and, though he knows each step can only lead him farther astray, cannot help running backwards and forwards.
>
> It was this that was terrible, this which to get free from I was ready to kill myself. I felt a horror of what awaited me; I knew that this horror was more terrible than the position itself, but I could not patiently await the end. However persuasive the argument might be that all the same something in the heart of elsewhere could burst and all would be over, still I could not patiently await the end. The horror of the darkness was too great to bear, and I longed to free myself from it by a rope or a pistol ball. This was the feeling that, above all, drew me to think of suicide. (pp. 36–37)

In addition to its meaning in ordinary speech the term depression has a technical meaning in psychiatry, describing a category of mental disorder, and a psychiatrist would be tempted to describe Tolstoi as suffering from a depressive disorder when he wrote this. Yet many of us feel despair at the inevitability of death at times.

WHAT IS MADNESS?

Part of Tolstoi's greatness as a writer lay in his ability to express his awarness of this dark side of life. He wrote a great short novel, *The Death of Ivan Illyich,* which is an artistic re-creation of the mood of the passage quoted above, and an attempt to answer the question asked there: "Illness and death would come . . . if not today, then tomorrow, to those whom I loved, to myself, and nothing would remain but stench and worms. . . . Why then busy one's self with anything? How could men see this and live?" Not everyone suffers such deep despair at the thought of death as Tolstoi did, but why reduce it to an illness?

Still, there are kinds of behavior that we recognize as being so far from the ordinary, involving so much distortion of reality, so much suffering to the individual, and so much danger to himself and others, that we feel there is some real meaning in the concept of "mental disorder" or "madness," no matter how hazy its boundaries are. The recognition of madness as a distinct kind of behavior, requiring both an explanation of its origin, and treatment to relieve the suffering it causes, is as old as recorded history.

HISTORY

The madness of King Saul, a theory of its cause, and the treatment used for it, are described in the first book of Samuel:

> And whenever the evil spirit from God was upon Saul, David took the lyre and played it with his hand; so Saul was refreshed, and was well, and the evil spirit departed from him. (I Samuel 16:23)

The therapy was not always effective:

> An evil spirit from God rushed upon Saul, and he raved within his house, while David was playing the lyre, as he did day by day. Saul had his spear in his hand; and Saul cast the spear, for he thought, "I will pin David to the wall." But David evaded him twice. (I Samuel 18:10–11)

Greek and Roman physicians recorded many cases of insanity. The physician Galen reported one of his patients having hallucinations of flutes being played incessantly. He also described depressed melancholics who thought of themselves as like Atlas, bearing the weight of the world on their backs. There are Roman records of men who murdered their mothers in what were clearly recognized as fits of insanity.[2] The Greek physician Soranus of Ephesus wrote a treatise in which a condition called melancholy (melancholia) is described: it is characterized by "mental anguish and distress, dejection, silence, animosity towards members of the household, sometimes a desire to live and at other times a longing for death, suspicion on the part of the patient that a plot is being hatched against him, weeping without reasons, meaningless muttering, and again, joviality. . . ." Soranus also distinguished a second disease, mania, characterized by excitement and hallucinations rather than despair.[3]

Treatment of the Insane

In the Middle Ages, the practice was begun of confining the seriously insane in hospitals, first in general hospitals, then later in special institutions. The insane were not treated badly in medieval times, but by the eighteenth century conditions had changed:[4]

> In England those who were unfortunate enough to be interned at Bethlehem Hospital, which during the medieval period had treated psychotics with some degree of kindness, had reason during the eighteenth century to regret their commitment. Bethlehem—or Bedlam, as it was called—was a favorite Sunday excursion spot for Londoners, who came to stare at the madmen through the iron gates. Should they survive the filthy conditions, the abominable food, the isolation and darkness, and the brutality of their keepers, the patients of Bedlam were entitled to treatment—emetics, purgatives, bloodletting, and various so-called harmless tortures provided by special paraphernalia. Conditions in Paris at the Bicêtre, which became a part of the General Hospital in 1660, which housed "madmen," and at the Salpêtrière, where "madwomen" were chained, were certainly no better.
> The innumerable contemporary descriptions of the miserable lot of the insane all testify to this fear. Their cruel segregation and restraint was described by Johann Christian Reil (1759–1813), one of the most advanced psychiatrists of his era: "We incarcerate these miserable creatures as if they were criminals in abandoned jails, near to the lairs of owls in barren canyons beyond the city gates, or in damp dungeons of prisons, where never a pitying look of a humanitarian penetrates; and we let them, in chains, rot in their own excrement. Their fetters have eaten off the flesh of their bones, and their emaciated pale faces look expectantly toward the graves which will end their misery and cover up our shamefulness." . . . Excited patients were locked naked into narrow closets and fed through holes from copperware attached to chains. Beatings were common and defended by shallow rationalizations. Strait jackets and chains attached to walls or beds were used to restrain patients, since the theory was that the more painful the restraint, the better the results, particularly with obstinate psychotics. The attendants were mostly sadistic individuals of low intelligence who could not find any other employment. "The roar of excited patients and the rattle of chains is heard day and night," says Reil, "and takes away from the newcomers the little sanity left to them." . . . The unsanitary conditions, lack of nourishment, wounds inflicted by the chains, and application of drastic skin irritants to increase the torment killed a large number of these patients. (pp. 114–116)

Does the Treatment of the Mad Make Them Madder?

By the end of the eighteenth century and the beginning of the nineteenth, a spirit of humaneness in the treatment of the insane was reborn. It arose in a number of different places at about the same time: in Tuscany under the Grand Duke Pietro Leopoldo; in Revolutionary France with Phillipe Pinel; and in England among the Quakers, one of whom, the tea merchant William Tuke, founded a hospital called

WHAT IS MADNESS?

the York Retreat. His work was continued by his son Henry and his grandson Samuel. Samuel who wrote a book on the hospital called *Description of the Retreat,* which was published in 1813. Some quotations from this book will give an indication of the spirit of this new movement:[5]

> Bleeding, blisters, evacuants, and many other prescriptions, which have been highly recommended by writers on insanity, received an ample trial; but they appeared to the physician [who directed the Retreat] too inefficacious, to deserve the appellation of remedies, except when indicated by the general state of the habit. As the use of antimaniacal medicines was thus doubtful, a very strong argument against them arose, from the difficulty with which they were very frequently administered; as well as from the impossibility of employing powerful medicines, in a long continuance, without doing some injury to the constitution. The physician plainly perceived how much was to be done by moral, and how little by any known medical means. He therefore directed, with his usual humanity and modesty, that any medicine which he might prescribe, by way of experiment, should not be administered, where the aversion of the patient was great; unless the general health strongly indicated its necessity; well aware, that otherwise, the probable good would not be equal to the certain injury. . . . (p. 480)
>
> The difficulty of obtaining sleep for maniacal patients, and the unpleasant effects frequently produced by the use of opium, are well known to medical practitioners. It occurred, however, to the sensible mind of the superintendent, that all animals in a natural state, repose after a full meal; and, reasoning by analogy, he was led to imagine, that a liberal supper would perhaps prove the best anodyne. He therefore caused a patient, whose violent excitement of mind indisposed him to sleep, to be supplied freely with meat, or cheese or bread, and good porter. The effect answered his expectation; and this mode of obtaining sleep, during maniacal paroxysms, has since been very frequently and successfully employed. In cases where the patient is averse to take food, porter alone has been used with evident advantage, always avoiding, in all cases, any degree of intoxication. . . . (pp. 483–484)
>
> Hence, also, the idea seems to have arisen, that madness, in all its forms, is capable of entire control, by a sufficient excitement of the principle of fear. This speculative opinion, though every day's experience decidedly contradicts it, is the best apology which can be made for the barbarous practices that have often prevailed in the treatment of the insane.
>
> The principle of fear, which is rarely decreased by insanity, is considered as of great importance in the management of the patients. But it is not allowed to be excited, beyond that degree which naturally arises from the necessary regulations of the family. Neither chains nor corporal punishments are tolerated, on any pretext, in this establishment. . . . (pp. 494–495)

In the above, we can see the recognition that at least part of the behavior of the insane is determined by how they are treated. The screams, wails, and maniacal laughter that made Bethlehem Hospital a source of entertainment to the Londoners of the eighteenth century decreased with the abolition of chains and whips (see Figure 8-1).

FIGURE 8-1. "Bedlam Hospital," engraving by William Hogarth, published in 1735. Hogarth portrays satirically two ladies of London, visiting the hospital for amusement, who are intrigued with the nakedness of the "King" in the back room. Apparently, Hogarth also perceived madness as something comical. (Reproduced with permission of the Pennsylvania Academy of Fine Arts.)

CLASSIFICATION AS THE STARTING POINT OF SCIENCE

Classification

We have mentioned that all of us, as part of everyday experience, encounter individuals who show forms of behavior we recognize as distorted, bizarre, and frightening. In this recognition, we are dividing people into categories—the sane and the insane. And we make further distinctions among those we think of as insane: the mild eccentrics who talk to themselves, the dangerously violent, the sex criminal, the suicide "who had everything to live for," and so on. These distinctions have important practical consequences: they determine whether we feel afraid or not, whether we call for the police or a psychiatrist, or whether we just smile and walk on. The process we go through in making these observations and coming to our conclusions is carried out in a subjective and imprecise way, but it is very natural for us to do this—as natural as to make the subjective and imprecise distinction between hot and cold that was the starting point for the scientific study of heat. This is how we organize most of our experience of the world. Yet what we have done, in recognizing some problem in the external world, observing some facts relevant to this problem, and organizing or classifying these facts into some coherent pattern, corresponds to the way any scientific endeavor begins.

In Chapter 4 on cholera we made the point that before the question of how cholera is transmitted could be answered, it was necessary to be able to distinguish cholera from other diseases with similar symptoms. This was not easy, in spite of the fact that cholera is a disease with characteristic symptoms and a characteristic course that most cases follow. To recognize and distinguish mental disorders is usually more difficult than to distinguish physical diseases, but we will see that the manner in which we do this is intimately connected with our success or failure in understanding them.

Facts and Their Classification

We have stressed in an earlier chapter that there are popular misconceptions about the role of facts and their classification in science. Among these misconceptions is the belief that the beginning stage in science is the collection of facts, that only after the facts are collected and subjected to some systematic arrangement do we look for a theory to explain them, and that theories are judged as correct only if they explain all the facts.

The truth is that this is a world in which there is an infinite number of facts available, and we have to begin with some feeling for which are important and which are not. The feeling may come from some well-defined theory we hold, or it may be a much more vague, intuitive feeling about what should and should not count, but unless we have some such concept, we cannot begin our study of the facts.

A well-known quotation from Darwin makes the point concisely:[6]

> About thirty years ago there was much talk that geologists ought only to observe and not theorize; and I well remember some one saying that at this rate a man might as well go into a gravel pit and count the pebbles and describe the colors. How odd it is that anyone should not see that all observation must be for or against some view if it is to be of any service! (pp. 194–195)

It is probably obvious to the reader that the individual weights, shapes, and colors of the pebbles in a gravel pit, although "facts," are not necessarily informative facts for understanding the history of the earth.

As varied as the pebbles are, the variety of facts about any human being whose state of sanity we want to study is enormously larger. Which events of his past life or circumstances of his present existence are important for his disorder and which are not? If he was spanked at the age of six by his mother for misbehaving at his sister's birthday party, is it more or less important than having had mumps at the age of four? Is it worth noting that, before the patient was born, an uncle of his father died in a mental hospital? We choose our facts in accordance with criteria of importance that may be based on theories we hold, or more often on vague subjective feelings that certain things are worth paying attention to and others not. There are obvious risks. We may choose the wrong things, and overlook the ones that provide the key to some important discovery. But there is no way to avoid this difficulty.

Just as the facts we select are determined by our preconceptions and theories, so the way we order these facts and describe the patterns in them depends on similar subjective criteria.

We will see in this chapter that classification in science is not a static thing that precedes discovery of theory, but is intimately intertwined with it. Good classifications make discoveries possible, and, in turn, discoveries change our ways of classifying the things we study.

We have learned in school that whales and bats are classified together as mammals; whales are not classed with fishes, nor are bats classified with birds. This is done not because there is some absolute objective proof that a whale is more like a bat than it is like a herring—in some ways the whale and bat resemble each other and in others they do not. It is because biologists feel that the ways whales and bats resemble each other are more important than the ways they differ. Biologists used this system of classification long before the theory of evolution was formulated, and their use of it helped make the discovery possible.

Many systems have been used in classification of diseases, and, depending on the degree of knowledge of the disease, different criteria may be employed at one time for different diseases. There was a time, for example, when the main classification was based on symptoms: all diarrheas were a single disease. More careful examinations of symptoms and the recognition of groups of symptoms occurring together enabled distinctions to be made, for example, among cholera, typhoid, and food poisoning. The development of the germ theory of disease and the use of microscopic methods of examination permit us now to identify some diseases by the specific organisms causing them. Other diseases not known to be caused by germs or viruses are classified by other criteria: leukemia and breast cancer are very different in symptoms, course of disease, and method of treatment. However, they are both characterized by a rapid proliferation of microscopically identifiable abnormal cells, and are classed together as cancer. Sometimes it is convenient if diseases are classed together by origin. There appear to be no common features shared by cataract, cleft palate, deafness, and mental retardation, but any of these may be the consequence to a child whose mother had German measles during her pregnancy.

In our discussion of the scientific study of mental disorders, we will focus on the two most common classes of severe disorders: schizophrenia and the depressive psychoses. These two are responsible for the great majority of admissions to mental hospitals, they have been considered definite and distinct disorders by most psychiatrists for many years, and they occur throughout the world. We will discuss how we decide whether a person has one of these disorders, and we will see that the way we do this cannot be separated from the theories we hold about the disorders.

THE KINDS OF MENTAL DISORDERS

In Greek and Roman times, three kinds of mental disorders were distinguished: mania, melancholy, and dementia.

Mania was a state of excitement, physical activity, ebullience, delusions, and hallucinations.

Melancholy was a state of depression, despair, inactivity, and loss of interest in food and sex.

Dementia was a state of being "without mind," inert, blank, unspeaking, torpid, and motionless.

To classify someone as suffering from one of these disorders, it was not necessary that all the symptoms be present. Some persons diagnosed as suffering from mania, for example, could be merely excited and hyperactive, without hallucinations.

By the end of the nineteenth century it was concluded that other classifications were more useful. One reason was that the advance of medicine had led to the identification of certain disorders as having a definite basis in disease or in some other malfunctioning of the body. Such disorders are said to have an *organic* origin.

Syphilis, Severe Retardation, Senile Dementia

For example, paresis, a disorder that made up a major portion of the cases in mental hospitals of the nineteenth century, was shown to be a consequence of syphilitic infection of the central nervous system. Paresis, which begins with delusions of grandeur, a state that would have been classified as mania, is accompanied by characteristic symptoms of nerve deterioration—certain reflexes are lost, and eventually paralysis and death occur. The organism of syphilis can be found by microscopic examination of the brains of its victims. But there were many other individuals who also had the symptoms of "mania," but did not have syphilis and did not become paralyzed, and were therefore presumed to be suffering from something other than paresis.

There are types of severe mental retardation which show clear evidence of an organic origin, such as Down's syndrome (formerly and misleadingly called *mongolism*), cretinism, and hydrocephalus. In all three, the disorder is present from birth, and the physical appearance of the patient characterizes the disease, allowing the three types to be distinguished both from each other and from other types of mental retardation. These were at one time all classed as "dementia." However "dementia" was also often observed as the end-product of disorders in which people of initially normal intelligence and normal behavior first suffered breakdowns to states of mania or melancholy, and then underwent a progressive deterioration. No organic origin could be found in patients following this pattern.

There was also the "dementia" of old age, again afflicting people of normal intelligence, a disease at one time believed to result from a hardening of the arteries of the brain, but now associated with microscopically detectable changes in brain cells (Alzheimer's disease).

Another pattern often shown by some patients was an alternation between states of mania and states of melancholy, often with normal intervals. One could have said about such a patient that he had recovered from "mania" only to fall ill of "melancholy," but it was felt more useful to regard this alternation as reflecting a single disorder.

On the basis of these observations and discoveries a new system of classification was gradually developed toward the end of the nineteenth century.

Organic versus Functional Disorders

First, there were those disorders with a clear indication of organic origin, as in paresis or the types of mental retardation described above. Then there were other common mental disorders, for which no evidence of organic origin was available. (These account for the majority of patients in mental hospitals in modern times.) The term functional disorder was used for these, to distinguish them from the disorders of known organic origin. The possibility that the functional disorders might ultimately be found to have an organic origin was not ruled out. But since no evidence for this had yet been found, the question was left open. More recent evidence, some of which will be described later in this chapter, suggests that some of the major disorders previously classed as functional may have an organic basis.

Cyclical versus Deteriorating States of Mind[7–10]

The German psychiatrist Kraepelin, working at the end of the nineteenth century and the first quarter of the twentieth, proposed that classification be based on the whole course of the disease, rather than on the collection of symptoms shown by the sufferer at one moment of time. He recognized two main categories of functional disorders, which he called "dementia praecox" and "manic-depressive psychosis."

Dementia praecox means "early dementia." (The Latin word *praecox* also gives us "precocious.") In contrast to the dementia of old age, which, as mentioned previously, has an organic origin, it usually afflicts the young, and manifests itself initially in manic or melancholic states often associated with hallucinations or delusions. In Kraepelin's view it led in most cases progressively to complete loss of memory, incoherence of mind, and the inability to function as an independent being—dementia as severe as seen in the most severe types of mental retardation or senility.

*Manic-depressive psychosis,** on the other hand, was defined by Kraepelin as a disorder characterized by a cyclical rather than a deteriorating course, involving alternating episodes of mania and melancholy, usually without hallucinations, and with long intervals of normal behavior.

The Risks of an Improved Classification

Kraepelin's distinction is still used, although the classifications have undergone refinements. It was a great advance to cut across the old categories of mania, melancholy, and dementia and to look at the whole course of a disease rather than at separate episodes. A system of classification, however, is not a purely neutral arrangement of facts. It also imposes a certain definite way of looking at problems. This way may be useful or it may not; more likely, it will have both advantages and drawbacks.

Kraepelin defined dementia praecox by its ultimate outcome—a final demented state which may not be reached for 20 or 30 years. But this is an awkward way to make a diagnosis—we do not wish to wait that long in order to be more certain we are correct. We want to treat the patient now, not 20 years from now. We would rather risk using the wrong treatment than wait until we are sure, by which time it would be too late to help.

Another Way of Looking at the Same "Collection of Facts"

Further, by defining dementia praecox in this way, we are excluding the question: does anyone ever recover? Obviously, by a narrow interpretation of Kraepelin's definition, if anyone recovered, he could not have had dementia praecox in the first place, no matter how closely his symptoms may have resembled other "genuine" cases of the disorder.

E. Bleuler, a contemporary of Kraepelin, noticed that there were many patients who began by showing the typical onset and symptoms of dementia praecox as defined by Kraepelin, but whose disorder did not follow the same hopeless and deteriorating course. They recovered, and were able to live normal or nearly normal lives without further breakdowns. Bleuler felt it more useful to consider this the same disorder as dementia praecox, in spite of the different outcome. Thus, concluding that the name "dementia praecox," with its grim implications of complete and inevitable destruction of the mind, would no longer be appropriate, Bleuler coined the name "schizophrenia," from the Greek words for "splitting" and "mind." It is this name which we use today.

---

* The word *psychosis* is defined by the *American Heritage Dictionary* as: "Any severe mental disorder . . . characterized by deterioration of normal intellectual and social functioning, and by partial or complete withdrawal from reality."

SCHIZOPHRENIA AND DEPRESSIVE DISORDERS

Description of Schizophrenia

Schizophrenia most commonly begins in youth, between puberty and the early thirties (*praecox*). The patient begins by showing an exaggerated concern with what the rest of us would class as trivial and easily solved problems—an obsession with cleanliness or money, or an excessive worry about the opinions other people hold about him. Things or events that others would regard as accidental or unimportant have an exaggerated significance for the patient (see, for example, the "referential mania" described by Nabokov[14] in the section "The Experience of Madness," below, p. 201). He might have delusions—beliefs that there is some sort of conspiracy afoot to harm him ("L. Percy King's" narrative,[12] below, pp. 199–200); or hallucinations—hearing voices that accuse him of monstrous crimes. General behavior becomes bizarre: mannerisms, grimaces, pointless motions or acts, sometimes repeated over and over again. The emotions displayed are no longer appropriate to the objective circumstances. Anger, upset, or suspiciousness may be expressed, but most commonly there is a "blunting of affect"— coldness, apathy, and withdrawal. Sequences of words with no relation to one another are uttered; new words are invented by combining several together; sometimes there is an obsessive concern with words that sound alike or rhyme. In other cases speech stops or is reduced to a minimum. Questions asked of the patient either are repeated verbatim instead of being answered, or are answered inappropriately.

Kraepelin distinguished several types of schizophrenia, and his categories, again with some modifications, are still with us. *Paranoid* schizophrenia is characterized by delusions of persecution, often quite elaborate and complex. The *catatonic* type, after an initial excited stage at the onset, develops into a state of immobility and stupor; the patient no longer dresses, washes, or feeds himself. There is a *hebephrenic* or *disorganized* type, harder to characterize, where there are delusions or hallucinations, but, in contrast to those of the paranoid, these are not formed into an elaborately structured system, and seem more pleasing in content to the patient. The speech disorders described earlier are common in this form of schizophrenia. The patient neglects habits of personal care, and severe deterioration of personality often occurs. A *simple* or *residual* type of schizophrenia has been identified, in which delusions, hallucinations, disorders of speech, and other striking symptoms are absent, but in which there is a deterioration of personality characterized by a preference for inactivity and withdrawal. The victim cannot hold a job, eventually stops looking for one and stays home, and sometimes cannot manage even the minor chores of the house, such as making the bed or washing the dishes.

From this description of schizophrenia, the reader might well wonder if we are describing a single disorder, or a whole collection of them which may or may not be related. Our description is necessarily brief and oversimplified, but the question is a very real one, and different specialists in the putative disorder have given conflicting answers. It may well be that what is called "schizophrenia" is several similar disorders with different causes. We do not know yet. The designation "the group of schizophrenias" is often used to express the uncertainty. Again, we are faced with the question, where do we draw the lines?

Description of Depressive Disorders

Kraepelin defined the manic-depressive disorders by their cyclical course, with returns to normality, in contrast to the continued deterioration of schizophrenia. The modern view is somewhat different and more complicated. Some patients show frequent depressed states without ever having a manic state in between; most of these patients recover fully. Only a minority of those who undergo depressive periods exhibit the manic-depressive cyclical pattern described by Kraepelin. According to data cited by Arieti, nearly 60% of depression victims experience only one such attack in their lifetimes and another 25% only two.[8] Again, as with schizophrenia, we may be looking at a group of disorders with similar symptoms, rather than a single disorder. We will use the term *depressive disorders*.

The onset of a depressed period is more often attributable to a specific precipitating event than is the onset of schizophrenic episodes. Such events may include the death of someone close, disappointment in a close relationship or in one's career, but need not always be one of bereavement or disappointment: the marriage of a son or daughter, or the let-down a writer feels after completing a book.

A certain amount of depression after bereavement is natural. It may be hard to tell the difference between a normal and an excessive reaction. In an excessive reaction the mood of the depressed person is gloomy and obsessed; nothing is worthwhile; life has lost its meaning. There is a strong sense of guilt and self-blame. The person complains that he cannot concentrate; what he reads does not stay in his mind. His speech slows down or stops. Hallucinations are less common than in schizophrenia, but, unlike in schizophrenia, if they occur it is usually at night, when competing stimuli are absent or less vivid. Often there are physical complaints: digestive disturbances, insomnia, loss of appetite. The variation of intensity of depression from patient to patient is great: it ranges from simple depression, hard to distinguish from moods any normal person might have, to depressive stupor. Suicide is a serious risk in depressed states.

The manic phases also vary enormously in intensity. They may range from cheerful, gregarious, and talkative states, which again are hard to distinguish from normal moods, to states of delirious mania, in which death from exhaustion or heart failure is a possibility. In general the dominant mood is one of elation and excessive activity; happy ideas rush pell-mell through the patient's mind without any logical order. Talk is rapid, but is not concentrated for long on any one topic. Jokes, puns, and some of the speech disorders described for schizophrenia also appear: ideas are associated by sounds of the words used to express them. The patient has an exalted opinion of himself—he is rich, famous, or has solved a problem that humanity has been struggling with unsuccessfully for thousands of years.

Comparison of the Two Groups

One can recognize that even within specific categories of the disorders, symptoms will vary from case to case, and also that the same group of symptoms can be seen in different patients suffering from different disorders. It is useful to have some concise way of describing the difference between the two groups, the schizophrenias and the depressive disorders, even if it is oversimplified and naive.

The characteristic pattern in schizophrenia is sometimes summed up by the phrase "thought disorder." Aside from the vivid hallucinations and complex delusions, there is some basic failure in the processes we call "thinking"—the ability to follow logical patterns and to recognize similarities or differences between objects or situations in the "normal" way.

In contrast, the group of depressive disorders are regarded as being disorders of mood and emotion rather than of thought. The technical term "affect" is essentially a synonym for feeling or emotion, and the depressive disorders are sometimes called the affective disorders.

However, in schizophrenia the emotions are affected, as well as the ability to think logically, and in the depressive disorders there are delusions, occasional hallucinations, and other symptoms of thought disorder. Therefore the boundaries between them are not sharp, and their classification presents problems.

There are as yet no laboratory tests, like those available for cholera or syphilis, that can distinguish between them. In recent years, certain drugs have been found effective in treating schizophrenia and certain others in treating depressive disorders, but their effectiveness has not been so universal as to enable us to diagnose the disorders by seeing which drugs help the patient.[10] We still rely on diagnosis by a psychiatrist interviewing the patient, and where a particular psychiatrist will draw the line may depend on his prior training, his theories of the disorders, or other factors.

DIAGNOSIS

How does a psychiatrist interviewing a patient make a diagnosis? One can always find what are called "textbook" cases, in which an individual shows a pattern of symptoms so clearly characteristic of one particular, well-recognized disorder, that no reasonable doubt exists. For example, most psychiatrists reading the quotation from "L. Percy King" (below, pp. 199–200) would be very tempted to diagnose a case of paranoid schizophrenia without even examining the patient.

But not all cases present such a classic pattern. The variety of symptoms reflects the tremendous varieties of human personality. It may take prolonged investigation to reach a conclusion. Even then the diagnosis may be in doubt, and the case record may show some qualifying term such as "probable paranoid schizophrenia." And although we have concentrated our discussion on the two most common "functional" disorders, schizophrenia and the depressions, many of the

WHAT IS MADNESS?

same symptoms are seen not only in other disorders, but also in types of human behavior that do not belong to the categories of mental disorder at all. One might well wonder how often two psychiatrists examining the same patient would agree on the diagnosis, and this is a question we will be concerned with in what follows.

Pattern Recognition—Art or Science?

The process of diagnosis in mental disorder, like the process of disease diagnosis by a physician, is an example of something one can learn to do and may do fairly consistently while at the same time finding it difficult to explain just how it is done. There is a technical term for this—*pattern recognition*.

It is like the process by which we recognize that a brother and sister look alike. But what do we mean by "look alike"? It is very hard to make the process of recognition precise. Although we can make statements like "they have the same eyes, the same chin," we cannot really spell out how much "the same" is really the same. In fact, we often make the judgment of similarity almost instantly, at a glance, before we have time to notice specific features—nose, eyes, hair—and compare them one by one.

The process is far from infallible, and we do not always agree. How often has the reader heard exchanges like this: "The man over there looks just like Harry." "Who? That one? Like Harry? Not in the least!" There is also the cliché of the tourist from a Western country, visiting the Orient for the first time, saying, "All Chinese look alike to me." Chinese do not resemble each other any more than Caucasians resemble each other, but the criteria we have learned to use in distinguishing among people whose features we are generally familiar with do not always apply to those who look quite different. We have to learn new ways of looking when we encounter new kinds of people.

The textbook descriptions of the mental disorders are necessary and helpful, but the beginning psychiatrist learns by working with more experienced colleagues until he has acquired the skills. Once having done so, he makes use of an intuitive sense of judgment in which not every step of his reasoning can be spelled out or logically justified: he sees the pattern as a whole. Much of our functioning, not just as psychiatrists but as human beings, depends on our ability to perceive patterns in this way.

However, there are problems with such a procedure. First is the problem of *reliability*. How well do different observers agree about what they see? If they repeat the observation at another time, under different conditions, do they get the same result? Second is the more difficult problem of *validity*. Do the diagnostic categories adequately reflect reality? Are these particular categories the most useful, and the most fruitful for treatment and for further research? Do the diagnoses remain stable, so that a patient diagnosed as schizophrenic today will not be shifted to the category of depressive disorder a few weeks from today, and to yet a third diagnosis after several months?

It is natural for us to trust our own judgment, and even more natural when it agrees with the judgment of others. But there are too many examples from the history of science where what all observers agreed on was simply not true. At one time, all observers saw the sun as moving around the earth. At one time, even dispassionate and reasonable observers saw certain individuals as witches. Our surroundings are sufficiently complex to allow many different patterns to be seen in them. However, once we have been conditioned by training or experience to see things one way, it is hard to break these habits and see other patterns that might be more fruitful and more significant.

The problems of reliability and validity in the diagnosis of mental disorders are key subjects in this chapter, to which we shall return.

THE EXPERIENCE OF MADNESS

The descriptions given above may not convey much to the reader about what mental disorders are like, or how they differ. It will help make them more vivid and real if we give examples of the feelings of individuals suffering from the manic and depressed phases of manic-depressive psychoses and from various types of schizophrenia. In subsequent sections we will describe the two disorders in more detail.

The first quotation reports the thoughts and feelings of John Custance, an English physician who suffered from a manic-depressive psychosis. The following was recorded by him in a hospital during the manic phase of the disease:[11]

> The first thing I note is the peculiar appearances of the lights—the ordinary electric lights in the ward. They are not exactly brighter, but deeper, more intense, perhaps a trifle more ruddy than usual. Moreover, if I relax the focusing of my eyes, which I can do very much more easily than in normal circumstances, a bright star-like phenomenon emanates from the lights, ultimately forming a maze of iridescent patterns of all colours of the rainbow, which remind me vaguely of the Aurora Borealis.
>
> There are a good many people in the ward, and their faces make a peculiarly intense impression on me. I will not say that they have exactly a halo round them, though I have often had the impression in more acute phases of mania. At present it is rather that faces seem to glow with a sort of inner light which shows up the characteristic lines extremely vividly. Thus, although I am the most hopeless draughtsman as a rule, in this state I can draw quite recognisable likenesses. This phenomenon is not confined to faces; it applies to the human body as a whole, and to a rather lesser degree to other objects such as trees, clouds, flowers and so on. . . .
>
> Connected with these vivid impressions is a rather curious feeling behind the eyeballs, rather as though a vast electric motor were pulsing away there. . . . (p.31)
>
> When in a depressive period I have an intense sense of repulsion to lavatories, excreta, urine, or anything associated with them. This repulsion extends to all kinds of dirt. I loathe going to the lavatory, using a chamber-pot, or

touching anything in the least bit dirty. With this repulsion is associated extreme terror, in my case terror of eternal punishment in Hell. It is also associated with repulsion to fellow-creatures, repulsion to self, repulsion in fact to the whole universe. Finally it is associated with a sense of intense guilt.

In the manic phase repulsion gives place to attraction. I have no repulsion to excreta, urine and so on. I have no distaste for dirt. I do not care in the least whether I am washed or not, whereas I am terrified of the slightest speck of dirt and continually wash my hands like Lady Macbeth when in a state of depression. At the same time I feel a mystic sense of unity with all fellow-creatures and the Universe as a whole; I am at peace with myself; and I have no sense of guilt whatsoever. . . . (p.42)

[Dr. Custance now describes his first attack, in 1938, when he was 38 years old.]

The first symptoms appeared on Armistice Sunday. I had attended the service which commemorates the gallant dead of the "War to End Wars." It always has an emotional effect upon me, partly because my work has had a good deal to do with the tragic aftermath of that war in Europe. Suddenly I seemed to see like a flash that the sacrifice of those millions of lives had not been in vain, that it was part of a great pattern, the pattern of Divine Purpose. I felt, too, an inner conviction that I had something to do with that purpose; it seemed that some sort of revelation was being made to me, though at the time I had no clear ideas about what it was. The whole aspect of the world about me began to change, and I had the excited shivers in the spinal column and tingling of the nerves that always herald my manic phases.

That night I had a vision. It was the only pure hallucination I have ever experienced; though I have had many other visions, they have always taken the form of what are technically known as "illusions." I woke up about five o'clock to find a strange, rather unearthly light in the room. As my natural drowsiness wore off, the excited feelings of the day before returned and grew more intense. The light grew brighter; I began, I remember, to inhale deep gulps of air, which eased the tension in some way. Then suddenly the vision burst upon me.

How shall I describe it? It was perfectly simple. The great male and female organs of love hung there in mid-air; they seemed infinitely far away from me and infinitely near at the same time. I can see them now, pulsing rhythmically in a circular clockwise motion, each revolution taking approximately the time of a human pulse or heartbeat, as though the vision was associated in some way with the circulation of the blood. I was not sexually excited; from the first the experience seemed to me to be holy. What I saw as the Power of Love— the name came to me at once—the Power that I knew somehow to have made all universes, past, present and to come, to be utterly infinite, an infinity of infinities, to have conquered the Power of Hate, its opposite, and thus created the sun, the stars, the moon, the planets, the earth, light, life, joy and peace, never-ending. (p. 45)

For an example of the feelings of a severely depressed person, the reader may return to the quotation from Tolstoi on page 184.

The following quotation is from a letter received by Dr. R. W. White, a professor of abnormal psychology at Harvard, who published parts of it in his book *The Abnormal Personality*.[12] The name "L. Percy King" is a pseudonym. The passage is a typical representation of paranoid schizophrenia.

I, LPK, had a few days to spend with Long Island relatives before returning to work for the War Dept., Wash., D.C. One day I went to reconnoitre in N.Y. City's East Side. Being a stranger I was surprised to hear someone exclaim twice: "Shoot him!," evidently meaning me, judging from the menacing talk which followed between the threatener and those with him. I tried to see who the threatener, and those with him were, but the street was so crowded, I could not. I guessed that they must be gangsters, who had mistaken me for another gangster, who I coincidentally happened to resemble. I thought one or more of them would try to shoot me so I hastened from the scene as fast as I could walk. These unidentified persons, who had threatened to shoot me, pursued me. I knew they were pursuing me because I still heard their voices as close as ever, no matter how fast I walked. As I rushed along, I tried to be lost in the street crowd. I tried to allude [sic] the "pursuers" by means of the "L," and a surface car. . . .

Days later while in the Metropolis again, I was once more startled by those same pursuers, who had threatened me several days before. It was nighttime. As before, I could catch part of their talk but, in the theatre crowds, I could see them nowhere. I heard one of them, a woman say: "You can't get away from us; we'll lay for you, and get you after a while!" To add to the mystery, one of these "pursuers" repeated my thoughts aloud, verbatim. I tried to allude [sic] these pursuers as before, but this time, I tried to escape from them by means of subway trains, darting up and down subway exits, and entrances, jumping on, and off trains, until after midnight. But, at every station where I got off a train, I heard the voices of these pursuers as close as ever. The question occurred to me: How could as many of these pursuers follow me as quickly unseen? Were they ghosts? Or was I in the process of developing into a spiritual medium? No! Among these pursuers, I was later to gradually discover by deduction, were evidently some brothers, and sisters, who had inherited from one of their parents, some astounding, unheard of, utterly unbelievable occult powers. Believe-it-or-not, some of them, besides being able to tell a person's thoughts, are also able to project their magnetic voices—commonly called "radio voices" around here—a distance of a few miles without talking loud, and without apparent effort, their voices sounding from that distance as tho heard thru a radio head-set, this being done without electrical apparatus. This unique, occult power of projecting their "radio voices" for such long distances, apparently seems to be due to their natural bodily electricity, of which they have a supernormal amount. Maybe the iron contained in their red blood corpuscles is magnetised. The vibration of their vocal chords, evidently generates wireless waves, and these vocal radio waves are caught by human ears without rectification. . . . (pp. 133–140)

Anton Boisen described his schizophrenic experiences in a book, *Out of the Depths*.[13] After his recovery he became a chaplain in several mental hospitals.

In Ward 2 I was given first the little room in the south east corner. I was tremendously excited. In some way, I could not tell how, I felt myself joined onto some super-human source of strength. The idea came, "Your friends are coming to help you." I seemed to feel new life pulsing all through me. And it seemed that a lot of new worlds were forming. There was music everywhere and rhythm and beauty. But the plans were always thwarted. I heard what seemed to be a choir of angels. I thought it the most beautiful music I had ever heard. Two of the airs I kept repeating over and over until the delirium ended. One of them I can remember imperfectly even now. This choir of angels kept hovering around the hospital and shortly afterward I heard something about a little lamb being born up-stairs in the room just above mine. This excited me greatly. . . .

The next night I was visited, not by angels, but by a lot of witches. I had the room next to the one I had occupied the night before. There was, as I remember it, nothing in it but a mattress on the floor. It seemed that the walls were of peculiar construction. There was, it seemed a double wall and I could hear a constant tap-tapping along the walls, all done according to some system. This was due, it seemed, to the detectives in the employ of the evil powers who were out to locate the exact place where I was. Then the room was filled with the odor of brim-stone. I was told that witches were around and from the ventilator shaft I picked up paper black cats and broomsticks and poke bonnets. I was greatly exercised, and I stuffed my blanket into the ventilator shaft. I finally not only worked out a way of checking the invasion of the black cats, but I found some sort of process of regeneration which could be used to save other people. I had, it seemed, broken an opening in the wall which separated medicine and religion. I was told to feel on the back of my neck and I would find there a sign of my new mission. I thereupon examined and found a shuttle-like affair about three-fourths of an inch long. (pp. 89–91)

The last of the examples of schizophrenic thinking is not by a patient at all, but an imaginative writer's recreation of the experience. It is from a story, "Signs and Symbols," by Vladimir Nabokov.[14]

"Referential mania," Herman Brink called it. In these very rare cases the patient imagines that everything happening around him is a veiled reference to his personality and existence. He excludes real people from the conspiracy—because he considers himself to be so much more intelligent than other men. Phenomenal nature shadows him wherever he goes. Clouds in the staring sky transmit to one another, by means of slow signs, incredibly detailed information regarding him. His inmost thoughts are discussed at nightfall in manual alphabet, by darkly gesticulating trees. Pebbles or stains or sun flecks form patterns representing in some awful way messages which he must intercept. Everything is a cipher and of everything he is the theme. Some of the spies are detached observers, such are glass surfaces and still pools; others, such as coats in store windows, are prejudiced witnesses, lynchers at heart; other again (running water, storms) are hysterical to the point of insanity, have a distorted opinion of him and grotesquely misinterpret his actions. He must be always on his guard and devote every minute and module of life to the decoding of the undulation of things. The very air he exhales is indexed and filed away. If only the interest he provokes were limited to his immediate surroundings—but alas it is not! With distance the torrents of wild scandal increase in volume and volubility. The silhouettes of his blood corpuscles, magnified a million times, flit over vast plains; and still further, great mountains of unbearable solidity and height sum up in terms of granite and groaning firs the ultimate truth of his being. (pp. 69–70)

THEORIES OF THE CAUSES OF MENTAL DISORDERS

It is not always true that knowing the cause of a disease will help find a cure. There are many illnesses that we know how to treat or prevent, but whose cause we do not know. We can treat diabetes with insulin, but we understand little about why people become diabetic. Snow discovered how to prevent cholera epidemics before the organism causing cholera was discovered. However, one way to find methods of treatment or prevention is to look for causes.

We have succeeded with some of the mental disorders. Paresis can be prevented by treating syphilis in its early stages; a psychosis of advanced stages of pellagra, a nutritional disorder, can be prevented by the same means used to prevent pellagra itself—an adequate diet. But the causes of the functional psychoses–the schizophrenias and the depressive disorders—have so far remained mysteries. There have been several general approaches to studying them.

One approach has been to look for the explanation of the disorder in the life experiences of the sufferer—his upbringing, how his parents treated him, his relationships with brothers and sisters. We will refer to theories of this type as *psychogenic*. A second, similar approach may be called *psychosocial*.[15] Here the focus is not just on the family environment, but on the whole social environment—which may include poverty, discrimination, and unemployment for some, or the stresses of striving for social and financial success in a fiercely competitive society for others.

Still another approach is a *biological* one—the explanation is sought in terms of the functioning of the body: one looks for substances in the blood or urine of the sufferer that are not found in normal people, one asks whether microscopic examination of the brain shows differences, or whether there is evidence for inheritance of the disorder.[16,17,18]

Our classification into these approaches should not be taken to express a view that only one explanation can be correct. Rather, it describes how most scientists studying the problems have actually proceeded, without ruling out the possibility that psychological and social stresses may combine with biological factors to produce the disorder.

Psychogenic Theories

Treatment of the insane with kindness, as advocated by Pinel and Tuke, was based on the belief that it could overcome the effects of the painful experiences that caused the disorder in the first place. This is one example of a psychogenic approach.

The most famous of psychogenic approaches is that based on the ideas of psychoanalysis, proposed originally by Sigmund Freud and extensively developed by others. These ideas have become part of the intellectual climate of our time, and most of us have at least superficial familiarity with them. We know of such things as the concept of an unconscious mind, the tendency to repress from the conscious mind memories or feelings that are painful or frightening, the possibility that these repressed experiences can remain influential in our lives, the importance of early childhood as the occasion when many of these processes occur, the fact that many repressed feelings are sexual in nature, and so on.

Those who have used psychoanalytical approaches in studying the functional psychoses have therefore focused on early childhood experiences, looking for those that might have produced such severe conflicts within the individual that the normal defense and mechanisms of dealing with unpleasant and painful events are overwhelmed by them. Although many theories of the origin of schizophrenia have been proposed, in the opinion of S. Arieti no one theory has won the support of a majority of psychogenically oriented investigators. The only point most of them agree on is that *"in every case of schizophrenia* studied, serious family disturbance was found" (italics Arieti's).[9] This disturbance is often, but not exclusively, schizophrenia, or borderline schizophrenialike behavior, in the parents. Arieti notes, however, that many families show similar family disturbances and yet fail to produce outright schizophrenia in the children. He concludes therefore that the family disturbance is presumably *necessary*, but not sufficient, to cause schizophrenia. Some other factor must be present also.

Theories of Social Causation

Some social scientists have suggested that one kind of stress possibly causing or contributing to mental disorder is that resulting from specific social conditions. People in the lowest social classes of modern societies are forced to live with unemployment, inadequate nutrition, and poor housing and are often the victims of crime and racial discrimination. Although people in higher social classes are spared these miseries, they may hold highly competitive and emotionally demanding jobs. Modern urban life produces its own stresses, in contrast with what is perceived as an idyllic and emotionally wholesome rural past.

Massachusetts, 1855, and Chicago, 1939

Indeed studies of the rates of mental illnesses among different social classes have shown striking differences. The first such report was by a physician, Edward Jarvis, who wrote a report in 1855 on the prevalence of mental disorders in Massachusetts.[19] This was before the modern era of diagnosis, and Jarvis described only two such disorders, "insanity" and "idiocy." Jarvis found that "The pauper class furnishes, in ratio of its numbers, sixty-four times as many cases of insanity as the independent class." Jarvis however did not explain this discrepancy as a consequence of the stresses of extreme poverty. He offered a reverse explanation:

"The mentally ill are poor because they are mentally ill. . . . They are unsuccessful in life; their plans of obtaining subsistence . . . often fail; and they are consequently poor. . . . The cause of . . . their mental derangement lies behind, and is anterior to, their poverty."

In 1939 a study of psychiatric disorders in Chicago was published by R. Faris and H. Dunham,[20] who found that hospital first admission rates for most psychoses, including schizophrenia, were highest in the central slums of the city, and were much lower in middle and upper-class suburban areas. These authors blamed the social disorganization of slum life for the excess rates.[15]

Further studies have confirmed a significant excess of schizophrenia in the lowest social class of American society, although depressive disorders appear to be higher in higher social classes.

Which Causes Which?

Now we have two possible explanations for the excess of schizophrenia in the lower class: that poverty causes schizoprenia (Faris and Dunham), or that schizophrenia causes poverty (Jarvis). Which is correct?

That poverty could cause schizophrenia is quite plausible. It is common knowledge that the general health of the poor is worse than that of other classes: infant mortality is higher among them, as well as death rates from many other diseases, and at all ages. There is little reason to doubt that this excess mortality is the result of—not the cause of—poverty, with its attendant problems of poor housing, crowded living conditions, poor nutrition, limited access to medical care, and so on. Further, it is obvious that the very poor are exposed to psychological and social stresses of a kind and severity few others suffer.

On the other hand Jarvis's view that mental disorders could cause people, initially of higher social class, to sink to the bottom level of society is also plausible. This hypothesis has been refined by the additional conjecture that in a society characterized by considerable social mobility, members of the lowest social class who suffer from schizophrenia or other severe psychopathology will fail to move upward. Either this, or Jarvis's hypothesis, or both acting together, could account for an excess of schizophrenia among the very poor.

It sounds like it should be an easy question to resolve, but attempts to do so ever since Faris and Dunham's original study have failed to resolve it to everyone's satisfaction.

One question that has been investigated is the social origins of lower-class schizophrenics: Were their parents and grandparents also lower-class, or from higher social classes? Others have studied immigrant groups arriving in the United States from more impoverished countries, and living initially in conditions of extreme poverty. If the social mobility hypotheses were correct, such groups would be expected to have *less* schizophrenia than would be inferred from their position in the social scale, as in the first generation there would not have been time for upward mobility to have sifted out the mentally healthy and better adjusted individuals. Unfortunately, results from different ethnic groups have led to conflicting conclusions.

It should be obvious that reliable information about social status and mental disorders over several generations is needed, and information of this kind is hard to come by. Reviews of current research on the problem have been published by Bruce Dohrenwend and Barbara S. Dohrenwend, who have made important contributions to this field.[15]

Biological Theories[16-18]

Biologically oriented scientists have studied mental disorders in two ways: they have looked for biochemical, neurologic, metabolic, or cellular differences between schizophrenics and normal people, and they have studied the families of schizophrenics for evidence of genetic origin. In fact, the occurrence of disturbance in the families of schizophrenics, which psychogenically-oriented psychiatrists regard as causing schizophrenia, has also been interpreted as evidence for genetic factors. For example, if a schizophrenic parent has a schizophrenic child, it may be the result of a common genetic factor; the parent may not have induced the disorder in the child by his or her behavior at all. We will discuss the genetic studies later in this chapter.

Progress in finding a biochemical basis for schizophrenia has been slow. Many laboratories have reported detecting unusual substances in the blood or urine of schizophrenics. However, it has often happened that other laboratories have tried to repeat these experiments and have failed to detect the substance in question. Occasionally, the unusual substances can be traced not to the disorder, but to the conditions under which schizophrenic patients in mental hospitals live—overcrowding leading to unsanitary living conditions and frequent digestive disorders, inadequate diet, insufficient exercise—or to the drugs used to treat the disorder. For example, one study found higher levels of a class of compounds known as urinary phenolic acids in the urine of schizophrenics than in the urine of normal test subjects. It turned out that these substances are produced by the body from substances in coffee, and that the schizophrenics in this study drank more coffee than the normal subjects.

Looking for the cause of a disorder by comparing two groups of people, those who have the disorder and those who do not, and by seeing how these groups differ in other ways, may sound simple, but it is not. The problem is that the number of ways in which the two groups may differ is enormous. Although most of the differences are irrelevant to the disorder, there is no automatic way to decide which are the relevant ones.

Many of the experimenters searching for a biological difference overlooked the fact that the schizophrenics differed not only in that they were schizophrenics, but also in that they were hospitalized for their disorder. They were subject, therefore, to all the changes in physical and emotional state that confinement in a mental hospital implies. Although these extra factors are not causative of the disorder, they are present in patients suffering from the disorder, and not in normal individuals. Scientists searching for biological differences do their best to control for these secondary factors, but it is hard to be sure that every possible source of confusion has been eliminated.

In recent years, certain drugs have been found effective for schizophrenia. These drugs have also been shown to influence the chemical processes that take place during the transmission of impulses from one nerve to another. Specifically, transmission of nerve impulses from one nerve cell to an adjacent one may take place either by electrical means, or by the secretion of a chemical substance by one cell and its absorption by an adjacent one. Some cell junctions (synapses) operate electrically, others chemically, using one of a limited number of chemical "transmitters." Among the chemical transmitters is a substance called dopamine. There is good evidence that "antipsychotic" drugs, introduced in the 1950s and found to be effective for schizophrenia, interfere with the action of dopamine-controlled synapses, both from studies on human beings and animals *(in vivo)*, and in laboratory experiments on extracts of living tissue *(in vitro)*. For example, these drugs often produce as an undesirable side effect muscular tremors which resemble those in Parkinson's disease, a malady known to be associated with decreased functioning of these synapses. The connection is additionally supported by the observation that certain drugs such as cocaine and amphetamine are on the one hand known to increase dopamine activity and on the other to produce schizophrenialike psychotic syndromes which can be alleviated by the same antipsychotic drugs that are used to treat schizophrenia itself. This is regarded as an important clue to a possible biochemical basis for the disorder, and is under active investigation at present.[21]

Interaction of Biological, Psychogenic, and Social Factors

The discovery of a definite biochemical difference between schizophrenics and normal people would not of itself prove that the disorder has an organic rather than a psychogenic cause, because the body is known to produce certain chemically detectable substances in response to emotions—epinephrine, for example, when we are excited or frightened. Even if schizophrenia had its origin in certain painful childhood experiences, it would not be surprising to find differences in the body chemistry between schizophrenic and normal people. But we would still have to find which comes first, the change in body chemistry or the disorder itself. Can we show, for example, that the chemical difference is inherited from a parent, or is the result of some purely physical occurrence such as an illness? Unless we can do this, no convincing case for a biological cause can be made.

AN EPIDEMIOLOGICAL STUDY

United States and British Rates of Mental Disorder—A Clue to Causes

In Chapter 4 on cholera, we saw that the differences in cholera rates in different parts of London played a key role in Snow's proof that the disease was transmitted through the water supply. In fact, time and again in the history of medicine important clues to the causes of diseases have been obtained by observing that one group of people got the disease more often than another. Once such differences in rates are found, we can start asking the question: In what ways do the two groups differ that might have a bearing on the difference in rates?

We might hope to use this epidemiological approach to study mental disorders also, and indeed it was noticed by psychiatrists in the 1950s and 1960s that the United States and Great Britain showed striking differences in their respective rates of hospital admissions for schizophrenia and for the depressive disorders.[22] Schizophrenia seemed to be more common in the United States, depressive disorders much more common in Great Britain. To take one example, *the rates for hospital admissions for manic-depressive psychosis in the age group 55–64 years were 20 times higher in England and Wales than in the United States.* This is as dramatic a difference as any that Snow ever observed for cholera in London. Conversely, in a comparative study of hospital admissions, *schizophrenia was found to be twice as common in New York as in London.* While not as large a difference as that seen in admission rates for manic-depressive psychosis cited above, this is still quite impressive. Such striking differences may provide important leads to the origins of the disorders.

Explanations

What explanations can be found? Certainly Great Britain and the United States are countries with much in common. Their language, political and social system, degree of industrialization, and standard of living are similar. Yet certain differences are easily noticed, even on a casual comparison. The pace of life is different, children are brought up differently, and so on. Further, even the small differences in social system and degree of industrialization may have a significant influence on the individual personality. Some psychosocially oriented psychiatrists, noticing differences in mental-disorder rates between countries, or changes over a period of time in one country, have suggested that there might be a close relationship between the type of personality favored by a society, and the types of mental disorders individuals in that society might be liable to contract. It has been suggested, for example, that an economically affluent society might place less stress on personal ambition and the competitive striving for success, and more on the social graces and personal adjustment.[8] These differences, which would be expressed in different practices of child rearing, could lead to differences in rates of the various disorders.

Caution! Discovery or Artifact?

There is another possible response to the surprising facts about rates of mental disorder in the two countries—one might ask, are they "facts" at all? Is it possible that these findings do not mean what they seem to, that they are the result of some error or misconception, and give a misleading picture of the true rates of mental disorders in the two countries?

A scientist confronted with a startling observation, one that conflicts with some well-established theory or even a less well-defined preconception, is in something of a dilemma. In the history of science the great majority of startling observations have meant nothing. They have been the result of some accidental circumstance—a mistake of the observer, a misunderstanding of experimental results, or, on rare occasions, outright fraud. But a small number of startling observations *are* significant, and represent major discoveries. Scientists may ignore the new "fact" on the grounds that it is implausible, and is probably a mistake or misinterpretation, but then they run the risk of missing an important new discovery. Or they may accept the "fact" and look for a new theory to explain it, but then they run the even greater risk of wasting time over their own or someone else's blunder.

There is no rule to use in such situations; scientists must follow their intuition and take the risks that go with it. To say "Be sure of the facts" is useless advice; the number of facts a scientist uses is enormous, and most of them have been discovered by others. There is no way to be sure of all of them. It takes time to be "sure" of anything, and life is too short for much of this.

However, there are experimental observations, not easily explained away as the result of some blunder, that would be very important if true. Yet they do not carry conviction—one feels there is something wrong even though one cannot easily explain just what it is.

Some scientists aware of the reported British-American differences felt this way. While recognizing that the differences would be important if real, they wanted to examine them more closely before trying to use them to build theories of mental disorders. There were two possible factors that could make these rates misleading: (1) Were the rates of *hospital admission* for schizophrenia or depressive disorders the same as the rates of *occurrence* for these disorders in the respective countries? (2) Did the psychiatrists in the two countries diagnose patients the same way?

1. Who goes to the hospital? The data quoted earlier suggested that more people in Great Britain have depressive disorders while more people in the United States have schizophrenia. However, they did not prove this, because they were rates of *hospitalization,* not rates of *occurrence* of the disorders. It is really this latter rate that is of interest, and that might provide a clue to the origins of the disorders. The technical term for this rate is *incidence.*

The two rates, one of hospitalization for a disorder, the other of developing the disorder, need not be the same. Not everyone who suffers mental disorder sees a doctor. Certainly many are not admitted to a hospital.

Ideally, what we would want for a comparison of British and American rates of the disorder is the true incidence: the rate at which people develop the condition, regardless of whether they go to a hospital. But how are we to get it? We cannot have every person in England or the United States interviewed by a psychiatrist once a month to evaluate his or her mental health. Rates of mental disorders have occasionally been studied by surveys in which the investigators interviewed a random sample of a population and estimated the rates from this smaller sample,[23] but even this is a slow and expensive procedure.

It is simpler to assume that, even though not every person who undergoes mental breakdown goes to a hospital, the proportion of those who do is the same in the two countries. We may not know what this proportion is—it may be 50% or 90%. But if the assumption is correct the *differences in rates of hospitalization* for a particular disorder would tell us that there are differences in the rates at which that disorder occurs in the two countries. There are risks in making such an assumption. We need to know the factors that determine whether an individual is hospitalized in each country. There are many reasons why a patient or doctor may decide against it, and the two countries may differ in the extent to which these reasons influence decisions. In the 1950s and 1960s, when drug therapy and electroshock were being introduced, some psychiatrists were administering these new treatments in their offices instead of sending patients to hospitals. This was particularly so for depressive disorders, which were considered on the whole less serious than schizophrenia, and for which the new treatments were more effective. If American psychiatrists were more willing than British psychiatrists to use office treatment instead of hospitalization, we would indeed find a higher rate of *hospitalization* for manic-depressive disorders in Great Britain, although the true rate of the disorder itself might not differ in the two countries. There could be other social factors that might operate this way: the British are considered more accepting of eccentricities than Americans; they might cheerfully tolerate mild schizophrenics whom Americans insist on hospitalizing.

The problem of distinguishing discovery from artifact, though common to all sciences, occurs more frequently in the social sciences. To test some theory, to understand some phenomenon, we need to measure something; but we find that what we want to measure is impossible, difficult, or expensive to measure directly. So we decide to use something else that can be measured more easily, something we hope will come close to representing the same thing. We have to be careful, though, to consider possible sources of error that might make what we can measure a poor indicator of what we really want to know.

2. *How reliable are the diagnoses?* Even if we concluded that the rates of hospitalization in the two countries really do measure the incidence of mental disorders, there is another possible source of error that could make the figures misleading. It had occurred to a number of scientists that perhaps part of the differences in rates of disorder arose because psychiatrists in the two countries, without being aware of it, were diagnosing the two major disorders differently. We

have pointed out in our discussion of diagnosis how difficult and possibly unreliable a procedure it is. We have mentioned in our descriptions of these disorders that there is some overlap of symptoms: a pervading sense of guilt so intense that it may be classed as a delusion rather than just a mood, and occasionally hallucinations, do occur in patients who would be diagnosed by some psychiatrists as depressives; yet delusions and hallucinations are also common symptoms of some forms of schizophrenia. Could it be that the terms "schizophrenia" and "depression" meant different things to psychiatrists in Great Britain and the United States?

Previous Studies of Reliability of Diagnoses

One of the factors that led scientists to question British–United States differences was previous studies of just this problem of the reliability of psychiatric diagnoses. One was a study made on patients admitted to a well-staffed university hospital in the United States. It was the practice of the hospital to assign the patients, as they arrived, at random to one of three wards. Only after assignment did the patients see the psychiatrist who made the diagnoses for that ward. The question investigated was, did the three psychiatrists diagnose in the same way? Since the patients were assigned randomly among the wards, it was reasonable to expect that if the psychiatrists used the same criteria for diagnosis they would have found about the same percentages of schizophrenics in each of the three groups of patients. In fact, it was found that one psychiatrist diagnosed 66% of the patients as schizophrenics, while the other two psychiatrists diagnosed between 20% and 30% as having this disorder. The authors of the study were able to show that these variations in diagnostic practices not only were of statistical interest but also were the basis for differences in treatment.[22]

The fact that patients with mental disorders are often hospitalized more than once was used in a British study. On going into a hospital for the second or third time, a patient is likely to see a different psychiatrist—he may enter a different hospital, or someone else may be on duty when he is taken in. A group of 200 patients were selected who had first been admitted to a hospital in 1954 and who had had at least four subsequent readmissions within the next two years. When their case records over this period were examined, it was found that only 37% were given the same diagnosis on all five admissions. Now this variability in diagnosis may not have been the fault of the psychiatrists in the various hospitals. Perhaps people suffering from mental disorders change in symptoms so much from one time to another that their diagnoses according to the classification system we have described really would change. If this were so, it would make us wonder whether the classifications we have been using are sensible, useful ones. Maybe they are really no better than the old categories of mania, melancholia, and dementia.

To test this, the scientists making the study took from the case records the notes made by the various psychiatrists who saw the patient on each admission to a hospital, and had all the cases rediagnosed by a single psychiatrist on the basis of these notes. This meant that the psychiatrist making the rediagnosis did not actually examine the patients, but only the records. The results were striking. The number of patients getting the same diagnosis on all five hospital admissions rose to 81%, implying that the differences in diagnoses did not arise from changes in the patients from one hospitalization to the next, but rather from differences among the psychiatrists who saw the patients on the different occasions.

If psychiatrists within one hospital or one city can disagree as much as this, it would not be so surprising if psychiatrists in one country, as a group, might differ from psychiatrists in another country. A study providing some evidence for this is described in the following quotation.[22]

> A filmed psychiatric interview was shown to 42 American and to 32 British psychiatrists: the patient was an attractive young woman in her middle twenties with a variety of fairly mild symptoms of anxiety and depression, who also complained of difficulty with inter-personal relationships and the frustration of her ambition to be an actress. In spite of one-third of the American psychiatrists making a diagnosis of schizophrenia, none of the British put this forward as the primary diagnosis. The predilection of American psychiatrists for symptoms related to or suggestive of schizophrenia was shown by the use of the terms schizophrenia and schizoid personality by nearly half of them, whereas over half of the British psychiatrists used diagnoses with a more affective connotation such as depressive neurosis or emotionally unstable personality. Another experiment reported in the same paper showed how wide a variety of diagnoses can be given to a single patient; 44 American psychiatrists between them used 12 different diagnoses after all viewing the same film. (pp. 11–12)

A Study of Diagnostic Practices

A project to study this question was organized in 1963. Known as the United States–United Kingdom Diagnostic Project (which we will refer to as "the Project"), it was a cooperative endeavor involving scientists from both countries.

From the studies on the reliability of diagnoses described above, one can see that there are two ways to find out how accurately the differences in rates of mental disorder represent real differences between the patients in the two countries, and to what degree they might reflect differences between the diagnostic practices of the psychiatrists: (1) have *two groups of psychiatrists* from each of the two countries examine the *same group of patients* and see how much the diagnoses of the two groups differ, and (2) have *one group of psychiatrists* examine *two groups of patients*, one from each of the two countries, and see how much the diagnoses of the two groups differ.

The Project scientists did studies of both kinds, but put special emphasis on a study of the second kind. However, the method of their study went somewhat further in its purposes. Feeling that psychiatric diagnosis had been shown to be too subjective and inconsistent, they decided to develop methods of diagnosis that would be more reproducible and objective, and to use these methods on groups of patients in both countries to determine if the rates were really different. The problem they faced in doing this is much like inventing an instrument—the thermometer—to measure quantitatively the property that gives rise to the subjective sensation of hotness.

A Thermometer for Mental Disorder?

One of the advantages of having some quantitative instrument is that it is objective. It gives the same answer regardless of whoever happens to use it.

A disadvantage is that it may not really measure the property or quality for which it was invented. As we pointed out in Chapter 6, the thermometer does not always agree with our feelings: a cold piece of metal feels colder than a cold piece of wood, yet a thermometer will indicate that they are at the same temperature. In looking at a much more complex question, one involving human beings, their feelings, and their behavior, one should expect much greater difficulties in devising a quantitative objective procedure that everyone concerned with mental disorders will accept. One might even question the possibility of such a procedure.

Psychiatric diagnoses are usually based on interviews with the patient, and sometimes also with members of his family. The psychiatrist will ask a large number of questions to determine just which symptoms of various possible disorders are present, and which are absent. Further, an unspecified part of the information comes across not only in the patient's answers to questions, but also in his behavior during the interview: how he answers, whether and how he looks at the interviewer, and other clues. Although such things cannot easily be made quantitative, most psychiatrists would agree that they are an important part of the process of diagnosis.

The Project Diagnosis

Tests based on questionnaires where the patient must answer "yes" or "no" to each question, and which could be scored as precisely as a true-false test in a history class, had already been devised for research on mental disorders, but one could not expect all psychiatrists to be willing to trust a purely mechanical procedure for such subtle questions requiring experience and judgment. The Project therefore chose a compromise between the objectivity and rigidity of a test, and the more sensitive and intuitive, but also more fallible judgment of the psychiatrist.

The test was based on a questionnaire, with a prescribed series of questions which had to be asked in the order given, but the psychiatrist interviewing the patient was free to ask additional questions and to interpret or disbelieve the patient's answers. The interviewer was required to select his diagnosis from a list based on a new, internationally accepted list of mental disorders, but he was left free to choose which diagnosis on the list he felt was appropriate.

The psychiatrists, both British and American, went through a period of training on the use of the procedure until they became proficient in its use and tended to agree on the results. As a further precaution, the written records of each interview were examined by two other psychiatrists also trained in the procedure, and if they disagreed with the diagnosis of the psychiatrist who had conducted the interview, that diagnosis could be changed.

Once this procedure for diagnosis had been worked out, its *reliability* was tested. It is obviously important to test the testers and find out how well two different testers agree on the same patient. One can see that 50% of the time would not be good enough, but it would have been unreasonable to expect agreement 100% of the time. The procedures for checking reliability are complicated, and we will not describe them, but the procedure, which we call "the Project Diagnosis," met as high standards of reliability as any other known procedure of psychiatric diagnosis.

The Results

The study required selecting groups of patients admitted to mental hospitals in both countries. It would have been inconvenient and expensive to study patients from hospitals all over the two countries, so only hospitals in the New York City and London areas were used. Two hundred patients admitted to hospitals in each city during a certain period of time were selected at random for the comparison. These patients had, on admission, been diagnosed in the usual way by the psychiatrists working at the hospitals. They were then given an additional diagnosis by the Project psychiatrists, using the procedure developed by the Project.

The rates found by the *hospital psychiatrists,* given in Table 8-1, show the usual striking differences betwen the two countries. We see from the third column, "Ratio London/New York," that schizophrenia appears to be twice as common in New York hospitals, and manic depressive disorders five times as common in London hospitals. The results of the diagnosis of the same 200 patients by the Project procedure are shown in Table 8-2. If the rates for the two types of disorders in the two populations were in reality exactly the same, the ratios in the third column of Table 8-1 would both be 1.0; in fact, they differ from 1.0 only by 6%–10%. The great differences found in the hospital psychiatric diagnoses, of twofold in

Table 8-1
Diagnoses of 200 Patients by Hospital Psychiatrists[a]

	New York	London	Ratio London/New York
Schizophrenia	76.6%	35.3%	0.46
Manic-depressive disorders	6.5%	32.3%	5.0
Other diagnoses	16.9%	32.4%	

[a] Hospital admissions for alcoholism and drug-related conditions have been eliminated from this table. The figures are percentages diagnosed as schizophrenia or manic-depressive disorders among all patients admitted to the hospital who were *not* diagnosed as suffering from alcoholism or drug addition.

schizophrenia and fivefold for the depressive disorders, have nearly disappeared. *The differences in rates between Great Britain and the United States, which looked like such a promising start for a study of the causes of the disorders, are not real. The conclusion is that it is not the rates of mental disorders that differ between the two countries, but the diagnostic practices of the psychiatrists.*

How Good Is the Project Diagnosis?

As we pointed out, there are two questions we can ask about any quantitative instrument we introduce in science: (1) Is it reliable? In other words, does it give the same answer all the time when applied in the same situation? The Project Diagnosis procedure met this test. (2) Is the Project Diagnosis valid? This is a harder question. In essence we are asking if a psychiatrist using this procedure will always diagnose schizophrenia if the patient really has schizophrenia. But when we put the question this way we are assuming that there is an objective reality to the concept "schizophrenia," and further, some objective procedure for showing whether a patient is afflicted with it, independently of the method of diagnosis we are using. Unfortunately we cannot make either claim.

Validity, in the social sciences, has been defined in many different ways, leading to different questions: Can the diagnosis be used to predict correctly the patient's future behavior? Does the diagnosis fit in well with the theories about mental disorders that are generally accepted at the time? Is the system of diagnosis fruitful when used in research on the disorders?

The same questions could be asked about the validity of I.Q. testing. If by validity we mean: "Does the I.Q. test successfully predict a child's performance in school?"—the answer is, fairly well, but not perfectly. If we mean: "Do I.Q. tests fit in with current theories of intelligence?"—the answer is that they fit in with theories held by some psychologists, but not with theories held by others, and that the two groups are at each other's throats. If we mean: "Have I.Q. tests been useful for research?"—the answer is that they have been the basis for the research programs of large numbers of psychologists, but there has been considerable argument as to the quality of this research.

The questions about the validity of psychiatric diagnoses are not easy to answer. They will be addressed repeatedly in this chapter, and we hope the concept of validity will become clearer as we proceed.

Table 8-2
Diagnoses of the Same Patients by Project Procedure

	New York	London	Ratio London/New York
Schizophrenia	39.4%	37.0%	0.94
Manic-depressive psychosis	34.5%	30.9%	0.90
Other diagnoses	26.1%	32.1%	

WHAT IS MADNESS?

We will have more to say about validity of diagnosis later on, but it should be pointed out here that if one wants to know whether the United States–British differences are real, one need be concerned only with the reliability of the Project Diagnosis, not its validity.

This can be understood when we consider that the problem is one of drawing some dividing line in the fuzzy area that distinguishes schizophrenia from the depressive disorders. We are asking whether American psychiatrists draw that line in about the same place as British psychiatrists do; we are not asking where the line *should* be drawn. We have found that American psychiatrists diagnosing American patients find more schizophrenia and less depressive disorder than British psychiatrists diagnosing British patients. Is it because the patients are different, and there really is more depression and less schizophrenia in Great Britain, or is it because the psychiatrists draw the lines in different places?

We could have answered this by having American psychiatrists examine British patients: do they find the same rates as the British psychiatrists did on the British patients, or do they find the same rates that they themselves did on American patients? If the first were the case, we would conclude that the rates really differ between the two countries; if the second were the case, we would conclude that diagnostic criteria differ. We would have the answer to our questions, even though we would not have established that the American psychiatrists' definitions of the disorders are in any way more or less valid than the British definitions. Obviously, we could have used British psychiatrists rather than American psychiatrists for this experiment, and it would have served the purpose equally well.

The reason for developing a system of Project Diagnosis was not only to answer the crucial question of the experiment, but also because the Project scientists felt that psychiatric diagnosis in general was unreliable, and it would be a useful by-product of the study if more reliable and consistent methods could be developed.

Other Studies

The scientists on the Project were led by their results to do other studies as well, but no longer on patients:[22]

> As it slowly became clear that the diagnostic differences we had set out to elucidate were being generated by psychiatrists rather than by patients we changed our plans and shifted the main focus of our investigation away from the patients and onto the psychiatrists. (p. 123)

They did studies similar to one described earlier (p. 211) as a test of diagnostic reliability. They made videotapes of diagnostic interviews on eight patients (five from England and three from the United States), three of whom were "typical" or "textbook" examples of one or another of the disorders, and the rest of whom were deliberately chosen to represent borderline cases difficult to diagnose. The tapes were shown to over 700 psychiatrists in the two countries. On the three textbook cases the psychiatrists of the two countries were, in the main, in agreement. For three of the borderline cases some disagreement appeared, with some British psychiatrists preferring a diagnosis of depressive psychosis to schizophrenia. For the last two cases there were striking differences. The majority of American psychiatrists diagnosed them as schizophrenia, while the majority of British doctors preferred other diagnoses not always involving serious psychoses. The American psychiatrists tended to see more pathology than the British ones—they perceived more of the patients' behavior as evidence of mental disorder. The conclusion of the Project scientists was as follows:[22]

> In general, the results of these videotape studies confirm and amplify those of the earlier hospital admission comparisons. Both indicate that the concept of schizophrenia held by psychiatrists in the New York area is much broader than that held by London psychiatrists and embraces substantial parts of what the latter would regard as depressive illness, neurotic illness or personality disorder, and almost the whole of what would be regarded in London as mania. (p. 124)

The Schizophrenia Epidemic in New York State

About the same time that the apparent differences in rates of mental disorders between Great Britain and the United States were noticed, other researchers had become aware that over several decades a dramatic rise in the relative proportion of schizophrenia among hospital admissions, roughly threefold, had been taking place in the United States. Specifically, the rate of all mental hospital admissions in New York State due to this disorder in the early 1930s was about 25%, similar to the English rate of 20% at that time. But then the rate in New York began a steady rise which peaked at 80% in 1952, followed by a slow fall to 50% by 1970. The English rate remained constant over this whole period at about 20%.

During the same period the rates of manic-depressive disorder had been falling significantly. These changes in rates of the disorders might offer important clues to their causes: one would now be able to study what social changes taking place in the United States during this period might be responsible.

The Psychiatrists Again

Of course now we must ask whether psychiatrists were changing their diagnostic criteria during this period; the rates of the disorders need not have been changing at all. Scientists associated with the Project decided to study this also.[24]

Since the case records of patients admitted to hospitals in both the 1930s and 1950s were available, these case records were reexamined by a single group of psychiatrists, who made new diagnoses to determine whether the relative proportion of schizophrenics had changed over the years.

They selected at random 64 case records of patients admitted to hospitals during the decade 1932–41. Of these 64 patients, 28% had been labeled schizophrenic from the hospital psychiatrists who saw them when they were admitted. Another 64 case records were similarly chosen for the decade 1947–56, on the

Table 8-3
Number of Diagnoses of Schizophrenia Out of 64 Case Records

	Original hospital diagnosis	Rediagnosis made in 1973
1932–41 decade	18 (28%)	27 (42%)
1947–56 decade	49 (77%)	30 (47%)

average 15 years later. Among these patients, 77% had received the diagnosis of schizophrenia from the hospital psychiatrists. All these case records were then edited to eliminate the original diagnoses as well as any other clues to what the diagnosis had been, such as the type of treatment the patient had received during his stay in the hospital.

These 128 case records were then submitted to a panel of 16 American psychiatrists. The members of this panel had not been trained in any specific diagnostic procedure such as that used by the Project; they may be assumed therefore to represent a cross-section of American psychiatric practice in 1973, the year the study was done. What matters is that a single group of psychiatrists was diagnosing both groups of patients.

The results of the 1973 rediagnoses are compared with the original hospital diagnoses at the time of admission in Table 8-3. Again the difference in rates of schizophrenia had just about disappeared. It was the psychiatrists who had changed, not the patients.

What Have We Learned?

One has the right to be disappointed with what we have learned from this United States–Great Britain comparison. We thought we were on to something interesting—a possible clue to the causes of schizophrenia and the depressive disorders—and we found out it was just a question of psychiatrists in two different countries using words differently. It is natural to react impatiently—"All right, let them get together and agree on what their terms mean and then get on with doing something about really solving the problem."

But it is not so simple. The differences in classification are not just differences about an arbitrary definition, like the differences between the Celsius and Fahrenheit temperature scales. They represent different philosophies about what is the best way to look at the problem.

What had been happening in the United States was the gradual growth of a belief about the functional mental disorders: that the distinctions made between schizophrenia and the depressive disorders were unimportant; that we were merely seeing individual responses to unbearable emotional stresses. American psychiatrists had come to regard these responses as just different symptoms of the same disorder. Although those holding this view did not abolish the classification "manic-depressive psychosis," they tended to define it very narrowly, using it only on patients showing "textbook" features of the disorders. They diagnosed most other patients as having schizophrenia, which they regarded as the more important and more comprehensive category.

In recent years diagnostic preferences among American psychiatrists have swung back towards a narrower definition of schizophrenia, more in accord with practices in other countries. In addition, recognition of the unreliability of psychiatric diagnosis has led to attempts to develop more objective and reliable methods. The World Health Organization has been publishing successive editions of a manual, *The International Classifications of Diseases*, which includes sections on mental disorders. The American Psychiatric Association publishes its own manual, *Diagnostic and Statistical Manual of Mental Disorders*. In the third edition, known familiarly as DSM-III, published in 1980 (from February 1980 to May 1982 eight printings, totalling about 250,000 copies, have appeared), the criteria for diagnosis was changed to be in better agreement with the *International Classifications of Diseases*.[25]

It is to be hoped that the establishment of internationally recognized criteria for diagnosis will lead to a greater degree of reliability, even though the ultimate source of a diagnosis will still be the individual psychiatrist relying on his training, his experience, and his intuitive judgment.

SECTION II: THE MENTAL HOSPITAL

DETERIORATION IN SCHIZOPHRENIA

We mentioned earlier that in a search for biological differences between schizophrenics and normal people, the schizophrenics were found to have higher concentrations of urinary phenolic acids. However, this turned out to have nothing to do with the disorder but only with the fact that, as mental patients with little to occupy their time in the hospital, they were drinking more coffee. Thus the excess urinary phenolic acids were a "result" of the disorder only in the sense that the disorder led to hospitalization, and the hospitalization led to more coffee. Although this anecdote may not seem to be of great scientific importance, the question it raises is a significant one: *How much of what we observe in a person suffering from a disorder is really part of the disorder, and how much is an accidental result of the fact that he has the disorder in a particular time and in a particular social setting?*

We will see this question from a deeper point of view if we return to the time when the original classification of the mental disorders was worked out by Kraepelin. A progressive and hopeless deterioration was regarded by him as the hallmark of dementia praecox. Descriptions by Kraepelin of severely deteriorated cases are as follows:[7]

WHAT IS MADNESS?

> [The patient] lay in bed dully, with a vacant expression, without occupying himself or paying attention to what went on around him. He alleged as the reason for his attempted suicide that he was ill; his brain had burst out a year before. Since then he could not think by himself; others knew his thoughts, spoke about them, and heard if he read the newspaper.
>
> The patient is still in the same condition to-day. He stares apathetically in front of him, does not glance round at his surroundings, although they are strange to him, and does not look up when he is spoken to. Yet it is possible to get a few relevant answers by questioning him urgently. He knows where he is, can tell the year and month and the names of the doctors, does simple sums, and can repeat the names of some towns and rivers, but at the same time he calls himself Wilhelm Rex, the son of the German Emperor. He does not worry about his position, and says he is willing to stay here, as his brain is injured and the veins are burst. . . . When told to give his hand, he stretches it out stiffly, without grasping.
>
> You will understand at once that we have a state of imbecility before us. . . . The complete loss of mental activity, and of interest in particular and the failure of every impulse to energy, are such characteristic and fundamental indications that they give a very definite stamp to the condition. . . . Together with the weakness of judgment, they are invariable and permanent fundamental features of dementia praecox, accompanying the whole evolution of the disease. (p. 26; Courtesy of the New York Academy of Medicine Library)

Figure 8-2 shows schizophrenic patients in a German hospital early in this century.

A similar description of severe deterioration in the mentally disordered is given in a book (published in 1976) by the psychiatrist Russel Barton:[26]

> A disease characterized by apathy, lack of initiative, loss of interest . . . in things and events not immediately personal or present . . . a deterioration in personal habits, toilet, and standards generally, a loss of individuality. . . . Occasionally the passive, submissive cooperation of the patient is punctuated by aggressive episodes. . . .
>
> The patient adopts a characteristic posture . . . the hands held across the body or tucked behind an apron, the shoulders dropped, and the head held forward. The gait has a shuffling quality, movements at the pelvis, hips and knees are restricted. (p.12) [Figure 8-3]

But Barton is not describing schizophrenia:

> It is only in the last years that the symptoms described above have been recognized as a separate disorder from the one which brought the patient into the hospital: *that the disease is produced by methods of looking after people in mental hospitals, and is not part of the mental illness preceding and sometimes existing with it*. (p. 13)

Dr. Barton is thus asserting his belief that a large part of the symptoms that have been regarded as characteristic of schizophrenia, and in particular the severe deterioration regarded as the hallmark and ultimate end-state of the disorder, may be a consequence of *what is done to* the patient in the mental hospital, rather than of the disorder itself.

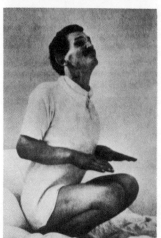

FIGURE 8-2. Photographs of schizophrenics from E. Bleuler's *Lehrbuch der Psychiatrie* (11th ed.). (Reproduced with permission of Springer-Verlag, Inc.)

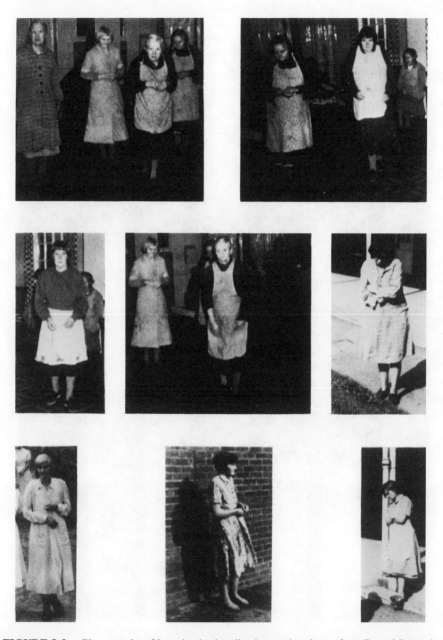

FIGURE 8-3. Photographs of long-institutionalized mental patients, from Russel Barton's *Institutional Neurosis*. (Reproduced with permission of John Wright and Sons, Bristol.)

FIGURE 8-3. (Continued)

Institutional Neurosis or Schizophrenia?

Dr. Barton was not the first to suggest that the hospital could be playing a role in producing the symptoms that had been attributed to schizophrenia, but his book, *Institutional Neurosis,* gives a graphic and moving description of the process. Barton quotes from a number of psychiatrists who were sensitive enough to the atmosphere of mental hospitals to perceive how the experience of hospitalization could be harmful to the patients and capable of contributing to a severe deterioration. His earliest quotation is from A. Meyerson, an American psychiatrist, who in 1939 described long-hospitalized mental patients as suffering from a "prison psychosis," the term for a type of behavior seen in inmates of prisons after long incarceration: apathy, submissiveness, a shambling gait, a lack of interest in getting out.[27] But the idea goes back to Pinel and Tuke. It was rediscovered by British psychiatrists during World War II, when mental hospitals had to be evacuated from the bombed cities.

How the rediscovery was made is described in an article by the American psychiatrist, E. M. Gruenberg:[28]

> Some British psychiatrists had been impressed by their surprising experiences at the time of forced hospital evacuations and severe staff shortages during World War II. . . . Not only did patients not take advantage of ample opportunities to escape, but many responded positively to war-time crises by helping less able patients. Some of the psychiatrists knew that during the first part of the 19th century many mental hospitals, inspired by the humanism of Pinel and Tuke, had completely given up physical restraints and operated as unlocked institutions. . . . Three British mental hospital directors began quietly experimenting to see in what ways how many of their patients could be cared for without the use of locked doors. (p. 130)

Barton's book describes the results of the changes in hospital practices and methods of treatment that started with these wartime observations.

Barton notes that the patients suffering from institutional neurosis are still suffering from schizophrenia as well: they may still be having the hallucinations and delusions that brought them to the hospital in the first place. Further, he does not make the claim that severe deterioration, at least in some cases of schizophrenia, could not have occurred regardless of the method of treatment or atmosphere of the hospital.

As evidence that institutional neurosis is distinct from schizophrenia though it often occurs with it, Barton cites the facts that (1) the same pattern of symptoms occurs in long-term patients in mental hospitals who have disorders other than schizophrenia—for example, organic disorders like arrested paresis and senile dementia (Alzheimer's disease)—and (2) the same pattern of symptoms is sometimes found in people without mental disorders who have been confined for long periods in such institutions as prisoner-of-war and concentration camps, orphanages, prisons, and tuberculosis sanatoriums.

The Origin of Institutional Neurosis

Barton attributes institutional neurosis to seven factors:

1. Loss of contact with the outside world
2. Enforced idleness
3. Bossiness of medical and nursing staff
4. Loss of personal friends, possessions, and personal events
5. Overuse of drugs
6. Ward atmosphere
7. Loss of prospects outside the institution

One should not think that these seven factors were individually discovered one at a time—they represent instead Barton's attempt to make systematic a flash of insight into the whole oppressive hospital atmosphere and what it does to the patients confined in it.

Barton illustrates each of the seven factors in detail, often with revealing anecdotes.

Under the first, loss of contact, he notes that at the very beginning of the illness the confinement within the locked and barred doors of the hospital is itself a shattering experience. Leave is grudgingly given, and when granted, is done so on humiliating terms. Nurses, doctors, and welfare workers only rarely encourage relatives to write letters; pen, paper, and a quiet place to write a return letter are not easily come by in the hospital. The hours of visiting are not always convenient for the average employed person who may want to come, hospitals are often located in isolated places requiring long trips by the visitor, and so on.

Barton attributes the enforced idleness of patients to an uncritical acceptance by doctors and nurses of the notion that the mental disorders are like physical diseases where bed rest may be really necessary. He describes the tendency of nurses to make the beds of patients and help them wash and dress when they are capable of doing these personal jobs themselves. We quote his description of the idleness of a patient's day:[26]

> After breakfast patients may be herded into the day room or garden and left to sit. A few may indulge in desultory occupational therapy such as knitting or rug making. If a patient gets up she may be told to sit down; if she asks to go out she may be snubbed or kept waiting. Individual activity of almost any sort may make the nurse afraid of imminent aggression. The nurse's behavior may actually cause an aggressive act which may be countered by sedation. Towards the end of the morning it is lunch time; the patient has no hand in purchasing, choosing, or preparing the raw materials or serving the finished product.
>
> After lunch a few patients may help clear away or a queue may file past a table putting dirty crockery and cutlery on it. The regular ward workers then set to and wash up, supervised by a nurse or ward orderly.
>
> The afternoon often presents another arid vista of idleness—nothing to do, no one to talk to, nowhere to go, the only event to look forward to being tea and, later, a sedative and bed.
>
> In some wards tea may be laid and prepared as early as 3.0 P.M., the bread

placed on the table already buttered, and nurses may spread jam on it. Preparations for bed may start as soon as tea has finished at 4.0 P.M., and often patients are in bed by 6.0 P.M. (pp. 17–18)

Barton describes the "bossiness"—the authoritarian atmosphere of the mental hospital—as follows:

> There is a tendency for . . . nurses to decide which clothes, shoes, and aprons a patient must wear, if and when and how their hair is dressed, where they must sit at table, which bed they must sleep in at night, what personal possessions they can have, if any, how much pocket money and "extra comforts" they can appreciate, if and when they can leave the ward, and so on. In some wards three doors had to be unlocked to enable patients to go from the day room to the lavatory. Needless to say, the incidence of incontinence was high. (pp. 18–19)

While it is clearly impossible for a patient in a hospital to have the same degree of personal privacy and sense of individuality as at home, Barton believes that mental hospitals could allow more of each than they have actually done. He remarks that prisoners in concentration camps were often pathetically attached to small personal possessions they were able to keep, such as a diary, a photograph, or a chess set. All the personal events that make up the pleasures of daily life outside are lost: family celebrations and meals together, going to the movies or a pub, and even watching television with friends or family.

> In the mental hospital institutional events exist, but the patient plays no part in ordering or altering them; they are largely impersonal. There is little to look forward to, and little to look back on. (p. 20)

In regard to the overuse of sedatives, Barton notes that when patients are expected to go to sleep at 7:00 P.M., itself a consequence of the absence of any worthwhile or challenging program of activity for them, it is natural to give them sedatives. When they wake up 8 hours later, having had sufficient sleep, it is only 3:00 A.M. They are then given another sedative to get them back to sleep, and then waken at 6:00 A.M. as part of the normal hospital routine, after which they are groggy and apathetic for the rest of the morning.

Ward atmosphere includes both the physical qualities—drabness, dinginess, smells, lack of views from the windows, and noise—and the more intangible aspects of atmosphere: attitudes of the nurses and ward attendants, posture and activities of the patients.

With respect to loss of prospects, Barton states:

> After admission to a mental hospital, as time goes by the prospects of finding a place to live, a job to work at, and friends to mix with diminish rapidly. It is difficult to persuade patients that the tremendous effort to re-enter the world outside is worth the gain. Many patients say they never wish to leave hospital: "I'm quite happy jogging along here, Doctor," or "Leave me alone, Doctor, I'm not well enough to manage outside," or "Nobody wants me—they haven't got room." Similar difficulties have been encountered with patients who have spent a long time in tuberculosis sanatoria. (p. 22)

The Cure

We have stated earlier that knowledge of the cause of a disease may or may not suggest a treatment. If "institutional neurosis" is real, the course of treatment is obvious, and if the treatment helps, we have provided confirmatory evidence for the hypothesis about its cause. If we change the depressing and dehumanizing atmosphere and practices of the mental hospital to more positive ones, based on the belief that the patients are still human beings and will react to the manner in which they are treated, and if we find that the symptoms described by Barton tend to disappear or diminish, and more patients recover well enough to leave the hospital, we have made it plausible that there is such a disorder as institutional neurosis, separate from schizophrenia or other mental disorders.

In fact, changes in hospital practices began before a clear-cut experimental test of the hypothesis was possible. This was as it should be, because as soon as it was recognized that mental hospitals were being run in an inhumane and depersonalizing way that *might* have prevented the inmates from recovering, the hospitals had to be changed. It would have been morally wrong to wait for convincing evidence. However, the need for such evidence does remain. If chaining patients to the wall and whipping them really cured them, most of us would accept the necessity of doing this, however repugnant the thought of it might be, but we do need to know if the more humane methods of treatment really work, not only for the sake of the patient's recovery, but also for our understanding of schizophrenia.

Some of the evidence on the effect of changing hospital practices is given in a continuation of the quotation from Gruenberg, describing the effect in three British mental hospitals of adopting a new approach suggested by the recognition of the concept of institutional neurosis:[28]

> After a decade or more, several things were established. First, no doors needed to be locked except medicine cupboards and that holy of holies, the staff toilet. . . .
> Second, patients could only be kept in an unlocked hospital when there was a full activity program to engage their attention. Third, patients were best cared for outside of the hospital once the initial period of disturbance had ended, which care in turn depended upon the establishment of extended outpatient and consultation services into the fabric of comunity life. This community care required a most relaxed approach to rehospitalization, which made sure that everyone knew that marginally adapted patients could be rehospitalized on very short notice and without fuss and red tape. Fourth, there appeared to be a radical reduction in the rate which new cases of chronic deterioration developed after the new open-hospital and community-care program had come into being. . . .
> I first became aware of these developments in 1953. It took me several years, extended visits, and a chance to observe the reactions of other skeptics to overcome my own skepticism regarding these claims. But I became convinced that basically what I have described actually did occur—and it all happened before the advent of any of the tranquilizing types of drugs. (p. 130)

Barton in his book chooses not to describe in detail the results for the patients of changes he recommended in hospital practices, but he does give one description of an experiment on the use and nonuse of sedative drugs which is illuminating:[26]

In March 1957, all sedatives were discontinued in one ward. The number of draw-sheets sent to the laundry each day through urinary incontinence was significantly reduced and the nurses said the patients were easier to handle and to rouse in the morning. During April 1957, patients were given the same sedatives as before with a result that the number of draw-sheets soiled soon increased to old numbers and nursing staff were incensed by the return to the old regime. When the drugs were again withdrawn in May 1957, the improvements noted in March reappeared and persisted. . . .

A second attempt to assess the role of sodium amylobarbitone and chlorpromazine was made in the first three months of 1958. In a geriatric unit of 200 patients 60 were found to be having a suspension of chlorpromazine 50 mg. three times a day and tablets of sodium amylobarbitone gr. 3 each night.

For a period of one month dummy tablets dusted with quinine to make them bitter and white suspension flavoured with ascorbic acid and quassia were substituted. Five out of six ward sisters were unable to say in which month the substitution was made. . . .

It would be foolish to say that all patients should be without drugs, but it seems likely that many are better without them. When a patient is not sleeping a warm drink and a small snack, especially if they are allowed to get it for themselves in the ward kitchen, may be more effective than a medicine glass of paraldehyde. (pp. 43–44)

The reader may remember the quotation from Tuke's book of 1813 on the sedative effect of "a glass of porter and a meal in the evening." It is surprising that this elementary human fact, known to anyone who has ever eaten a midnight snack before going to bed after a stimulating evening, should have been forgotten, requiring rediscovery after 150 years (see p. 187).

Other observations tending to demonstrate the reality of the institutional neurosis concept are described by Gruenberg, but for a variety of reasons discussed by him a fully convincing proof was difficult to obtain. There is no doubt that the catastrophic course in schizophrenia common in Kraepelin's time is seen more rarely today, but which of the enormous changes since his time in hospital administration, social attitudes toward mental disorder, new drug therapies, etc., are responsible, and to what extent, is hard to determine.

Gruenberg expresses the view that the concept of community care plays a central role in this new approach to treatment. Community care involves easy hospitalization and easy discharge, with treatment within the community by outpatient clinics when necessary. The mental hospital, instead of being regarded as a terminal institution for long-term custodial care, is a place to be used only in periods of extreme disturbance, when there is no alternative. He reviews the evidence for the conclusion that this approach is responsible for the known decline in the number of severely deteriorated schizophrenics, and states: "I find the conclusion inevitable but recognize that the evidence for it is inadequate." This may seem "unscientific," but as we noted earlier, it is often necessary to act on the basis of beliefs for which the evidence is inadequate.

Gruenberg's conclusion, as qualified as he made it, is not shared by all psychiatrists. There are some who believe that severe deterioration does sometimes occur in schizophrenia regardless of the kind of hospital atmosphere and course of treatment, and they have suggested a new way of classifying schizophrenic disorders on this basis. We will return to this question at the end of the chapter.

The best we can conclude is that the question raised by Barton and others—*How much of the symptoms and course of schizophrenia are determined by the hospital and how much by the disorder itself?*—is an important one, and that, difficult as it may be to answer, it deserves our serious attention.

GENERALIZING A CONCEPT

Focus in Science

Science begins with the general recognition of a problem or an area to be studied. The next step is concerned with focusing on just what to study to solve the problem. What do we begin with? What should we observe, what facts should we determine, what data should we gather?

We cannot repeat often enough that one of the greatest misconceptions about science is that the "facts" are there, clear-cut, well defined, and inescapable, and that scientific investigation starts with them. The reality is quite different. Faced with any problem, we quickly realize that the number of facts we might consider is enormous, and we must rather begin with some choice as to which ones we will assume to be relevant and which not. If we do not begin with some preconceptions as to which are worth looking at, we cannot begin at all. And if it is pointed out that preconceptions might mislead us, we can only answer that this is a risk we must take. The most useless advice we can be given when starting on a problem is that we should rid our minds of all preconceptions. We should rid our minds only of the false ones, but which are they?

If we want to study madness, the obvious place to begin is with the mad, and the obvious place to find them is in madhouses. Nothing could be more straightforward than to go to the asylums and observe the patients closely, see how they act, what their symptoms are, how the disorder progresses. Yet we now realize that much of what we have observed was not the result of a disorder itself but the result of a complicated interaction between the disordered patient and the institution in which he was confined.

The reader may feel that this particular problem is easily dealt with. Since we find that prolonged hospitalization produces effects of its own that are not necessarily characteristic of either schizophrenia or depressive disorders, if we want to study these psychoses in their "pure" forms independent of the confusing effects of hospitalization, let us study the victims who are not yet in a hospital or who have been confined only a short time. However, this only shifts the problem to another place. When a person undergoes a psychotic breakdown, it is not like catching a cold or measles, a process whose course is pretty nearly independent of the social situation of the person in the days or weeks preceding the onset. Instead, the breakdown takes place in a social environment in which the victim is closely involved with his family, friends, neighbors, co-workers, and ultimately such figures

WHAT IS MADNESS?

as policemen and doctors. The question can be raised as to what extent the way these people act toward the victim, once they perceive that he is behaving strangely, influences the course and symptoms of his disorder. Is the effect of the environment on the mental patient limited to the mental hospital, or does it exert its influence long before?

The realization by psychiatrists that the mental institution is only one special example of the way the entire social environment of the mental patient affects him and his behavior led to the development of a concept more general than institutional neurosis, labeled "social breakdown syndrome." (*Syndrome* is the medical term for a group of symptoms that are usually found together, whether or not they are the symptoms of a particular well-identified disease.)

The syndrome, as defined by Gruenberg, consists of many symptoms earlier regarded as characteristic of the specific psychoses such as schizophrenia or depression:[29]

> The social breakdown syndrome can be manifested by a wide range of overt disturbed behavior. Withdrawal, self-neglect, dangerous behavior, shouting, self-harm, failure to work and failure to enjoy recreation are the main manifestations. . . .
> The onset is sometimes insidious, the course indolent, and the end the vegetative state described in textbooks. More commonly, onset occurs in a single, explosive leap, beginning with violent behavior or the sudden termination of all ordinary social roles, often accompanied by a confused or clouded state. (p. 703)

It is the kind of behavior that rapidly leads to confinement in a hospital, often against the wishes of the patient himself. Is this pattern of behavior intrinsic to the disorder, or is it the result of the interaction between a person undergoing a psychotic episode and his social situation? Gruenberg inclines to the latter opinion. He regards the behavior as emerging "as a result of a spiraling crescendo of interactions between the patient and his social environment."

It begins, in his view, when the person starts to feel that there is a discrepancy between what he thinks he is capable of doing and what he thinks others expect of him. Now perceiving such a discrepancy is not of itself a symptom of a mental disorder—rather it is a situation most of us get into more than once in life. One way or another, we resolve it without breakdowns: we work harder or quit our job, drop out of school, change careers, or find other friends. The person undergoing a breakdown may lack the resiliency and personal resources to develop constructive ways of dealing with the perceived discrepancy. He sees himself as trapped, unable either to handle it or escape from it. He tries desperately to develop new ways of behaving to compensate for his difficulties—withdrawal, anger, fantasy—but these do not always work. Those close to him may become frightened and upset by his behavior. At first, they may try to cope with the situation by themselves, but eventually they are driven to seek help or advice from outside sources. The consequence is often that the family or friends are advised to request hospitalization for the person, on the grounds that he is "mentally ill." By this time they have begun to treat him differently from before, both as a result of the concern and anxiety stimulated by his own behavior, and in response to the judgment of society that he is a mental case. Once hospitalized, the person, often feeling betrayed by those closest to him, is in an environment in which he is told that he is sick, in need of care, and no longer responsible for his behavior.

The patient is now in a different environment from the one in which his breakdown began. He may, depending on the institution, the mode of treatment, and the seriousness of his disorder, recover and be discharged, or he may remain for a prolonged stay, with the additional risk of developing "institutional neurosis." Both institutional neurosis and the social breakdown syndrome can be seen to be results of interactions between a disordered individual and a social environment. The environments, however, are quite different, and the two conditions are not the same, though one person may in the course of time suffer both.

How can we get rid of the influence of social interactions and really study schizophrenia or depression by themselves? There may be no satisfactory answer to this question. There are those who feel that there is no such thing as "schizophrenia by itself" or "depression by itself"; that these are disorders of "interpersonal relations" and not definable except in a situation of social interaction. Thus, they can be viewed more fruitfully as belonging more to sociology than to psychology.

Some of those who hold this view of schizophrenia have gone so far as to propose that there is no such disorder as schizophrenia, that it is a label we apply to certain types of marginally deviant behavior, and that the problem is created by this very labeling. This idea will be discussed later in the chapter.

We can see that the questions asked earlier, "What do we begin with? What should we observe, what facts should we determine, what data should we gather?" are crucial, and that the way we answer them will determine whether our research will lead to something of value, or be wasted on accidental and unimportant details. Yet there is no choice but to rely on preconceptions that may as easily lead us astray as direct us on the right path. There is no rule to follow, no objective source of guidance to tell us which way to go.

A New Discovery?

We would like to quote a passage from Tuke's 1813 book that is relevant to the concept of the social breakdown syndrome. It shows us once more that we have rediscovered what was known 150 years ago.[5]

> A patient confined at home, feels naturally a degree of resentment, when those whom he has been accustomed to command, refuse to obey his orders, or attempt to restrain him. We may also, I conceive, in part, attribute to similar secondary causes, that apparent absence of the social affections, and that sad indifference to the accustomed sources of domestic pleasure of which we have just been speaking. The unhappy maniac is frequently unconscious of his own disease. He is unable to account for the change in the conduct of his wife, his

children, and his surrounding friends. They appear to him cruel, disobedient, and ungrateful. His disease aggravates their conduct in his view, and leads him to numerous unfounded suspicions. Hence, the estrangement of his affections may frequently be the natural consequence, of either the proper and necessary or of the mistaken conduct of his friends towards him. (p. 492)

TOTAL INSTITUTIONS

A crucial step in the discovery of institutional neurosis was recognition that the apathetic demeanor of long-hospitalized mental patients resembled that of long-confined prisoners—"prison psychosis." Barton, in his book, drew the parallels between such patients and inmates of prisons, concentration camps, sanatoriums, and so on.

The sociologist Erving Goffman made a detailed and searching analysis of such parallels in an essay, "On the Characteristics of Total Institutions," reprinted in his book *Asylums*. A "total institution" is defined by Goffman as follows:[30]

> A basic social arrangement in modern society is that the individual tends to sleep, play, and work in different places with different coparticipants, under different authorities, and without an over-all rational plan. The central feature of total institutions can be described as a breakdown of the barriers ordinarily separating these three spheres of life. First, all aspects of life are conducted in the same place and under the same single authority. Second, each phase of the member's daily activity is carried on in the immediate company of a large batch of others, all of whom are treated alike and required to do the same thing together. Third, all phases of the day's activities are tightly scheduled, with one activity leading at a prearranged time into the next, the whole sequence of activities being imposed from above by a system of explicit formal rulings and a body of officials. Finally, the various enforced activities are brought together into a single rational plan purportedly designed to fulfill the official aims of the institution. (pp. 5–6)

Goffman's book is an exploration of the significant features all these institutions have in common, features that shape the inmates into a pattern regarded by the institution or society as desirable. He does not make the claim that all total institutions are identical in every way; he is well aware that they differ in many significant features. Goffman divides total institutions roughly into five different types:

> First, there are institutions established to care for persons felt to be both incapable and harmless; these are the homes for the blind, the aged, and orphaned, and the indigent. Second, there are places established to care for persons felt to be both incapable of looking after themselves and a threat to the community, albeit an unintended one: TB sanitaria, mental hospitals, and leprosaria. A third type of total institution is organized to protect the community against what are felt to be intentional dangers to it, with the welfare of the persons thus sequestered not the immediate issue: jails, penitentiaries, P.O.W. camps, and concentration camps. Fourth, there are institutions purportedly established the

better to pursue some worklike task and justifying themselves only on these instrumental grounds: army barracks, ships, boarding schools, work camps, colonial compounds, and large mansions from the point of view of those who live in the servants' quarters. Finally, there are those establishments designed as retreats from the world even while often serving also as training stations for the religious; examples are abbeys, monasteries, convents, and other cloisters. (pp. 4–5)

Of course, in some of these situations the inmate is present voluntarily and in others not; in some the inmate is in for a specified time, while in others the stay is indefinite. Such factors must make important differences to the inmates.

What Do Convents and Concentration Camps Have in Common?

However, there are certain features shared by such diverse institutions as a convent and a concentration camp, and these are what Goffman focuses on. He is interested in the ways these institutions reshape the individual: change him from someone who enjoyed considerable freedom of choice and a sense of personal privacy and worth, into someone who follows a deindividualized pattern of behavior that serves the institutional purpose.

We cannot summarize Goffman's analysis in detail, but we can give its flavor by some quotations describing what might be called the "initiation" of the inmate:

> The recruit comes into the establishment with a conception of himself made possible by certain stable social arrangements in his home world. Upon entrance, he is immediately stripped of the support provided by these arrangements. In the accurate language of some of our oldest total institutions, he begins a series of abasements, degradations, humiliations, and profanations of self. His self is systematically, if often unintentionally, mortified. . . . In many total institutions the privilege of having visitors or of visiting away from the establishment is completely withheld at first, ensuring a deep initial break with past roles and an appreciation of role dispossession. A report on cadet life in a military academy provides an illustration:
>
>> This clean break with the past must be achieved in a relatively short period. For two months, therefore, the swab is not allowed to leave the base or to engage in social intercourse with non-cadets. This complete isolation helps to produce a unified group of swabs, rather than a heterogeneous collection of persons of high and low status. Uniforms are issued on the first day, and discussions of wealth and family background are taboo. Although the pay of the cadet is very low, he is not permitted to receive money from home. The role of the cadet must supersede other roles the individual has been accustomed to play. There are few clues left which will reveal social status in the outside world. . . . (pp. 14–15)

The inmate, then, finds certain roles are lost to him by virtue of the barrier that separates him from the outside world. The process of entrance typically brings other kinds of loss and mortification as well. We very generally find staff employing what are called admission procedures, such as taking a life history, photographing, weighing, fingerprinting, assigning numbers, searching, listing personal possessions for storage, undressing, bathing, disinfecting, haircutting, issuing institutional clothing, instructing as to roles, and assigning to quarters. . . . (p. 16)

[T]here is a special need to obtain initial co-operativeness from the recruit. Staff often feel that a recruit's readiness to be appropriately deferential in his initial face-to-face encounters with them is a sign that he will take the role of the routinely pliant inmate. The occasion on which staff members first tell the inmate of his deference obligations may be structured to challenge the inmate to balk or to hold his peace forever. Thus these initial moments of socialization may involve an "obedience test" and even a will-breaking contest: an inmate who shows defiance receives immediate visible punishment, which increases until he openly "cries uncle" and humbles himself.

An engaging illustration is provided by [the Irish writer] Brendan Behan in reviewing his contest with two warders upon his admission to Walton prison:

> "And 'old up your 'ead, when I speak to you."
> "Old up your 'ead, when Mr. Whitbread speaks to you," said Mr. Holmes.
> I looked round at Charlie. His eyes met mine and he quickly lowered them to the ground.
> "What are you looking round at, Behan? Look at me."
> I looked at Mr. Whitbread. "I am looking at you." I said.
> "You are looking at Mr. Whitbread—what?" said Mr. Holmes.
> "I am looking at Mr. Whitbread."
> Mr. Holmes looked gravely at Mr. Whitbread, drew back his open hand, and struck me on the face, held me with his other hand and struck me again.
> My head spun and burned and pained and I wondered would it happen again. I forgot and felt another smack, and forgot, and another, and moved, and was held by a steadying, almost kindly hand, and another, and my sight was a vision of red and white and pity-coloured flashes.
> "You are looking at Mr. Whitbread—what, Behan?"
> I gulped and got together my voice and tried again till I got it out. "I sir, please, sir, I am looking at you, I mean, I am looking at Mr. Whitbread, sir." . . . (pp. 17–18)

One set of the individual's possessions has a special relation to self. The individual ordinarily expects to exert some control over the guise in which he appears before others. For this he needs cosmetic and clothing supplies, tools for applying, arranging, and repairing them, and an accessible, secure place to store these supplies and tools—in short, the individual will need an "identity kit" for the management of his personal front. He will also need access to decoration specialists such as barbers and clothiers.

On admission to a total institution, however, the individual is likely to be stripped of his usual appearance and of the equipment and services by which he maintains it, thus suffering a personal defacement. Clothing, combs, needle and thread, cosmetics, towels, soap, shaving sets, bathing facilities—all these may be taken away or denied him, although some may be kept in inaccessible storage to be returned if and when he leaves. . . . (p. 20)

The impact of this substitution is described in a report on imprisoned prostitutes:

> First, there is the shower officer who forces them to undress, takes their own clothes away, sees to it that they take showers and get their prison clothes—one pair of black oxfords with cuban heels, two pairs of much-mended ankle socks, three cotton dresses, two cotton slips, two pairs of panties, and a couple of bras. Practically all the bras are flat and useless. No corsets or girdles are issued.
>
> There is not a sadder sight than some of the obese prisoners who, if nothing else, have been managing to keep themselves looking decent on the outside, confronted by the first sight of themselves in prison issue. (pp. 20–21)

At admission, loss of identity equipment can prevent the individual from presenting his usual image of himself to others. After admission, the image of himself he presents is attacked in another way. Given the expressive idiom of a particular civil society, certain movements, postures, and stances will convey lowly images of the individual and be avoided as demeaning. Any regulation, command, or task that forces the individual to adopt these movements or postures may mortify his self. In total institutions, such physical indignities abound. In mental hospitals, for example, patients may be forced to eat all food with a spoon. In military prisons, inmates may be required to stand at attention whenever an officer enters the compound. In religious institutions, there are such classic gestures of penance as the kissing of feet, and the posture recommended to an erring monk that he

> lie prostrate at the door of the oratory in silence; and thus, with his face to the ground and his body prone, let him cast himself at the feet of all as they go forth from the oratory.

In some penal institutions we find the humiliation of bending over to receive a birching. (pp. 21–22)

The Institution and the Condition

The relation both of Goffman's general concept of the total institution, and of its methods of reshaping the inmate, to the specific case of the mental patient reduced to the "institutional neurosis" state described by Barton, can be easily seen.

Since Goffman's concept applies to all kinds of total institutions, it loses sight of the special features that distinguish mental institutions from the others. Most of the inmates of mental institutions are there because of behavior disorders, disorders in responding to the stresses of ordinary life outside. That this is so implies that their responses to those characteristics of the mental hospital shared by other total institutions are likely to differ in significant ways from the responses of political prisoners in a jail, or young men attending a military academy.

WHAT IS MADNESS?

Of course, by making this distinction we are implying that there really is such a thing as schizophrenia (even if the course it takes may be determined in large part by the social situation in which it occurs). Goffman is one of those who have suggested that it is primarily created by our tendency to pin labels on socially unacceptable types of behavior. There is no question that prison psychosis would not exist without prisons. Would schizophrenia exist without either asylums, or the social response to deviant behavior characteristic of our culture? This is a crucial question in our attempt to understand mental disorders scientifically, more general than the question whether the rates of different psychoses in the United States and Great Britain are the same, but encompassing that question also. We will try to answer it in a later section of this chapter.

Belief and Evidence

Like the concept of institutional neurosis, the concept of the total institution itself has the ring of truth. We believe it almost as soon as we hear it. It makes sense—institutions really are like that. Goffman's demonstration proceeds by quoting example after example of incidents that support his idea. It is a process that resembles the listing of incidents by Snow in the beginning of his book to support the idea that cholera is transmitted by ingestion of the excreta of victims of the disease. Such a way of providing evidence for a belief is often called "anecdotal." Snow, having given evidence for his view on the transmission of cholera, went beyond the anecdotal evidence to a controlled experiment—a comparison of the rate of cholera with the degree of contamination of the water supply. However, the nature of Goffman's hypothesis is such that it does not lend itself easily to test by a controlled experiment.

Yet there are sciences, and disciplines we do not ordinarily regard as sciences, that must proceed this way. Goffman offers us a belief, an insight, a theory, and gives us certain objective evidence that supports it. One may find his evidence unconvincing, or come up with contradictory evidence that makes it less plausible. But one must recognize that it is much harder to find decisive or critical ways to test such theories than in those sciences in which controlled experiments are possible.

Geology is a science with similar problems in testing theories. There is very little we can do in the way of laboratory experiments to answer questions about what has been happening on the earth over millions of years. We cannot control at will the enormous forces that shape the earth's surface—the glaciers, the volcanoes, the drift of the continents—nor can we wait millions of years for the outcome of such an experiment. We must form our geological theories and test them on what we find when we go out and observe nature. We will, of course, select some facts as significant and disregard others, while people who hold opposing theories may build on the facts we overlook. This is not a claim that Goffman's sociology is as much a science as geology, but only a recognition that both are "observational" sciences rather than "experimental" ones.

What Have We Learned So Far?

One has a right to be a little disillusioned at this point. We seem to have promised much from applying scientific method to the study of the mental disorders, and gotten little.

We started with the exciting possibility that the rates of schizophrenia and the depressive disorders were dramatically different in Great Britain and the United States, only to discover that these differences were consequences of differences in psychiatrists' definitions of the disorders. Then we learned that the symptoms and course of schizophrenia, painstakingly studied and classified by psychiatrists over a period of fifty years, may have been determined in part by the way schizophrenics were treated in hospitals, rather than being intrinsic features of the disorder.

Having recognized the role of the social response in shaping the behavior and symptoms of the schizophrenic, we are led to ask, How far does this shaping go? How much of the syndrome we call "schizophrenia" has an existence in the individual sufferer and how much is created by the social environment?

However, we can see progress in these episodes. We have found it possible to develop methods of diagnosis for schizophrenia and depressive disorders that are at least reliable: different psychiatrists using these methods on a given patient will agree most of the time. And with the aid of these diagnostic methods we have shown that the rates of the disorders in Great Britain and the United States are really about the same, which, while less exciting than if they were different, is still useful knowledge. Further, our recognition of the way the social environment affects the course of schizophrenia not only gives us more insight into schizophrenia, but also may have played an important role in diminishing the catastrophic consequences of the disorder. And in this recognition we have been led to raise a significant question: does schizophrenia exist at all, or have we invented it?

So while it cannot be claimed that we have yet learned as much about the causes and prevention of mental disorders as was known a hundred years ago about the causes and prevention of cholera, we have gotten somewhere. We have learned to be critical of "facts" and of subjective impressions, and we have learned some humility in the face of the complex problems human beings present. We are at least asking better questions than in the past.

LABELING

We must now confront the question whether mental disorders generally, and schizophrenia specifically, are artificially created by the social context in which they occur. In this view, we produce schizophrenics by labeling individuals engaging in certain marginally abnormal types of behavior as "schizophrenics." Once so labeled, they are perceived as such by the rest of society, and the treatment they receive, by family, friends, psychiatrists, the staff of mental hospitals, imposes a pattern of behavior on them that fulfills the stereotype of the disorder. In this perspective, schizophrenia is not just a disease of civilization in the sense that cancer of the bladder resulting from industrial pollution is a disease of civilization—after all, the cancer is real. Instead, it is a label assigned to unlucky individuals by the society they live in, like the label of being a witch in Europe in the sixteenth or seventeenth century. Once the label is assigned, the individual must play the part—people accused of witchcraft or schizophrenia no longer behave like normal people. But it is the accusation and its social consequences, rather than a preexisting objective condition, that have produced the altered behavior.[31]

Once we appreciate the complicated interaction between the individual and his society that produces the social-breakdown syndrome, this theory of mental disorders has considerable plausibility. But there are ways of examining it critically. We have shown at length that the rates of schizophrenia and depression in the United States and Great Britain are about the same. This does not prove much about the "labeling" view of mental disorders, as the two cultures are perhaps not very different and labeling practices might be similar. But what if we examine cultures that are very different from these? Do we find that they have concepts of psychotic behavior at all? If so, are their concepts anything like our own concepts of schizophrenia and depression? What are the rates of these disorders? Is it possible that these cultures also produce mental disorders by labeling practices?

Eskimos and Yorubas

The anthropologist J. M. Murphy made a study of an Eskimo community in northwest Alaska and of the Yoruba tribe of rural Nigeria.[32] She found that both communities recognize a distinct type of behavior characterized by such symptoms as hallucinations, talking to oneself, refusing to eat for fear of poisoning, grimacing, violent or threatening behavior, tearing off one's clothes, laughing when there is nothing to laugh at. These are all symptoms of schizophrenia in our own society. In both cultures there are words in the language to describe this type of behavior. In both communities the treatment of individuals behaving in this fashion involves a combination of care for those who cannot care for themselves, laughter when the behavior appears foolish or incongruous, restraint when the individual threatens harm to himself or others. The Eskimos, for example, sometimes tie up individuals who act violently or tend to wander off in a way that endangers their own lives, or confine them to igloos with barred doors. The Yoruba healers who treat such people use shackles to restrain those who wander off, and herbal medicines for sedation.

The rate of occurrence in the Eskimo and Yoruba communities of this schizophrenialike behavior was estimated by Murphy as comparable to the known rates of schizophrenia in such Western countries as Canada or Sweden.

It is of course possible to argue that this research has only established that labeling produces "mental disorders" even in cultures disparate from our own. But the similarities of symptoms and the similarities of rates suggest that it is more plausible to conclude that schizophrenia really has an objective existence that does not depend on the type of society in which it occurs.

In general, studies of schizophrenia in different countries of the world have not supported the idea that there are differences in rates between more-developed and less-developed countries. This, however, has not closed the question of "labeling": while the *rates* at which people develop schizophrenia in different cultures is about the same everywhere, the *duration* of the disease seems to be shorter in less-developed societies, and this has been attributed to a greater degree of social acceptance of the schizophrenic in these cultures.[33]

SECTION III: GENETIC STUDIES

WHY DOES SCHIZOPHRENIA RUN IN FAMILIES?

We would like to close this chapter by describing some recent studies of the genetic aspects of schizophrenia that seek to answer the questions: Is schizophrenia a hereditary disorder? If there is a hereditary component, how important is it, compared to the influence of the family environment studied so intensively by psychodynamically oriented psychiatrists?

We quoted earlier a conclusion of the psychogenically oriented psychiatrist S. Arieti, that in families that have produced a schizophrenic child, there seems to be severe disturbance, of one form or another, in the parents. Examples of parental behavior often seen—sometimes incredibly domineering, other times rejecting or loveless, and still other times just plain bizarre—are given in the following quotations:[34]

> One therapist tells the story of parents who for months had begged their hospitalized schizophrenic daughter to come home for a weekend. Then one day she asked whether she could make the visit soon. Father and mother continued their conversation as if nothing had been said until the therapist broke in, unbelieving, to ask whether they had not heard. The daughter asked again.
>
> Apparently delighted, her parents asked formal permission, and then for twenty minutes they debated the arrangements for driving the girl to and from the house, both having cars and available time. Their daughter spent the weekend in the hospital, and . . . she will no doubt continue to do so. The hospital provides her only available protection against such conflicting statements and the real rejection inherent in their leaving her there. (p. 141)

The following is a description of relationships in the family of identical twins who later developed schizophrenia:[35]

> Mrs. Nebb insisted that the twins were geniuses whose activity must not be inhibited by any restrictions. She believed she had failed with their brother because she had been punitively controlling, and she would succeed with twins by using obverse methods. She categorically eschewed any discipline, but she was nonetheless extremely controlling and intrusive in other areas, particularly in all matters that threatened her serious phobia concerning contamination. This

applied to bowel habits, bathing, food fads, and abhorrence of animals, where any breach in the practices she dictated produced severe rebuff. She had banished her husband from her bedroom and bathroom when he had a mild transient fungus infection of the groin, and thereafter he could only use the basement lavatory. The older brother and father were labeled "dirty," and the brother "dirty-minded" for refusing to let mother bathe him and examine his stools after he was fourteen.

The twins were raised as a unit. They were dressed identically until nine or ten, and they were practically indistinguishable in appearance. Often they were not differentiated by the parents and both were punished or praised for the deeds of one. . . .

Often, when she was ill and received medicine, she also gave it to the twins. At one time, the boys could not keep awake in school, and learned that Mrs. Nebb had been placing the sedative she had received for herself in their breakfast food. They were both bowel trained starting at three months. Suppositories and enemas were liberally used, and enormous amounts of parental affect and energy were concentrated on bowel functions. Later, indeed until late adolescence, mother would give both twins enemas together, often because one was angry with her, which meant to her they were constipated. The enemas were administered according to a ritual in which both boys lay naked on the floor, and mother lubricated their anuses with her finger and inserted the nozzles with water as hot as they could stand. The twin more dilatory in getting into position would have to dash to another floor to the toilet. She continued to bathe them and to wash their genitals until they were fifteen. The mother constantly praised the twins to her friends and strangers and asked them to perform for guests. Friends were often amazed to learn that Mrs. Nebb had another son. (pp. 209–210)

Another example:[35]

After some months of intensive therapy, the patient improved considerably. She repeatedly expressed her hopelessness that her parents would ever listen and understand her unhappiness over her school and social problems. As the patient could not be kept in the hospital for a long period and as the psychiatrist who was interviewing the parents found it impossible to get them to focus upon any meaningful problems that might be upsetting, a therapeutic experiment was undertaken with great trepidation. The patient and the parents would meet together and, with the help of both psychiatrists, would try to speak frankly to one another. The daughter carefully prepared in advance what she wished to convey, and we tried to prepare the parents to listen carefully and to reply meaningfully. The patient, to the surprise of the psychiatrist, freely poured out her feelings to her parents and in heart-rending fashion told them of her bewilderment and pleaded for their understanding and help. During the height of her daughter's pleas, [the mother] offhandedly turned to one of the psychiatrists, tugged at the waist of her dress and blandly remarked, "My dress is getting tight. I suppose I should go on a diet." The mother had fallen back upon her bland equanimity. The next day the patient relapsed into incoherent and silly behavior. (p. 182)

It is easy to conclude, as did the psychiatrists reporting these cases, that the cruelty, rejection, or smothering love of the parents described in these examples is responsible for the schizophrenia of the children.

Those who have argued for a hereditary component have been quite aware of the existence of distorted family relationships in families of schizophrenics, but they have suggested another way of looking at the observations. Could it be that the bizarre behavior so often seen in the parents of schizophrenics, and usually believed to play a causative role, is rather an expression of a common hereditary factor shared both by schizophrenics and by their parents? Is the parents' behavior in some sense a borderline or unrecognized form of the same disorder? It is clear that both points of view could explain a tendency of schizophrenia to run in certain families.

The distinction between the two approaches was put graphically by Paul Wender: the children of red-haired parents tend to be red-haired; the children of Chinese-speaking parents tend to speak Chinese. How do we tell whether the children have the trait in question because they inherited it from their parents, because of the way their parents brought them up, or because of a combination of both?[36]

Still a third possible explanation for deviant behavior in the parents of a schizophrenic should be mentioned. Schizophrenia is a disorder of early adult life, and it is often insidious in its onset. Many years of disturbed behavior, gradually increasing in severity, may precede a real breakdown and the psychiatric identification of a case of the disorder. In the course of time, parents, who initially may have treated their child in the same way other parents do, may develop altered responses as a reaction to the child's disturbance. A psychiatrist, who generally sees the family for the first time only after the child has become schizophrenic, might be led to interpret the parent's behavior as the cause of the disorder in the child, when in reality it is an effect of it.

INHERITANCE

Introduction[37]

We present here a discussion of some basic facts about genetics. It will, we hope, enable the reader not only to follow our summary of some genetic studies of schizophrenia, but also to appreciate the difficulties and controversies that have arisen.

We will begin with a description of the basic Mendelian laws which govern the transmission of inherited traits from parents to children. Then we will point out some complications that arise when we try to apply these laws to specific situations.

Mendel's Work

Gregor Mendel discovered the laws named after him in the course of a study of pea plants. In 1865 he published a paper reporting his conclusions, which was overlooked by biologists for the next 35 years.

WHAT IS MADNESS?

In the simplest case—inheritance governed by one Mendelian factor or "gene"—the laws work somewhat as follows. An inherited trait may be classed as *dominant* or *recessive*. If the trait is dominant, an individual can have it only if at least one parent had it also. If recessive, the individual may have the trait even if neither parent did, but only if both parents were *carriers* of the trait.

We have thus distinguished three types of individuals when a particular inheritable trait is under discussion—those who possess the trait, those who do not possess it but can under certain circumstances transmit it to their children (we use the term *carrier* for these), and those who neither possess it nor can transmit it.

Given parents, one or both of whom may possess a trait or be carriers, the question of which if any of their children will have the trait is not a strictly determined matter, but depends on chance. We cannot usually predict *which* child will have it, but rather how many children *on the average* will. Table 8-4 gives some examples of such predictions.

The derivation of the above rules from the laws of probability is given in the Appendix.

Using these rules it might seem to be an easy matter to identify which traits in human beings are inherited, and whether they are dominant or recessive. All you would need to do is find enough cases of parents who possess or are carriers of particular traits, and count the numbers of possessors of the trait or carriers among the children. If the fractions correspond to those given in the table, the conclusion is easily drawn. This is in fact how Mendel was able to identify seven or eight traits of pea plants as either dominant or recessive. Unfortunately it is not always that simple, especially in human beings.

Table 8-4
Patterns of Inheritance

Type of trait	Parent 1	Parent 2	Fraction of offspring		
			Having trait	Carriers	Neither
Dominant[a]	Has trait	—	$1/2$	0	$1/2$
Recessive	Has trait	—	0	1	0
Recessive	Has trait	Carrier	$1/2$	$1/2$	0
Recessive	Carrier	Carrier	$1/4$	$1/2$	$1/4$

[a] If a trait is dominant, it is impossible to be a carrier—if you can transmit the trait to your children, you must have the trait to begin with. The fractions in this case were calculated on the assumptions that the trait is rare in the population, so the possessor of the trait is virtually certain to be what is called "heterozygous" (see Appendix). This is almost always the case when the trait is a disadvantageous one, such as a disease or malformation.

ficulties

Among the many factors that make it quite difficult are the following:

1. We do not often have direct means of determining whether an individual is a carrier, but can only infer it if the individual has been the parent of a child having the trait.
2. Many human inherited characteristics are determined by the simultaneous operation of several genes instead of one ("polygenic" inheritance). Examples are such familiar characteristics as eye and hair color, height, and skin color (as in interracial matings).
3. An individual may have the genetic makeup to have a trait, and yet not show that trait in a detectable way. Often the extent to which an individual will show a trait depends upon a combination of an inherited predisposition and environmental factors. In other cases traits do not manifest themselves, or vary greatly in the intensity with which they do, for reasons as yet not fully understood.
4. Spontaneous alteration of a gene in one particular body cell or germ cell of an individual may occur, and then be transmitted to a child who then develops a genetic trait not previously known in his family. Such a spontaneous alteration is called a mutation.
5. It is not only mental disorders that present problems of classification. Diseases of an unquestionably biological nature may also. Diabetes is known to run in families (and to be more common in certain ethnic groups as well), but there are tremendous uncertainties in predicting the risk to children from knowledge of the family history—to such an extent that the disease has been called a "geneticist's nightmare." Lately several types of this disease have been distinguished. Juvenile onset versus adult onset is one classification, and another is by severity—whether the patient can be treated by diet alone, or requires insulin. These different categories seem to show different patterns of inheritance, and may best be regarded as different diseases.
6. Last but not least, proving inheritance by consideration of family pedigrees alone is risky, as was pointed out earlier. Not only may traits run in families for nongenetic reasons, but their pattern may closely resemble that for a genetic trait. Professor A. M. Lilienthal of Johns Hopkins University did a "genetic" analysis of the trait "Attended Medical School" among students at the University of Buffalo (N.Y.) Medical School. The students, all of whom clearly possessed the trait in question, were asked whether either of their parents had attended Medical School. The resulting data were consistent, by the usual statistical tests applied in cases of Mendelian inheritance, with the hypothesis that the trait in question was determined by a single recessive gene which occurred in the general population with a frequency of about 6%.[38]

ADOPTION STUDIES

We said that some scientists who have studied the clustering of cases of schizophrenia in families have become convinced of the involvement of genetic factors. Others have considered psychogenic or psychosocial factors responsible: there is clustering in families because the family interactions induce the disorder in susceptible children. Both parties to the conflict are aware that not all children of a given set of parents become schizophrenic, even if one of the children does. The genetically oriented scientist is aware that not all children born to a given pair of parents are genetically identical, and is also aware that not all individuals possessing the genetic makeup for a trait will necessarily show that trait; the psychogenically oriented scientist explains the fact on the grounds that parents do not treat all their children exactly alike, with the result that children of the same parents often differ strikingly in personality. Further, the geneticist does not necessarily deny the possible role of stress in producing the disorder in genetically susceptible individuals; nor does the psychogenically oriented scientist always deny the possibility of some genetic components in an individual's susceptibility to the stresses and conflicts of the family situation. One feels that both could be right, even though they emphasize different aspects of the problem. But still there is room to argue the question of *how much?* Just how much of schizophrenia is genetic and how much is stress?

Fortunately, there is a clear-cut and decisive experiment which, in principle, can provide an answer. We quoted earlier the remark of Paul Wender that the children of red-haired parents are often red-haired and the children of Chinese-speaking parents tend to speak Chinese, but without further experiments we cannot decide whether this is a question of genetics or of environmental influence.

We are of course sure that speaking Chinese is not an inherited characteristic. We know that when Chinese children are adopted at a sufficiently early age by people who speak only English, they grow up speaking only English. While genetically they remain Chinese, with the characteristic appearance and other traits of their biological parents, they speak the language of their adoptive parents. So we have concluded that when it comes to language, environment is everything. This suggests that the relative importance of genetic factors and of stresses within the family environments for causing schizophrenia might be determined by a similar procedure. We can look at children adopted at an early age, so that the *biological* parents, from whom the child receives his genetic heritage, and the *nurturing* parents, who bring him up, are different people.

We do not have the right to take children away from families with schizophrenic traits just to satisfy our curiosity, any more than Snow would have had the right to add deliberately the excretions of cholera victims to the water supply of half of London to satisfy his curiosity. But, like Snow, we can look for a natural experiment. Can we find children, born to parents of whom one is schizophrenic, who were adopted at an early age by normal parents? Can we find children of normal parents who were adopted by parents who had schizophrenic traits, or who later developed schizophrenia?* Which plays the larger role in determining whether a child becomes schizophrenic—the traits inherited from his biological parents, or the upbringing given him by his adoptive parents?

The Danish Folkeregister

The experiment is simple in concept, but in practice was found difficult to carry out. Some pioneering experiments were done by the American psychiatrist L. Heston,[39] and a comprehensive study was made by the Americans S. Kety, D. Rosenthal, and P. Wender, together with the Danish psychiatrist F. Schulsinger.[40,41] The latter group did most of their studies using cases of adoption in Denmark, because the records of birth and other vital statistics there (called the *Folkeregister*) are more detailed and comprehensive than the United States statistics. It was possible, for example, to locate almost all the children born during a certain period of time, and adopted at an early age, who were later admitted to mental hospitals with diagnoses of schizophrenia, a task that would have proved impossible in the United States. Out of 14,500 children born from about 1925 to 1950 and adopted shortly after birth, 74 were found to have later developed clear-cut cases of schizophrenia.

Schizophrenia affects about 1% of the total population, regardless of country. But the statement that schizophrenia runs in families implies that members of families with a schizophrenic member will have a higher rate. A brother or sister of a known schizophrenic has a 10% chance of sharing the disorder, a parent 5%. These figures by themselves, as we have indicated, do not enable us to tell whether psychodynamic or genetic factors are important, but they do tell us what to look for in the experiment. The adopted child who later becomes schizophrenic has two sets of "relatives": biological relatives—his real mother and father and their blood relatives—from whom he was separated at an early age and who thus played very little part in bringing him up, and adoptive or nurturing relatives, who are not his blood-relatives but who provided the environment in which he grew to maturity.

If schizophrenia is induced by family interaction, we would expect to find excess cases of the disorder among the adoptive relatives. If it is primarily a genetically transmitted disorder, we would expect to find an excess among the biological relatives.

It was among the biological relatives, who shared genetic traits with the schizophrenic children, that the excess appeared. Among the adoptive relatives there was no more schizophrenia than would be expected of any group of people picked at random.

---

* Parents actively suffering from schizophrenia are not likely to want to adopt children, nor are the agencies that handle adoptions likely to permit children to go to such families.

Another experiment, also using the Danish adoption data, was done under the supervision of D. Rosenthal. He and his associates searched the Folkeregister for people who had given out children for adoption and who themselves suffered from schizophrenia. The children showed a high rate of schizophrenia even though the people who adopted them did not show an abnormal rate of the disorder. This experiment was similar to the earlier study done in America by L. Heston on children born in state institutions to schizophrenic mothers and adopted at the age of one or two weeks; the conclusions of the two studies were similar.

In the studies of Kety and his associates, careful consideration was given to possible sources of error and to alternative hypotheses that might have accounted for these dramatic results.

"Blind" Classification

In the course of these studies it was necessary to divide the subjects into two classes: "schizophrenic" and "normal." In so doing, there is some risk that the bias of the scientist making the classification may influence his choices. If, for example, he is a strong believer in a genetic origin for schizophrenia, he may be more likely to class a borderline case as "schizophrenia" if he knows it involves a biological relative of a schizophrenic. That he may do so implies not that scientists are generally dishonest, but that they are human. To avoid this possibility of error, the classification was done "blind"—the scientists doing it did not know at the time whether the individuals they were examining were blood relatives of schizophrenics or not.

In addition, the reliability of diagnosis was tested by having several psychiatrists make the diagnoses independently, in ignorance of each other's conclusions. A case was considered "schizophrenia" if four psychiatrists independently agreed on the diagnosis.

The Control Group

One possible alternative hypothesis that could explain the high rate of schizophrenia among the biological parents of the adopted children who became schizophrenic is that *people with schizophrenic tendencies are more likely to give out their children for adoption than normal people are.* If this were so, adopted children in general, whether they were schizophrenic or not, would be more likely than nonadopted normal children to have schizophrenic biological parents. It would therefore follow that adopted children who *did* become schizophrenic would have a higher proportion of schizophrenic biological parents than the nonadopted children. This, however, would be the result of a particular social practice rather than a genetic factor. This kind of consideration, among others, led Kety and his associates to choose as a control group 74 adopted children who had *not* developed schizophrenia. No excess of schizophrenia among the natural parents and blood relatives of this control group was found, so the alternative hypothesis can be ruled out as an explanation.

The Uterine Environment

The adopted children in this study were not totally unexposed to some environmental influences from their biological relatives. While many of these children were born out of wedlock and may never have been seen by their real fathers, they did spend some weeks or months with their mothers, and of course nine months in the womb. Perhaps the critical period for inducing schizophrenia is just those early weeks. Or perhaps schizophrenia is neither psychogenic nor genetic in origin, but related to an abnormal environment in the uterus—abnormal for some reason we can only speculate about, but associated with schizophrenia or schizophrenic tendencies in the mother. This possibility represents a third hypothesis: an environmental origin, but not a psychogenic one. How can we test these possibilities?

If this hypothesis—influence either during the early weeks of life, or within the uterine environment as the causative factor—were correct, it would be the mothers rather than the fathers who would be the source of schizophrenia in the adopted child. One would therefore expect the mothers and their blood relatives to show a high rate of schizophrenia, but not the fathers.

We mentioned that many of these children were born out of wedlock. Their fathers were often married to or had lived with other women. It was possible to track down the children of these other relationships. These children were half-brothers and half-sisters of the schizophrenic adopted children; they shared only half as much of their genetic makeup (due to a common father) as full brothers and sisters do. More precisely, full siblings have, on the average, half their genes in common. Half-siblings have one quarter of their genes in common. But they had neither any common upbringing, even for a few weeks, nor a common uterine environment. They showed a 13% rate of schizophrenia, compared with a 2% rate found in a control group of half-siblings of nonschizophrenic adopted children.

Cross-Fostering Studies[42]

In a further experiment, children adopted by families in which one parent became schizophrenic after the adoption were followed through early adult life to see if they would show a greater tendency to contract the disorder. As noted earlier, the children of a schizophrenic parent, growing up in the same home with that parent, usually have about a 10% chance of becoming schizophrenic. This fact alone does not tell us whether genetic factors or upbringing counts most. But the *adopted* children in such homes were found to suffer no more risk of schizophrenia than children born to and raised by nonschizophrenic parents. If schizophrenia owes anything to psychological factors, then, the factors do not include exposure to a parent with the disorder.

TWINS[43]

Studies of twins have been useful in human genetics in general, and have contributed also to our understanding of genetic factors in schizophrenia.

There are two kinds of twins: identical and fraternal. Identical twins result when the fertilized ovum, early in the course of the repeated cell division that gives rise to the fetus, breaks up into two separate groups of cells, each of which develops into a fully formed individual. Since both such individuals have originated from a single fertilized ovum, they have an identical genetic heritage. They are always of the same sex and, of course, resemble each other physically to an extraordinary degree. Fraternal twins on the other hand result when there are two viable ova in the uterus at the time of conception, and each is fertilized by a separate sperm cell. Except for being of the same age, the two children do not resemble each other any more than other siblings, and share a common genetic heritage only to the same extent that siblings normally do. In terms of "genes," the basic unit of heredity contained in the microscopically visible chromosomes of the cells of the body, identical twins have *all* their genes in common, while fraternal twins and other siblings share, on the average, 50% of their genes.

Siblings of a schizophrenic have about a 10% chance of themselves becoming schizophrenic, a fact that is consistent both with a genetic and a psychological or psychosocial origin for the disorder. With fraternal twins, studies have shown that the probability is the same: if one twin is schizophrenic, the other has about a 10% chance of being schizophrenic as well. The rate of concordance among identical twins—the percentage of the time that when one member of a pair of identical twins is found to be schizophrenic, the other will be also—has been found to approach 50%.

If we assume that treatment of both members of a fraternal-twin pair is as much alike within the family environment as treatment of both members of an identical-twin pair—an important assumption, requiring justification—then the 50% concordance for identical twins, compared with the 10% concordance for fraternal twins, demonstrates two things: first, that genetic factors play an important role, and second, that there is a considerable environmental effect also. If the possession of a particular gene were sufficient to determine the presence of the trait, the concordance rate among identical twins would be 100%—but it is not.

One of the first questions that must be dealt with is the validity of the basic assumption: Are fraternal twins always treated so much alike that if one becomes schizophrenic the other should also, assuming a psychological origin for the disease? One thought that comes to mind is that fraternal twins need not be of the same sex, and in some families children of one sex are exposed to different and more deleterious treatment than children of the other sex. The scientists studying twins, aware of this possibility, have used fraternal twins of the same sex only for the comparison with identical twins. It has also been suggested that the higher concordance rate among identical twins reflects greater stresses of being an identical twin—problems of a sense of identity, for example. However this possibility is adequately ruled out by the fact that identical twins are no more susceptible to schizophrenia than the rest of the population.

Obviously this question could be resolved neatly by a strategy similar to the adoption studies. If identical twins are separated at an early age and reared in different homes, and one becomes schizophrenic, does the other also? What is the concordance rate then? The number of such cases that could be discovered by extensive search is very small—about a dozen. In this small group the concordance rate is about 50%, the same as in identical twins raised together, but the sample size is too small to carry much conviction.

Researchers have therefore turned their attention to the following problem: It is known that the children of a schizophrenic parent have a high (10%) chance of developing schizophrenia themselves, compared to a 1% chance for people in general. Suppose, in an identical-twin pair, one twin develops schizophrenia but the other twin remains normal; such a pair is called "discordant." What are the rates of schizophrenia among the two sets of children of the two twins? Are the children of the schizophrenic twin more liable to the disorder than their cousins, children of the normal twin? The answer is that the risk is found to be the same: 10% in both sets of children.

The identification of such a high risk among children of parents neither of whom is schizophrenic could only have been made on the basis of a genetic hypothesis, and it is strong additional evidence for something in the genetic heritage that predisposes to schizophrenia. However, it draws attention again to the observation made earlier on the basis of the 50% concordance rate among identical twins that while genes are important, they are not the whole story. What determines whether a person with a genetic predisposition to schizophrenia gets it or not? Is it some kind of extra psychological or social stress? Studies of discordant twins have failed to demonstrate any apparent differences in the emotional stresses to which the twins have been subjected that should cause one to develop schizophrenia and the other not. Scientists have therefore looked at physical differences: infections, physical trauma, differences in birth weight (identical twins usually differ in birth weight, which could reflect a different uterine environment). Again no plausible difference has been revealed.

So the conclusion of these studies was that although there must be some environmental factors—which might include both psychogenic factors and physical illnesses, either of the child, or of the mother when pregnant—that play a part in producing overt schizophrenia in a genetically susceptible individual, they have yet to be identified. This accords with the conclusions reached through adoption and cross-fostering studies.

THE CONCLUSIONS AND THE CONTROVERSIES

If the validity of these studies is accepted, the conclusions are quite dramatic: they imply that the kind of upbringing a child receives has less to do with his risk of developing schizophrenia than does his genetic background. It is the children born of parents who are schizophrenic or have schizophrenia in their own families who are most likely to develop schizophrenia, even if they are separated from these parents at birth or when very young, and raised in loving and stable adoptive families.

It is hardly surprising that not all psychogenically oriented psychiatrists have been convinced by these studies. A number have taken exception to them, and have made plausible-sounding criticisms, which have received plausible-sounding answers from the authors of the studies and others.[44-48]

The controversy is likely to continue for a long time, not only because of the purely scientific and clinical issues involved, but also because the question of the relative importance of genetic and environmental factors in schizophrenia is felt by some to be part of a larger and older question with social and political implications as well: the "nature versus nurture" question. How much of what human beings are or can achieve is limited by their biological make-up, and how much is determined by environmental and cultural influences? The controversy over the heritability of intelligence and the validity of the I.Q. test, which we referred to earlier (pp. 103–104, 214), has been intense and bitter because it is also, and more directly, concerned with this question. So is another acrid controversy which has just broken out over the anthropological research of Margaret Mead in Samoa.[49,50]

In this book we have tried to deal only with the scientific side of the dispute. We do so because we believe that the issues of genetics in schizophrenia and genetics in intelligence are independent questions, and each can be ultimately resolved only by the scientific process of making hypotheses and testing them, while continuing to subject both the hypotheses and the tests to critical examination.

We would like to note that the critiques of the genetic studies touch on issues which we have stressed as basic in science. Particularly central is the question of classification. One of the main points at issue in the genetic studies is just this question of whether the classifications—diagnoses—of the subjects were made properly or not, and, in turn, whether the results of the studies will lead to changes in diagnostic procedures.

As we have pointed out, severity of schizophrenic behavior shades off gradually from the extreme of disturbed patients who require hospital care, to individuals who, though they show some features of thought and behavior patterns which resemble the disorder, are able to function adequately enough in their daily lives to avoid breakdown. It was found in the genetic studies that more of these individuals with diagnoses of "latent schizophrenia" or "schizotypal personality disorder" turned up among the biological relatives of the schizophrenics than among the biological relatives of the control group. Thus, for the purposes of the genetic studies a "broad" definition of schizophrenia was adopted. Because the number of individuals in the study was small, it happens that without this the case for genetic factors would have been much weaker, and indeed Kety and his co-workers have been criticized on the grounds that they adjusted their definition of "schizophrenia" arbitrarily for the sake of producing data that would confirm their bias in favor of a genetic hypothesis. Kety has defended the procedure on the grounds that rather than being an arbitrary adjustment of the definition of the disorder, it was based on a hypothesis made originally by Bleuler that there was a group of disorders closely related to schizophrenia but less severe, which he had termed "latent schizophrenia."

Other criticisms based on issues of classification have concerned the reliability of the diagnoses made by the original research team. However several other teams of psychiatrists have reexamined the case records of the subjects, and confirmed the consistency of the original diagnoses with current diagnostic procedures as specified in DSM-III.

Other more technical points concern the surprising fact that, of the biological relatives of the adopted children who became schizophrenic, there were more schizophrenics among the half-brothers and half-sisters, than among the more closely related siblings and parents. This fact, though hard to explain by a genetic hypothesis, seems no easier to explain by any other hypothesis.

There have been other technical criticisms of the manner in which the tests of statistical significance were performed by the original research team.

Many of the criticisms would be resolved one way or another if more data were available. (See Chapter 9, pp. 271–272 for a discussion of this point.) Indeed, Kety and his associates are now gathering additional data from all parts of Denmark, rather than just the Copenhagen area, and we await with great interest the publication of their results in the near future. It is to be assumed that further studies will be forthcoming from other research groups as well—the impact of the genetic studies in the field of psychiatry has already been very great, to say nothing about their wider implications, and the question is too important to be left long without a clear-cut answer.

Kety in his turn has criticized the studies that have led to the conclusion that the familial environment is a major cause of schizophrenic disorders, in the following words:[48]

> [These studies are] flawed by one or more of the following deficiencies which do not permit that conclusion: absence of controls, inability to minimize subjective bias in the observations or diagnoses, inability to dissociate genetic from environmental contributions, or inability to remove the impact of rearing a schizophrenic individual on the personalities of the parents. . . . (pp. 726–727)

While his critics would probably answer that in spite of these safeguards he and his associates have failed to provide convincing evidence for a genetic factor, Kety's critical view of much prior research into the causes of mental disorders is justified.

Many psychiatrists and psychologists today believe that the case for genetic factors in schizophrenia is very strong, and further, that if there are any psychological factors involved, they have yet to be identified. But clearly this view is not yet unanimous.

Another Classification of Schizophrenia

In recent years some psychiatrists have challenged the usefulness of the classical division of schizophrenia into the subtypes of paranoid, hebephrenic, catatonic, and simple. They have observed that the onset and course of the disorder, *regardless of subtype,* tend to fall into one of two patterns which have been called *reactive* and *process* schizophrenia.[51]

In process schizophrenia, the victim has shown a characteristic schizoid personality long before the onset: shy, oversensitive, seclusive, avoiding competitive situations, eccentric with a tendency to daydream, and having poor social and sexual adjustment. The onset of the disorder is gradual, and there is no clear precipitating factor, no obvious personal situation or problem that places the victim under a severe stress.

In *reactive* schizophrenia, on the other hand, the onset is acute, and can clearly be related to personal situations producing stress. The victim does not have the "schizoid" personality to start with.*

An important distinction between the two categories is prognosis: The process schizophrenic has a much poorer chance of recovery than the reactive schizophrenic, who may have one such episode in a lifetime and be normal otherwise.

We may recall now the classification system of Kraepelin, who distinguished "dementia praecox" from other disorders by its deteriorating course, and the change in classification made by Bleuler, who found other patients with similar symptoms who did not deteriorate, and who felt that it was more useful to regard both kinds of patients as suffering from the same disorder. We see that Kraepelin's distinction between a disorder with a deteriorating course and disorders without such a course has been in a sense revived, and may after all turn out to be a more useful classification than Bleuler's more optimistic conclusion that they were one and the same.

This does not mean that every case Kraepelin regarded as true "dementia praecox" would have been classed as process schizophrenia today, nor does it mean that process schizophrenia today follows the same hopeless course as in his time.

---

* The DSM-III uses the term *schizophreniform disorder* in this case.

The role of the hospital, and more generally of the social situation of the patient, in influencing the symptoms and course of the disorder cannot ever again be overlooked. But it does mean that there has been some turning back to ideas that are closer to Kraepelin's concept than to Bleuler's.

The genetic studies have given some indication that when the cases of schizophrenia are divided up according to the categories of *process* and *reactive,* it is process schizophrenia, not the reactive type, that is inherited.

If these indications stand up to more careful scrutiny, it may be concluded that the process-reactive system of classification is more fundamental and more important than the earlier system, with its categories of paranoid, catatonic, and so on.

We thus see that genetic studies not only have strongly suggested a major role for genetic factors in determining whether someone gets schizophrenia, but also may shed new light on the question of classification of mental disorders. As we stated earlier, good classifications make discoveries possible, and, in turn, discoveries change our ways of classifying the things we study.

The "Myth" of Schizophrenia

We discussed earlier the hypothesis that schizophrenia itself may be a myth—an artifact of society's response to certain types of behavior. As evidence against this view, we then cited the cross-cultural study of J. M. Murphy, who showed that a disorder with the same pattern of symptoms occurs in primitive societies at about the same rates as in Western industrialized countries. It should be apparent that if schizophrenia, whatever it is, can be transmitted from parents to children by biological inheritance, then it must be real.

Implications for the Psychogenic Approach

Psychogenically oriented psychiatrists started with the hypothesis that the life experiences of schizophrenics, particularly the early experiences with their parents, are the significant causative factors. Starting from this concept, it was natural to look for the differences in family relationships between those families that have produced schizophrenics and those that have not. Such studies on schizophrenia can be thought of as controlled experiments with the second kind of family, the normal, as the control group. Differences were found between the two groups—the families of schizophrenics, on the average, are different from the normal families, and the intimate relationships within the first kind of family do often present disturbed and bizarre features. But if, as suggested earlier, the case for inheritance is correct, then these disturbed and bizarre features are not necessarily causative of schizophrenia, but may only be reflections of the common heritage of parents and children.

We have stated that in a controlled experiment the experimental and control groups should ideally be alike in every relevant factor except one, the factor whose influence the experiment is designed to test. But in a practical sense no two groups of human beings (or even white mice) are exactly equivalent. The best we can hope for is that the various ways they differ are not "relevant," but we have no infallible procedure for deciding what characteristics may or may not be relevant.

In experiments searching for psychological factors causing schizophrenia, the experimental group—parents with disturbed behavior and with schizophrenic children, and the control group—parents without disturbed behavior and with normal children, differed not only in their behaviors, but also in the genetic makeup that parents share with their children. This second factor was overlooked until the alternative theory of a genetic factor in schizophrenia was taken seriously enough to require experimental testing.

The debate over psychogenic factors in schizophrenia has not been put to rest by evidence for a genetic factor: the question of what sort of stresses or other environmental factors convert a genetic predisposition into an outright case of the disorder still remains open. We now know only that when we look for such factors, we must try to control the genetic variable.

A current view on the problem of the competing roles of nature and nurture— heredity and environment—in schizophrenia has been given by a leading sociologist, Bruce Dohrenwend:[(52)]

> There is nothing on the socioenvironmental side of the nature-nurture issue anywhere near so compelling . . . as the evidence on the genetic side that has been supplied by investigators using twin and adoption strategies in research on schizophrenia. In fact, one of the reasons some investigators believe that environmentally induced stress is indeed important in [causing schizophrenia] is that the quasi-experimental strategies used by the genetically-oriented investigators, while demonstrating a role for genetic factors, have not isolated the mechanisms involved and have left much of the variability still to be explained! This argument by subtraction, however, is a far cry from a demonstration of the etiological importance of environmentally-induced stress in its own right. The evidence we now have is simply not sufficient, for example, to permit investigators with socioenvironmental theoretical orientations toward [causation] to make confident nominations of "high risk" factors on the sociocultural and social-psychological side of the nature-nuture issue. (p. 386)

POSTSCRIPT

In this case study we have not attempted to present a comprehensive picture of current knowledge of mental disorders. We have chosen to discuss only limited areas of this subject, not necessarily because they are intrinsically important but rather because they illustrate important features of scientific method, and also because they can be explained to the reader who lacks an extensive background in the field.

We have said little about such important areas as drug therapies, biochemistry, and neurology, or about those aspects of human behavior for which psychodynamically oriented psychiatrists have claimed their greatest successes—theories of the human personality, the study of neuroses, and so on.

No one reading our previous case studies which deal with the medicine of 125 years ago and the physics of almost 200 years ago would think he was getting accurate accounts of the state of our knowledge today. And although the present study, in contrast, does deal with some areas of current active interest, it should not be taken as any more representative of the whole story than the others.

APPENDIX 8-1: GENETICS AND PROBABILITY[36]

We have described in our discussion of the genetics of schizophrenia some of the unpredictable workings of inheritance: brothers and sisters, children of the same parents, need not resemble each other or their parents in every characteristic. They may have different eye color, for example, or height. Yet these factors can be clearly shown to be determined, or greatly influenced, by inheritance. There are other genetically transmitted factors of more importance to the individual than eye color, such as certain diseases or predispositions to diseases. Schizophrenia is only one of many, but its mode of inheritance and the relative importance of environmental factors in producing it in a susceptible individual are not yet well understood. There are other genetic diseases that are more lethal: sickle-cell anemia, for example, which affects many Africans and their descendants in other countries. In this condition the red blood-cells of the body are deformed in shape and unable to fulfill their proper function of carrying oxygen in the bloodstream. The victims of this condition rarely survive to maturity. Children who suffer from it can be born to parents who do not have it, yet not all the children of these parents need have it. How can we predict who will get it and who will not? If a couple has a first child who suffers from the disease, should they risk having a second? What are the chances of the second child suffering from the same disease?

Four O'Clock Plants

To get some answers to these questions, we will begin with a discussion of inheritance in plants. There is a flowering plant called the "four o'clock" or, less commonly, the "marvel of Peru" (*Mirabilis jalapa*) that occurs in two pure strains, one with red flowers and one with white. Plants reproduce sexually: pollen plays a role analogous to the sperm of animals and fertilizes an ovum to produce the seeds of a new generation. It happens that the four o'clock is bisexual, so the pollen of a single flower can be used to fertilize the ovum of the same flower. When red plants are self-fertilized, the seeds produce only red plants; the self-fertilization of the white strain produces only white ones. If the ovum of a red plant is fertilized

WHAT IS MADNESS?

by the pollen of a white (or the ovum of a white one fertilized by the pollen of a red), plants with pink flowers are the result. This finding is no surprise: the surprise comes in the next generation. When pink flowers are self-fertilized, the new generation might be expected to be pink also, but it is not. All three types occur: red, pink, white. When the experiment is done a large enough number of times, so that many third generation plants can be counted, it is found that the proportions of red, pink, and white flowers are 1 : 2 : 1. One-quarter are red, one-half are pink, one-quarter are white.

Mendel's Hypothesis

What does this observation tell us about inheritance? What hypothesis can we make about the reproductive process to explain it? In a paper published in 1865, Gregor Mendel (1822–1884) put forward an explanation of this kind of experiment which is still regarded as correct.

We do not know how Mendel happened to think of his hypothesis, but it must have been suggested to him by some simple results from the theory of probability. If a coin is tossed 2 times, there are 4 *equally probable* outcomes:

$$
\begin{array}{c}
H\ H \\
H\ T \\
T\ H \\
T\ T
\end{array}
$$

This in turn means that in a very large number of trials in which a coin is tossed twice, 2 heads will occur $1/4$ of the time, 1 head and 1 tail $1/2$ of the time, and 2 tails $1/4$ of the time.

Mendel's conjecture was that most of the cells of each flower contain two factors related to the flower color, one from the male parent and one from the female parent. However, the cells involved in the reproductive act—the (male) pollen cells and the (female) ova—contain only one of the two factors. These reproductive cells are formed during the development of the organism by a splitting in two of the body cells, which contain both factors. Any one pollen grain may have either one—but only one—of the two factors of the individual cell that gave rise to it. The same is true of each ovum (see Figure 8-4).

Consider a pink four o'clock plant. Each body cell of the plant contains both a red factor (R) and a white factor (W), whose combined action produces the pink color. When this plant forms pollen and ova by a splitting of its cells, each pollen grain and each ovum formed contain either R or W, but not both. Since each cell of the original plant contained one R and one W, half the pollen grains contain R and half contain W; similarly, half the ova contain R and half contain W.

The *probability* that a pollen grain picked at random is R is $1/2$; the probability that an ovum picked at random is R is $1/2$. Both probabilities are the same as the probability of getting a head on one toss of a coin.

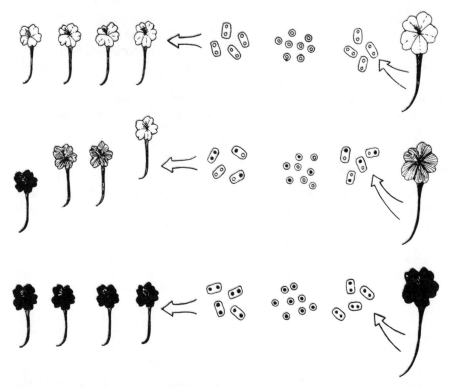

FIGURE 8-4. Mendelian inheritance. The body cells of the plant contain two factors (shown as black and white circles) for color. Plants whose cells have two red factors are red, those with two white factors are white, and those with one of each are pink. The male and female sex cells (gametes) are formed by the division of the body cells, and each contains only one factor. The sex cells of red and of white plants all have the factor for only one color, that of the parent plant. Half of the sex cells of the pink plant have the factor for red, the other half have the factor for white. On self-fertilization, the random mating of the sex cells of a red plant produces only seeds with two red factors, and of white plants produces only seeds with two white factors. Random mating of the mixture of sex cells of the pink plant produces seeds with either two red, one red and one white, or two white factors in the proportion 1 : 2 : 1.

Now we imagine fertilization—the pairing of pollen with ova—taking place at random. The outcomes are given in Table 8-5 with the equivalent problem of coin tossing. We can see that Mendel's hypothesis accounts for the results of self-fertilization of this plant.

Dominant and Recessive

Mendel actually did his experiments with peas rather than four o'clocks, and

Table 8-5
Breeding of Four O'Clock Plants and Corresponding Coin Tosses

Ovum	Pollen	Probability	Outcome: plant color	Net probability
R	R	1/4	red	1/4
R	W	1/4	pink ⎱	
W	R	1/4	pink ⎰	1/2
W	W	1/4	white	1/4

First toss	Second toss	Probability	Outcome	Net probability
H	H	1/4	2 heads	1/4
H	T	1/4	1 head, 1 tail ⎱	
T	H	1/4	1 head, 1 tail ⎰	1/2
T	T	1/4	2 tails	1/4

his results were different and slightly more difficult to interpret. But the behavior of flower color in peas happens to be more typical of inherited factors than that of color in four o'clocks. In peas, the flowers are either purple or white, but when pure purple and white strains are crossed only purple flowers result. There are no plants produced whose colors are intermediate between the extremes. The flowers that result *appear* to be the same as those that result when purple flowers of a pure strain are self-fertilized or are crossed with others of the same pure strain.

But the purple flowers resulting from the cross have not forgotten their parentage. This appears if they are self-fertilized or mated with each other. Both purple and white flowers result from the cross, in the ratio 3 : 1. The probability that a "child" of a purple plant of mixed parentage will be purple also is $3/4$; it has a $1/4$ probability of being white. Note that the trait of whiteness has skipped a generation: one of the grandparents possessed it, but neither parent did.

Mendel was led to the conjecture that in pea plants, in which the ordinary cells have *different* factors for color, the purple factor dominates, and the individual plant cannot be distinguished by its appearance from one in which both factors are purple. He coined the terms *dominant* and *recessive* to describe the two kinds of factors. In his study on peas he examined seven or eight characteristics of pea plants, all of which were governed by factors that could be classed as either dominant or recessive, unlike the simpler-to-understand relation between flower color and color factors in the four o'clock plant.

Table 8-6 shows the result of self-fertilization of the purple hybrids resulting from cross-breeding the pure purple strain with the white one. Body cells in which the two genes for the trait are identical (*PP* or *WW*) are said to be *homozygous* for that trait. When the two genes are different (*PW*), the cells are said to be *heterozygous*.

Table 8-6
Breeding of Pea Plants

Ovum	Pollen	Probability	Plant color	Net probability
P	P	1/4	purple	
P	W	1/4	purple	3/4
W	P	1/4	purple	
W	W	1/4	white	1/4

Human Genetics

Similar considerations apply to human beings. Since we do not do controlled mating experiments on human beings to test our theories of genetics, we cannot so easily identify traits as being governed by one or several genes, or as dominant or recessive, and so on. Yet much is known about human inheritance.

In sickle-cell anemia, the presence of the disease is determined by a single recessive gene. If we symbolize the sickle-cell factor as S and the normal factor as N, the affected individuals have body cells of SS type. Since until recently SS individuals have almost never lived long enough to have children of their own, most SS children must have been born to parents both of whom were of NS type—the mode of inheritance is the same as with the RW factors of flower color in the four o'clock plant or the PW factors of Mendel's peas. Note that not all children of two NS parents will be SS: only $1/4$ on the average will be. This answers the question we raised earlier about the chances that the younger siblings of a victim of the disease will suffer from it also: the probability is 0.25, independent of whether the first-born child was a victim or not.

There was one puzzle about this terrible condition: the fact that its victims die before having children of their own means that in time the S gene should be eliminated from the population. Why is it so common in certain parts of Africa? Why has it not yet been eliminated through attrition of its possessors?

The answer in this case appears to be that apparently normal individuals of the NS type are not identical biologically with the completely normal NN individuals. The S trait is not completely recessive: the situation is more like the pink flowers of the four o'clock plant. The NS individuals have somewhat abnormal blood cells which, for reasons not fully understood, are able to resist the ravages of malaria *better* than normal blood cells. The result is that the NS individuals have a competitive advantage over NN individuals in malarious areas even though a fraction of their children die before maturity. The end result is a population in which the proportion of the S gene remains stable over time. This hypothesis is supported both by direct clinical evidence, and by the fact that the sickle-cell condition is most common in those parts of Africa where malaria is most prevalent.

There is another mechanism through which harmful genetic traits can continue to appear in the population in spite of the tendency of individuals who possess them to die or fail to reproduce, to which we alluded earlier. Sometimes the hereditary factors possessed by individual members of a species undergo a spontaneous *mutation:* in one sperm cell or one ovum, the chemical molecule that determines the hereditary trait undergoes a change. This may occur because of high-energy radiation, toxic substances in the environment, or may simply be the effect of a collision with another molecule in the cell (remember, molecules are always in motion and colliding with each other). Most of the time these changes are harmful, and the individual that results from the altered ovum or sperm cell does not survive. However, once in a while the mutation is beneficial, and the individual born from the mutated germ cell has a trait that puts him at an advantage with respect to other members of his own species. As a result, the individual has a better chance of survival and of having children, and of transmitting the new gene to his descendants. Over the millions of years that life has existed on earth, many such events have occurred, and have allowed species to adapt to new conditions. This process is called *natural selection,* and has been proposed as the mechanism of evolution.

Today, we know more about genetics than in Mendel's time, when nothing was known about which parts of the cell contained the "factors" (genes) he hypothesized to account for heredity. Now not only have those parts of the cell that contain the genes (the chromosomes) been seen in the microscope, but also the chemical nature of the molecules of the gene has been identified (DNA: deoxyribonucleic acid), and the mechanism of their action is a topic of current active study.

The theory of probability can be applied to many of these more complicated problems also, and one cannot imagine research in modern genetics being carried on without it.

REFERENCE NOTES

1. Leo Tolstoi, *My Confession, and the Spirit of Christ's Teaching* (London: Walter Scott, undated).
2. George Rosen, *Madness In Society* (Chicago: University of Chicago Press, 1968).
3. Charles Ernest Goshen, ed., *Documentary History of Psychiatry: A Source Book on Historical Principles* (New York: Philosophical Library, 1967).
4. Franz Gabriel Alexander, and Sheldon T. Selesnick, *The History of Psychiatry: An Evaluation of Psychiatric Thought and Practice from Prehistoric Times to the Present* (New York: Harper and Row, 1966).
5. Samuel Tuke, *Description of the Retreat, an Institution near York, for Insane Persons of the Society of Friends* (1813; facsimile reprint, London: Dawsons, 1964).
6. Charles Darwin, *More Letters of Charles Darwin,* ed. Francis Darwin, vol. I (London: John Murray, 1903).
7. Emil Kraepelin, *Lectures on Clinical Psychiatry* (1904; reprint, New York: Hafner, 1968).
8. Silvano Arieti, "Manic-Depressive Psychosis." In *American Handbook of Psychiatry,* ed. Silvano Arieti, vol. I, pp. 419–454 (New York: Basic Books, 1959).
9. Silvano Arieti, ed., *Interpretations of Schizophrenia,* vol. I (New York: Basic Books, 1974).
10. Solomon H. Snyder, *Madness and the Brain* (New York: McGraw-Hill, 1975).

11. John Custance, *Wisdom, Madness, and Folly* (New York: Pellegrini and Cudahy, 1952). Copyright 1952 by John Custance, copyright renewed 1980 by John Custance. Reprinted by permission of Farrar, Straus & Giroux, Inc.
12. Bert Kaplan, ed., *The Inner World of Mental Illness: A Series of First-Person Accounts of What It Was Like* (New York: Harper and Row, 1964).
13. Anton Boisen, *Out of the Depths* (New York: Harper and Row, 1960).
14. Vladimir Nabokov, *Nabokov's Dozen* (Garden City, N.J.: Doubleday, 1958).
15. Bruce P. Dohrenwend, "Sociocultural and Social-Psychological Factors in the Genesis of Mental Disorders," *Journal of Health and Social Behavior* 16 (1975): 365; Bruce P. Dohrenwend, and Barbara Snell Dohrenwend, "Social and Cultural Influences on Psychopathology," *Annual Review of Psychology* 25 (1974): 417–452.
16. Snyder, *Madness and the Brain;* Paul H. Wender and Donald F. Klein, *Mind, Mood, and Medicine: A Guide to the New Biopsychiatry* (New York: New American Library, 1982).
17. E. R. Kandel, and James H. Schwartz, *Principles of Neural Science* (New York: Elsevier-North Holland, 1981).
18. Seymour S. Kety, "Biochemical Theories of Schizophrenia," *Science* 129 (1959): 1528–1532.
19. Dohrenwend, "Sociocultural and Social-Psychological Factors."
20. R. E. L. Faris, and H. Warren Dunham, *Mental Disorders in Urban Areas: An Ecological Study of Schizophrenia and Other Psychoses* (Chicago: University of Chicago Press, 1939).
21. Edward J. Sachar, "Psychobiology of Schizophrenia," in Kandel, *Principles of Neural Science.*
22. J. E. Cooper, et al., *Psychiatric Diagnosis in New York and London: A Comparative Study of Mental Hospital Admissions* (London: Oxford University Press, 1972).
23. Leo Srole, *Mental Health in the Metropolis: The Midtown Manhattan Study* (New York: McGraw-Hill, 1962).
24. Judith B. Kuriansky, W. Edwards Deming, and Barry J. Gurland, "On Trends in the Diagnosis of Schizophrenia," *American Journal of Psychiatry* 131 (1974): 402–408.
25. *International Classification of Diseases,* 3 vols., 9th rev. (Ann Arbor: Commission on Professional and Hospital Activities, 1978); *Diagnostic and Statistical Manual of Mental Disorders,* 3rd ed. (Washington, D.C.: American Psychiatric Association, 1980).
26. Russel Barton, *Institutional Neurosis* (Chicago: Year Book Medical, 1976). Quoted with the permission of John Wright & Sons, Bristol, U.K., and Dr. Barton.
27. Abraham Myerson, "Theory and Principles of the 'Total Push' Method in the Treatment of Chronic Schizophrenia," *American Journal of Psychiatry* 95 (1939): 1197–1204.
28. E. M. Gruenberg, "Hospital Treatment in Schizophrenia," in Robert Cancro, ed. *The Schizophrenic Reactions: A Critique of the Concept, Hospital Treatment, and Current Research* (New York: Brunner/Mazel, 1970).
29. E. M. Gruenberg, "The Social Breakdown Syndrome and its Prevention." In *American Handbook of Psychiatry,* ed. G. Caplan and S. Arieti, 2nd ed., vol. II (New York: Basic Books, 1974).
30. Erving Goffman, *Asylums: Essays on the Social Situation of Mental Patients and Other Patients* (Garden City, N.Y.: Doubleday, 1961) © 1961 by Erving Goffman. Reprinted by permission of Doubleday & Company, Inc.; Sanford M. Dornbosh, "The Military Academy as an Assimilating Institution," *Social Forces* 33 (1955): 317; The Holy Rule of St. Benedict; Brendan Behan, *Borstal Boy* (Boston: David R. Godine, 1982); John M. Murtaugh, and Sara Harris, *Cast the First Stone* (Westport, CT: Greenwood, 1978).
31. Thomas S. Szasz, *The Myth of Mental Illness: Foundations of a Theory of Personal Conduct* (New York: Harper and Row, 1974).
32. Jane M. Murphy, "Psychiatric Labelling in Cross-Cultural Perspective," *Science* 191 (1976): 1019.
33. Nance E. Waxler, "Is Outcome for Schizophrenia Better in Nonindustrial Societies? The Case of Sri Lanka," *Journal of Nervous and Mental Disease* 167 (1979): 144–158.
34. Don D. Jackson, *Myths of Madness: New Facts for Old Fallacies* (New York: Macmillan, 1964).
35. Theodore Lidz, L. Fleck, and R. Cornelison, *Schizophrenia and the Family* (New York: International Universities Press, 1967).
36. Wender and Klein, *Mind, Mood, and Medicine.*

37. Victor A. McKusick, *Human Genetics*, 2nd ed. (Englewood Cliffs: Prentice-Hall, 1969); F. Vogel, and A. G. Motulsky, *Human Genetics: Problems and Approaches* (Berlin: Springer Verlag, 1979); James S. Thompson, and Margaret W. Thompson, *Genetics in Medicine*, 3rd ed. (Philadelphia: W. B. Sanders, 1980).
38. Abraham M. Lilienfeld, "A Methodological Problem in Testing a Recessive Genetic Hypothesis in Human Disease," *American Journal of Public Health* 49 (1959): 199–204.
39. L. L. Heston, "The Genetics of Schizophrenia and Schizoid Disease," *Science* 167 (1970): 249–256.
40. David Rosenthal, and Seymour S. Kety, eds., *The Transmission of Schizophrenia. Proceedings of The Second Research Conference of the Foundations' Fund for Research in Psychiatry*, Dorado, Puerto Rico, June 26 to July 1, 1967. (New York: Pergamon Press, 1968). In particular: (a) Paul H. Wender, David Rosenthal, and Seymour S. Kety, "A Psychiatric Assessment of the Adoptive Parents of Schizophrenics," pp. 235–250; (b) Seymour S. Kety, David Rosenthal, Paul H. Wender, and Fini Schulsinger, "The Types and Prevalence of Mental Illness in the Biological and Adoptive Families of Adopted Schizophrenics," pp. 345–362; and (c) David Rosenthal, Paul H. Wender, Seymour S. Kety, Fini Schulsinger, Joseph Welner, and Lise Ostergaard, "Schizophrenic Offspring Reared in Adoptive Homes," pp. 377–392.
41. Seymour S. Kety, David Rosenthal, and Paul H. Wender, "The Biologic and Adoptive Families of Adopted Individuals Who Became Schizophrenic," in *The Nature of Schizophrenia: New Approaches to Research and Treatment*, ed. Lyman C. Wynne, Rue L. Cromwell, and Steven Matthysse (New York: John Wiley, 1978).
42. Paul H. Wender, David Rosenthal, and Seymour S. Kety, *et al.*, "Crossfostering: A Research Strategy for Clarifying the Role of Genetic and Experiential Factors in the Etiology of Schizophrenia," *Archives of General Psychiatry* 30 (1974): 121–128.
43. Irving I. Gottesman, and James Shields, *Schizophrenia: The Epigenetic Puzzle* (Cambridge: Cambridge University Press, 1982).
44. Lorna S. Benjamin, "A Reconsideration of the Kety and Associates Study of Genetic Factors in the Transmission of Schizophrenia," *American Journal of Psychiatry* 133 (1976): 1129–1133.
45. Theodore Lidz, Sidney Blatt, and Barry Cook, "Critique of the Danish-American Studies of the Adopted-Away Offspring of Schizophrenic Parents," *American Journal of Psychiatry* 138 (1981): 1063–1068.
46. Paul H. Wender, "The Danish-American Studies: Critiquing the Critique," letter to the editor, *American Journal of Psychiatry* 138 (1981): 1392–1393.
47. R. Abrams, and M. A. Taylor, "The Genetics of Schizophrenia: A Reassessment Using Modern Criteria," *American Journal of Psychiatry* 140 (1983): 171–175.
48. Seymour S. Kety, "Mental Illness in the Biological and Adoptive Relatives of Schizophrenic Adoptees: Findings Relevant to Genetic and Environmental Factors in Etiology," *American Journal of Psychiatry* 140 (1983): 720–727.
49. Margaret Mead, *Coming of Age in Samoa* (New York: William Morrow, 1973).
50. Derek Freeman, *Margaret Mead and Samoa: The Making and Unmaking of an Anthropological Myth* (Cambridge: Harvard University Press, 1983).
51. Jonathan H. Pincus, and Gary J. Tucker, *Behavioral Neurology* (New York: Oxford University Press, 1978).
52. Dohrenwend, "Sociocultural and Social-Psychological Factors."

SUGGESTED READING

Alexander, Franz Gabriel, and Sheldon T. Selesnick. *The History of Psychiatry: An Evaluation of Psychiatric Thought and Practice from Prehistoric Times to the Present.* New York: Harper and Row, 1966.

Goshen, Charles Ernest, ed. *Documentary History of Psychiatry: A Source Book on Historical Principles.* New York: Philosophical Library, 1967.

Kaplan, Bert, ed. *The Inner World of Mental Illness: A Series of First Person Accounts of What It Was Like*. New York: Harper and Row, 1964.
Rosen, George, *Madness in Society*. Chicago: University of Chicago Press, 1968.
Snyder, Solomon H. *Madness and the Brain*. New York: McGraw-Hill, 1975.
Wender, Paul H., and Donald F. Klein. *Mind, Mood, and Medicine: A Guide to the New Biopsychiatry*. New York: New American Library, 1982.

PROBLEMS

1. Differences in rates of physical illnesses are not subject to local variations in diagnosis the way mental disorders are, or are they? Do you know of any examples either now or in the past of differing practices of diagnosis among different groups of physicians?
2. In what ways is your school a "total institution" according to Goffman's terms? In what ways is it not?
3. Discuss the use of "blind" diagnoses in the adoption studies of schizophrenia. State who was "blind" and why this procedure was followed.
4. Give briefly the type of evidence that you would look for to test the theory that schizophrenia is a disease resulting from the stresses of life in modern industrialized societies.
5. As described in the text, studies of adopted children were used to test the hypothesis that schizophrenia is a genetic disorder. Describe briefly at least one experiment of this kind. Was it a controlled experiment? If so what were the "experimental" and "control" groups? What was the outcome?
6. As described in the text, some scientists raised the possibility that the great differences between the U.S. and Great Britain in the rates of schizophrenia and depressive disorders were due to differences in diagnostic practices in the two countries. Describe briefly one of the experiments used to test this. Was it a controlled experiment? If so what were the "experimental" and "control" groups? What was the outcome?

9
Statistics

THE PROBLEM TURNED AROUND

In the theory of probability we started with knowledge of the chances of an individual outcome—heads or tails—and used this to predict the average behavior of sequences of individual trials. Our assumption that the probability of heads is 0.5 was based both on long and varied experience in tossing coins and on our intuitive feeling for the physics involved in tossing flat disk-shaped objects whose opposite sides are not very different.

But we are much more often in the situation of not having such long and varied experience combined with physical insight to tell us what the probabilities are. Instead, we are forced to guess the probabilities from a limited number of trials. The limited number of trials can be thought of as a sample of the behavior, and we try to guess the general behavior from the characteristics of our sample.

We imagined earlier a large container filled with a mixture of black and white balls in a 4 : 1 proportion. What if we did not know the relative proportions of the two kinds of balls and wanted to find out? Now the surest way would be to empty out the container and count them. But this might not be possible—there might be too many to count, or it might take too long. What we can do is draw a small, randomly selected sample of balls from the container, and count the numbers of each kind in the sample. If we draw 10 and find, for example, 7 black balls and 3 white balls, we might conclude that 30% of the balls are white.

But can we be sure? Table 7-5 (p. 159) shows us that even if there are 20% white balls, we have a good chance of drawing 3 white and 7 black balls; the probability is 0.201, compared to a probability of 0.302 of drawing 2 white and 8 black balls, but it will happen fairly often. We can see that our estimate based on drawing 10 balls can be seriously in error. It is much less likely that the balls are 90% black and 10% white, but even this is possible.

Now it is obvious that we can have more confidence in our estimate if we draw 50 balls instead of 10, and even more if we draw 200. But we must stop somewhere, because we will eventually run out of time, patience, or money. So we choose a sample size that is the best compromise we can find between our lack of patience or time and our desire for certainty. How to make that choice is the kind of question statistics tries to answer.

Statistics is in this sense the reverse of the theory of probability. In the theory of probability, we use known probabilities of outcomes of individual trials to predict the results of sequences of trials. In statistics, we use the results of sequences of trials to estimate what the probabilities of the outcomes of individual trials are.

The necessity for statistics arises from the impracticality of doing our individual trials—we can now call them our *experiments*—an enormously large number of times. The only way to achieve certainty about the probabilities is to do just that, but it is not always possible.

How Tall Is the Average Person?

As an example, suppose we wish to know the average height of the inhabitants of a city of two million people. If we wanted to be sure, we would have to measure each of those two million people. But this is obviously impractical. Instead, we select a small sample of the population at random—say, 1000 people—and measure their heights. Now we recognize that this procedure may give us the wrong answer: the 1000 people in the sample may not have the same average height as the two million they are assumed to represent. We may have been unlucky and selected, by sheer accident, too many tall people. We expect that the chances of this happening are not large. But we would like to have some quantitative answer to the question, what are our chances of being wrong? What is the probability that even though the average height of our sample of 1000 is 5 feet, 9 inches, the average height of the whole two million is really 5 feet, 8 inches? Should we measure 2000 people? How much more confidence will we have if we do?

Is the Drug Effective?

Let us consider as another example a trial of a new drug for a serious, often fatal, but relatively rare disease. Our experience with the disease is that 50% of those who get it die. The rarity of the disease is such that we can find only 10 patients for our experiment. We administer the drug, and find that 2 patients die and 8 recover. From the fact that only 20% died rather than 50% we might be tempted to conclude that the drug is effective.

Again we recognize that we cannot be sure. The drug might be useless, and the fact that only 2 of the patients died rather than the expected 5 a chance result.

We could consider the problem as one of comparing two different hypotheses that might be used to explain the result of our experiment:

1. The drug is effective: If used widely, it will reduce the death rate from

50% to 20%. The probability of any one patient given the drug dying is 0.2, the same as the probability of drawing a white ball from the container holding 20% white balls.
2. The drug is useless: The death rate remains 50% on the average, but by chance only 20% (2 patients) died in this experiment. The chance of any one patient given the drug dying is 0.5, the same as his chance of dying if he didn't receive the drug.

We should note that the two hypotheses above do not exhaust all possibilities. For example, the drug might be worse than useless: the real chances of surviving when it is taken are reduced to 25%. Yet in our experiment we might, as a possible although not very likely outcome, have had only 2 deaths. For simplicity we will ignore all other hypotheses, though in proper statistical analysis they are taken into account.

In any event the result of our experiment is clearly consistent with either of the two hypotheses listed above. We cannot be certain on the basis of our experiment alone which is correct. But we can ask the following questions:

If hypothesis (1) is right, what is the probability that the experiment would have the outcome it did? The answer, from Table 7-5 (p. 159), is 0.302. If hypothesis (2) is right, what is the probability that the experiment would have the outcome it did? The answer, from Table 7-2 (p. 154), is 0.044. These two results give us an estimate of the probability that the drug is effective. There is a good chance that it is, but the chance is not overwhelming. The probability of hypothesis (1) is, on the basis of this calculation, about seven times greater than that of hypothesis (2).

The only way we could achieve a greater degree of certainty is to perform the experiment on many more patients. If we had 100 instead of 10, and again only 20% died, we would have a lot more confidence in hypothesis (1) than before.

However, there could be reasons that prevent us from doing this more extensive experiment. The disease is rare. We may have a number of other drugs of equal promise to test, and therefore have to limit the number of patients on whom we try any given drug. Perhaps we have very limited supplies of the drug because it is difficult to prepare, and we do not know yet whether it is worth the expense and effort to prepare more. In short, we may be forced to draw the most reasonable conclusion possible from a limited amount of data, and to hope we will do the right thing more often than not. Statistics helps us do the best we can with the data we have.

Random versus Nonrandom

We have said that statistics is concerned with the problem of inferring, from a small, randomly selected sample of a large population, what the characteristics of that large population are. And we recognize that because the sample is small, there is always a chance it will differ from the large population, just as 10 tosses of a coin need not always give 5 heads and 5 tails.

We have used the words "randomly selected," and it is time to ask what they mean. It may be easiest to convey this meaning through examples of studies that failed to select their samples randomly.

One of the classic examples was the poll conducted by a magazine, *Literary Digest*, prior to the 1936 United States presidential election when Franklin D. Roosevelt was running for reelection against the Republican Alf M. Landon. The *Digest* reported that the results of its poll predicted a clear victory for Landon, who had received 60% of the votes in the poll. As it turned out, Roosevelt won the election by a landslide, carrying every state except Maine and Vermont, while Landon got only 40% of the actual vote. The reason for the *Digest*'s astounding error emerged after the election. The *Digest* had drawn its sample from three sources: its own subscribers, lists of registered owners of automobiles, and telephone books. In 1936, having an automobile or a telephone or subscribing to the *Literary Digest* were privileges mainly of members of the wealthier and more politically conservative classes in American society. The non-auto- and non-telephone-owning working-class and lower-middle-class Americans, who made up the majority of the population, strongly supported Roosevelt and the Democratic party.

Bias in selection is not always so obvious. For example, surveys are often made by investigators visiting homes at random and asking questions of the occupants. In a certain proportion of homes, no one is there when the investigator calls. Can these be assumed to be purely random circumstances, and these homes be left out of the survey? It should occur to us that families with small children are more likely to have someone home at a given time than other families. This clearly will bias the sample toward such families, who are likely to differ in many characteristics from the average family—they will tend to be younger, for example, and their economic status, ethnic origin, or religious beliefs may differ from those of the average family.

Surveys taken by mail replace this problem with others. Not everyone answers a mail survey. Rates of return often represent a minority of the questionnaires sent out. Do the people who fail to return them differ from those who do? It is not easy to tell, and will certainly depend on the object of the survey.

Another sort of bias is shown in this quotation:[1]

> A house-to-house survey purporting to study magazine readership was once made in which a key question was: What magazines does your household read? When the results were tabulated and analyzed it appeared that a great many people loved *Harper's* and not very many read *True Story*. Now there were publishers' figures around at the time that showed very clearly that *True Story* had more millions of circulation than *Harper's* had hundreds of thousands. Perhaps we asked the wrong kind of people, the designers of the survey said to themselves. But no, the questions had been asked in all sorts of neighborhoods all around the country. The only reasonable conclusion then was that a good many of the respondents, as people are called when they answer such questions,

had not told the truth. About all the survey had uncovered was snobbery. (p. 16)

It should be clear that there is no easy way to cope with possible bias in the selection of a sample. Our best protection is the good judgment and experience of the sampler.

Another Meaning of "Statistics"

In the preceding discussion we defined statistics as concerned with the problem of inferring the properties of large groups from small samples.

There is another common usage of the term: to describe the properties of a large group when the individual members of the group differ.

There are situations where every individual case can be counted, and there is no need to limit our examination of a group to a sample. For instance, in the United States census, the attempt is made to count every individual. To the extent that this is successfully done, we can answer exactly such questions as: How many Indians live in Nevada? What is the number of divorced women living alone?

Now a complete census involves the tabulation of an enormous quantity of data. Each individual is listed, with his or her age, marital status, religion, race, type of house lived in, who else lives in the house, and so on. In fact, there is too much there to be useful when left in that form, even if it is stored in a computer.

The data become useful only when some specific question about the properties of the group as a whole is asked, instead of details about each individual member. What is the average family size? What is the average per-capita income? Has either changed since the last census? How do these compare with figures for France or Thailand? The numerical measures that attempt to give general information about a population without giving all the individual details are called *statistics*. Used in this sense, statistics is a plural noun.

How Good Is an Average?

It should be clear that a statistic in this sense of the word is a kind of distortion. We sacrifice a wealth of complex detail for a simpler general picture. But there is an element of judgment involved—is the statistical measure we use an accurate picture of the complex reality? How much have we lost by using it?

We are all familiar with the meaning of an average and know how to calculate it. We would expect, if we wanted to compare two groups, to use the average as the appropriate measure. Suppose we wanted to find out which of two communities is better off economically: we would be likely to compare the average family income in each. If one community had an average income of $15,000 and the second an average income of $10,000, we would conclude that the first community is better off. This is not a bad way to make the comparison, but there are times when it could be very misleading. The first community might have a few very wealthy people who bring the average up to a high value. If this small number of millionaires is excluded, the average income of the rest of the community might turn out to be only $8,000. Another example of how misleading an average can be is given by Darrel Huff and Irving Geis:[1]

> A corporation was able to announce that its stock was held by 3,003 persons, who had an average of 660 shares each. This was true. It was also true that of the two million shares of stock in the corporation three men held three-quarters and three thousand persons held the other one-fourth among them. (p. 129)

A person running a clothing factory who intends to manufacture 10,000 pairs of mens' pants may be satisfied with the average man's waist and leg-length sizes when he orders the cloth, but will surely need more detailed information about men's sizes than just the average when he actually makes the pants (Figure 9-1).

Again, the Average of the Unaverage

Obviously, to find out how well the average really represents the individual members of a group, we need some measure of how much the individuals differ from the average. Now it should be obvious that the average of a list of numbers will be larger than some numbers on the list and smaller than others. It follows that if we make a new list of numbers by subtracting the average of the list from

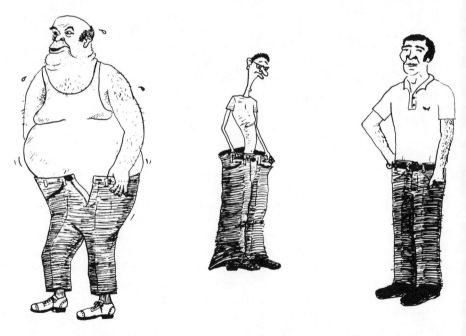

FIGURE 9-1. Various men wearing the average pair of pants.

Table 9-1
Results of a Chemistry Test Given to 12 Students

Student	Grades	Deviations: Grade minus average grade	Square of deviations
1	88	9	81
2	57	−22	484
3	93	14	196
4	91	12	144
5	75	−4	16
6	82	3	9
7	76	−3	9
8	85	6	36
9	80	1	1
10	72	−7	49
11	89	10	100
12	60	−19	361
Sum	948	0	1486
Average	79	0	123.8

r.m.s. of deviations = square root of 123.8 = 11.12

each individual number, some numbers on our new list will be positive and some negative. We might have hoped to use the average of this new list—the list of *deviations* of the individual from the average—as a measure of the representativeness of the original average, but we recognize that the presence of both positive and negative quantities will tend to lead to a cancellation. Indeed, worse than this, it can be proved that the cancellation will be *exact:* the average of the deviations can be shown—we will not attempt to do so—to be *zero* always. An example illustrating this behavior is given in Table 9-1.

The problem is one we encountered before, when we calculated what we called D: the average value of the number of heads minus the number of tails in a coin-tossing experiment (see Chapter 7, Appendix 7-1, p. 161). There we dealt with it by taking the root-mean-square of the outcomes; we do the same thing here.

Statisticians use the term *standard deviation* for the r.m.s. of the deviations.*

When the number of individuals in the group is large, it is often useful to represent the distribution of the individual scores by a histogram (see Figures 7-1 through 7-5). Very often the histogram will follow a Gaussian (bell-shaped) curve fairly closely. When this is so, it can be shown that the number of individuals who

---

* Sometimes the standard deviation is defined by a different formula: the numerical difference between the r.m.s. of the deviations and the standard deviation calculated by this different formula is negligible when the number of individuals in the sample is large. For the chemistry-test example with 12 grades, the standard deviation calculated by the other procedure is 4% larger, a difference hardly worth bothering about.

differ from the average by no more than one standard deviation is about $2/3$ of the group (68.2%, to be precise). In the chemistry exam results of Table 9-1, the histogram is not in the least like a Gaussian curve, but it happens that exactly $2/3$ of the grades (8 out of 12) differ from the average by one standard deviation or less.

STATISTICAL TESTS IN SCIENCE

The purpose of a statistical test should be clear from the preceding discussion. We are comparing two hypotheses. The first is the hypothesis of the scientist doing the experiment: that *something* we can vary will make a difference to the outcome of our experiment. In the example given, it was the use of a new drug on the outcome (death or life) of a serious disease. The second hypothesis (called the null hypothesis) is that the *something* has no effect whatever.

The experiment seemed to show an effect: 80% of the patients survived, instead of the expected 50%. It sounds encouraging, but could it be that the result is due to chance, and that the null hypothesis is true after all?

Now if we ask a foolish question, we get a foolish answer. If we ask "Could the result of the experiment be due to chance?" the answer is simple: it is always "Yes." Any result of an experiment *could* be due to chance alone. Toss a coin 10 times and there is a chance of getting all heads. Toss it 100 times and there is still a chance, though a smaller one, of getting all heads. If we demand absolute certainty that chance could not be responsible for the result of an experiment, we are asking for the impossible. We must be satisfied with less.

The proper question to ask is something like "What is the probability that the result of the experiment could have occurred by chance alone?" If the probability is small enough, we have a right to feel that the outcome of the experiment has favored the first hypothesis. That is the best we can do.

Here a slight complication arises. For reasons that cannot be gone into here, statisticians prefer to ask a different question: "What is the probability that the result of the experiment *or an even more favorable result* could have occurred by chance alone?" In the example we used, 8 patients survived. The probability of this occurring by chance alone was 0.04395. If 9 had survived, it would have been even more favorable to the first hypothesis: the probability of this, if the null hypothesis is correct, is 0.00977. The survival of 10 is more favorable still to the first hypothesis, since the probability that 10 will survive if the null hypothesis is true is only 0.00098. The probability of 8 *or more* surviving if the null hypothesis is correct is the sum of the probabilities of these three mutually exclusive outcomes, or 0.0547.

How small should the probability that chance alone might have produced the observed experimental result or a more favorable one be, before we agree to rule out the null hypothesis? In a matter of human judgment like this, there are no firm rules. There is however a common practice: if the probability is 0.05 or less, the result is called "statistically significant," and the null hypothesis is rejected (with the qualification that we used a 5% criterion); if the probability is 0.01 or less, the result is called "highly statistically significant."

It should be recognized that there are two kinds of risks in our procedure. There is the risk of accepting some hypotheses that are false, because by chance alone the experiment came out in a way that seemed to support them. There is also the risk that we may discard true hypotheses because, through the operation of chance, the experimental results did not seem to support them. Chance can produce either kind of mistake.

The rejection rule for the null hypothesis has been set at 5%, or even 1%, to lean away from the acceptance of false hypotheses, even at the price of increasing the risk of rejecting true ones. Assigning a 6% probability that the outcome of an experiment was a result of chance alone is logically equivalent to saying there is a 94% probability that the difference observed in the experiment was *not* due to chance. A 94% probability is fairly convincing, is it not?

The reasons for setting up such a tough test arise, in part, from a certain skepticism lying at the heart of the scientific process. Individual scientists *want* to believe their own theories. In spite of all the honest self-criticism they can bring to bear on themselves, they still find it hard to avoid accepting favorable evidence and disregarding the unfavorable. It is the job of the scientific consensus to act as a check to the individual scientist's enthusiasm for his own theory. That individual may often be right, but science as a whole benefits from caution, from the demand for more stringent evidence.

Let us apply this criterion to our experiment on the new drug. We have shown that if the rate of survival were actually unaffected by the drug, in an experiment on 10 patients there is a 0.0547 probability that 8 or more patients would survive. This is greater than 0.05. Hence the conclusion is, we cannot rule out the null hypothesis that the drug is worthless.

This conclusion would surely be a disappointment to the scientists who developed the drug, and who may have had good reasons for believing, prior to the experiment, that it *should* be effective. But research does not end here. It is always possible to do an additional experiment, if one feels strongly that the hypothesis is really correct, and to hope for a different answer.

In the experiment, 80% of the patients survived when a survival rate of 50% would have been expected. Yet our statistical test tells us the difference is not statistically significant. The problem is clearly that the size of our experimental group was too small. Suppose it were possible to study an additional 10 patients, and combine the result of this second experiment with the first (note that we cannot disregard the first experiment solely on the grounds that we were not pleased with the result). If we still had an 80% survival rate, this would mean 16 survivals out of 20, and the probability that 16 *or more* out of 20 would survive if the drug were useless is only 0.006—less than 1%, and highly statistically significant!

But remember, we have no guarantee that the survival rate would be 80% in the second group also. In fact, if the null hypothesis—that the drug is useless—

were correct, this is unlikely. It is more likely that about half of the second group will survive, so that the total number surviving in the whole experiment will be about 13. Although this is more than half, the probability of having 13 or more survivors out of 20 if the null hypothesis is true is about 0.13, much more than the 0.05 level we require to reject the null hypothesis.

There are several morals to this long story:

1. Use a large enough sample in your experiment to be able to detect a scientifically useful result if there is to be one. Get advice from a statistician on how large it should be.
2. Statistical tests are done on the *numbers* of individuals, not on the *percentages*.
3. Don't give up on a good hypothesis too easily, but don't stick with it too long (useless advice, of course).
4. If a statistical test rules out chance at a significant level, say at the 1% level, DO NOT make the mistake of thinking that this means the hypothesis being tested has a 99% chance of being correct. It doesn't mean that at all.

Moral 4 is important enough to deserve a little discussion. What the test has ruled out is a likelihood that the result was due to chance; the probability is thus 0.99 that the difference was *not* due to chance. That is all the statistical test can show.

But if it was not due to chance, why was the result obtained? One possibility among many is that the hypothesis the experiment was designed to test is, after all, correct. Among other possibilities, one is that the person gathering the data cheated, and reported the results falsely. This is not, so far as is known, common, but it happens. More commonly, there is an unconscious bias on the part of the experimenter that influences the way he evaluates the outcome of the experiment (see Chapter 13). More often the data are real enough, but there is a fortuitous relation between a variable *believed* to be causative and the true cause. We gave an example of this in Chapter 4: Dr. William Farr observed a very close inverse relation between cholera rates in districts of 19th-century London and the elevation of those districts above sea level. He concluded that this relation between cholera and elevation was a general rule, applicable everywhere. Though he did not perform a statistical test on the data from London, as such tests had not yet been developed, there is no doubt that if he had the observation would have proven highly statistically significant. Yet his hypothesis was wrong, as John Snow pointed out, merely the result of an accidental relation between elevation and degree of contamination of the water supply, in that city, at that time.

Statistics and Science

We can see that the problems of using statistics correctly are similar to the problems of doing science properly.

One problem is that of determining what the facts really are, as opposed to

STATISTICS

what they seem to be or what we think they are. We encountered this question in discussing the different rates of mental disorders in the United Kingdom and the United States. These rates were authoritative figures, found by presumably skilled and competent professional psychiatrists in the two countries, and reported by appropriate governmental agencies. Yet they did not reflect reality. A more trivial example is that provided by Huff: the figures on the numbers of families reading *Harper's* rather than *True Story* represented only what people were willing to admit to an interviewer, not what they actually read.

The second problem is more subtle. Scientific explanations cannot deal with everything in the world at one time, but must focus on a limited number of facts that are selected from the enormous number possible. Which are the important ones to use and which are trivial, peripheral, irrelevant? In the stock-shares example given by Huff, the average number of shares held by each stockholder really was 660. This is a fact: it is not in dispute. But if we are interested in the question of how widely the ownership of shares of stock in American corporations is distributed, it is the wrong fact to pay attention to.

APPENDIX 9-1: STATISTICS EXPERIMENT

The following experiment was used pretty much as given here in the required science course for non-science majors at Yeshiva College. (Yeshiva College is the men's college of Yeshiva University, while Stern College is the women's college. Yeshiva College is commonly referred to as YC.)

The Chi-squared Test on a 2 × 2 Table

Moshe, a student at YC, expresses his opinion that women are not as good in science as men. As evidence he offers his impression that there is a higher proportion of pre-meds and other science majors at YC than at Stern College. He justifies this impression by going through a list of his friends and acquaintances at both schools, and finds that of 22 YC students he knows, 13 are science majors, while of 10 Stern College students, only 3 are.

There are many criticisms we could make of this reasoning, but before doing so we should recognize that it is a type of reasoning all of us engage in much of the time, and that a large number of the things we believe are based on just such evidence.

We could argue against Moshe's conclusion in many ways: We could ask whether his personal circle of friends is representative of YC and of Stern generally. For example, we could note that since Moshe is himself a pre-med, he is more likely to have a high proportion of pre-meds among his friends at YC. We could also ask whether YC and Stern College are representative of men and women generally. Or we could ask whether the subject one chooses to major in at college reflects one's intrinsic ability alone, or various social and parental pressures as well.

All of these questions, and many others which could be asked, call for deeper insight, additional data, or both.

But there are a number of questions we can ask and answer without additional data: questions of a statistical nature. For example, assuming that Moshe's sample of 22 YC students was randomly selected, unbiased by his being a pre-med, does it give a reliable estimate of the proportion of science majors at YC? What about the even smaller sample of 10 Stern College students? How large a sample should we use to get a reliable estimate? What do we mean by "reliable" in this case?

We note that the number of men and the number of women in our sample are not equal. Does this matter? Is it possible to compare the two groups when the sample sizes are not equal? The answer to this is simple enough: "Why not?" Of course 22 YC students picked at random may not give as reliable an estimate of the average student at YC as 100 students, but it represents some degree of information, and permits some conclusions about YC students to be drawn. While a random sample of 10 Stern students is likely to give a less reliable estimate of the average Stern student, in the absence of a larger sample we would be willing to use it. If we can give some degree of credence to each of the samples individually we should be equally willing to give some degree of credence to a comparison between them.

It has not been claimed, we note, that the 22-to-10 ratio is the correct ratio of the numbers of YC-to-Stern student-populations. (By chance it is probably not far from it.) Does this invalidate the comparison? Again, by the same reasoning, why should it?

Moshe's samples were obtained first by picking the students by sex, then counting the science majors. The same group of 32 students might also have been produced by picking 16 science majors and 16 non-science majors at random from the combined student bodies of YC and Stern, and then counting the students by sex. If one asks whether exactly half the combined student bodies are science majors, the answer is, we do not know, and from the point of view of the question under investigation (not necessarily from the point of view of other interesting questions) we do not care either.

Moshe's samples can be summarized as:

	Science majors	Non-science majors	
YC students	13	9	Subtotal: YC students = 22
Stern students	3	7	Subtotal: Stern students = 10
	Subtotal: Science majors = 16	Subtotal: Non-science majors = 16	Grand total: Students = 32

The above results do show more science majors among the men than among women; could this result have arisen by chance alone? More precisely, what is the probability that it arose by chance, and that in reality the proportion of science and non-science majors is the same for the two schools?

The hypothesis to be tested, therefore, is that the two groups being compared are really the same, and that the difference we found in our experiment is a result of chance alone: this is called the *null hypothesis*.

Now there are two features of the above sample that are given, and that cannot be changed without getting a new sample.

1. There are 22 YC students and 10 Stern students.
2. There are 16 science majors and 16 non-science majors.

For a sample with these two features (because it is all we have to go on at the moment), is the excess of science majors among YC students the result of chance?

To visualize the probability problem (it is a little more complicated than tossing coins) imagine 32 squares in a row, 22 of which are labelled "YC" and 10 "Stern." We now take a pack of 32 cards, 16 of which are labelled "Sci" and 16 "Non-Sci." We shuffle the 32 cards and without looking at them place them in the 32 boxes, one to a box. Then we count the numbers of "Sci" cards in the "YC" boxes. We repeat this thousands of times (at least in thought). On the *average,* the proportion of "YC" boxes that get "Sci" cards will be the same as the proportion of "Stern" boxes that get "Sci" cards. In fact, since in our sample the numbers of "Sci" and "Non-Sci" cards happen to be the same, on the *average* half the "YC" boxes and half the "Stern" boxes will get "Sci" cards. But the average is not achieved each and every time. Some of the times, a higher proportion of "YC" boxes will get "Sci" cards, other times a higher proportion of "Stern" boxes will. So a greater proportion of science majors among the YC students than among the Stern students known to Moshe could occur by chance alone, even if there really is no difference between the proportions of science majors at the two schools. What is the probability of such an outcome? Is it so high that Moshe's argument is invalidated? Is it so low that we must concede that Moshe has a point? Is it in between?

There are a number of statistical tests which could be used to answer this question. We will describe one of the common ones, which is relatively simple to perform. It makes use of a number of assumptions and approximations that will not be described in detail; the interested reader should consult a standard text in statistics for more information. One warning: do not use this test if the total number of subjects is less than 20 (we have 32 here).

The test is called a chi-squared test. Chi (pronounced *"kai"*) is the English spelling of the Greek letter χ. We will explain here in "cook-book" fashion how to perform the test.

Refer to the 2 × 2 table.

1. Multiply the two numbers on each diagonal of the 2 × 2 table.

$$13 \times 7 = 91 \qquad 9 \times 3 = 27$$

2. Subtract the smaller from the larger: $91 - 27 = 64$
3. Subtract half the grand total from the result:

$$64 - \frac{32}{2} = 48$$

4. Square this number: $48 \times 48 = 2304$
5. Multiply by the grand total: $32 \times 2304 = 73728$
6. Multiply the four subtotals together: $16 \times 16 \times 22 \times 10 = 56320$
7. Divide the result of Step 5 by the result of Step 6:

$$\frac{77328}{56320} = 1.309$$

The result of Step 7 is the test statistic, χ^2 (chi-squared). So $\chi^2 = 1.309$ in this case. Next we look up in Table 9-2 the probability P associated with this value of χ^2.

P is the probability that the two groups in the sample being compared—in this example, the men at YC, the women at Stern—really are the same with respect to the property investigated—in this sample, the proportion of science majors—but the proportions in the sample could *appear* to be different, solely due to chance, by the amount observed, or more. We find that P lies between 0.30 and 0.20. The exact value could be found from a more detailed table, but we do not always need such an accurate value.

We see, then, that Moshe's results miss statistical significance by a considerable margin, and we would therefore reject his hypothesis about male scientific superiority in favor of the null hypothesis.

This means that even if there were *no* real differences between the percentage of science majors among YC and Stern students, differences at least as great as

Table 9-2
A Chi-square Table

χ^2	P
0.0158	0.90
0.0642	0.80
0.148	0.70
0.455	0.50
1.074	0.30
1.642	0.20
2.706	0.10
3.841	0.05
5.412	0.02
6.635	0.01
10.827	0.001

Moshe's have a 20%–30% probability of occurring in a sample of 22 YC students and 10 Stern students and in which there are 16 science majors.

What if Moshe had taken a large sample, say three times as great, *and found the same proportion?* This time the table would be:

	Science majors	Non-science majors	Subtotals
YC	39	27	66
Stern	9	21	30
Subtotals	48	48	96

$$\chi^2 = \frac{\left(39 \times 21 - 9 \times 27 - \frac{96}{2}\right) \times 96}{48 \times 48 \times 66 \times 30} = 5.87$$

and P would be close to 0.02 (2%). Moshe would have made a more convincing case, *assuming that the proportions had remained the same in his larger sample.* Of course if the difference really were due to chance alone, Moshe's larger sample would not have been at all likely to show the *same* proportions.

The experiment, as indicated above, is for you to find a worthwhile problem to which this statistical method can be applied. You must choose the problem from the subject in which you are majoring. A photocopy of the source of the data you use must be included with your lab report.

Here are some examples which may serve as a guide to the kinds of problems to which the test may be applied.

1. A news broadcaster announces that yesterday on Wall Street oil stocks generally went up, while heavy industry stocks went down. Choose some examples of each kind of company, find out from the financial pages whether each went up, stayed the same, or went down. Note that for the test given here you must "dichotomize" the outcome: you may lump "staying the same" together with "going down" or with "going up," but you must have only two categories of outcome.

2. You wish to compare the writing styles of two writers: you are willing to assume that "difficulty" of style is synonymous with the use of long words. Take a sample of each writer's prose, define a "long word" as, say, one with 12 or more letters, and count the "long" and the "short" words in each.

3. Do students who sit in the front of a lecture hall get the same grades as those who sit in the back? Define the first two rows as "the front," and A and B as "good" grades. Find out from as many students as you can where they usually sit, and what grades they received.

You will be graded on the *choice* of a problem—it should be a serious and interesting one, related to your own field of specialization.

REFERENCE NOTE

1. Darrel Huff and Irving Geis, *How to Lie with Statistics* (New York: W. W. Norton, 1954).

SUGGESTED READING

Tippet, L. H. C. *Statistics*. Oxford: Oxford University Press, 1968.
Huff, Darrel, and Irving Geis. *How to Lie with Statistics*. New York: W. W. Norton, 1954.
Freedman, David, Robert Pisani, and Roger Purves. *Statistics*. New York: W. W. Norton, 1978.

The book by Freedman *et al.* is a text for a statistics course requiring the minimum of prior mathematical training. It is clear and readable.

10

Science—The Search for Understanding

UNDERSTANDING AS A COMMON EXPERIENCE

In the first chapter, we described science as consisting of three essential elements:

1. It is a search for understanding
2. by means of laws or principles of the greatest generality
3. which are capable of experimental test.

In this and the next few chapters we will discuss each of these elements in detail.

An understanding of the world is the major goal of science. It is a goal, however, that is not unique to science but is shared with other fields of human activity as well: religion, the arts, philosophy. For all, the understanding sought is associated with the perception of an underlying order and unity in the chaotic world of experience. We have remarked earlier that it is not easy to spell out in precise detail what is meant by a sense of order and unity behind appearances, but fortunately such a perception is not limited to scientists or philosophers, but is a common enough experience of ordinary living to permit us to illustrate it by examples.

The following, an example of a child perceiving order in the world, is from Helen Keller's autobiography.[1] It describes an episode that occurred shortly after her teacher, Anne Sullivan, began to work with her. Helen was seven years old at this time. She had become blind and deaf after an attack of scarlet fever at the age of 18 months, just when she was beginning to learn to talk. Anne Sullivan was trying to teach her to speak and read using an alphabet in which the letters are spelled out by touch.

> Earlier in the day we had had a tussle over the words "m-u-g" and "w-a-t-e-r." Miss Sullivan had tried to impress it upon me that "m-u-g" is mug and that "w-a-t-e-r" is water, but I persisted in confounding the two. In despair she had dropped the subject for the time, only to renew it at the first opportu-

nity. . . . We walked down the path to the well-house, attracted by the fragrance of the honeysuckle with which it was covered. Someone was drawing water and my teacher placed my hand under the spout. As the cool stream gushed over one hand she spelled into the other the word water, first slowly, then rapidly. I stood still, my whole attention fixed upon the motions of her fingers. Suddenly I felt a misty consciousness as of something forgotten—a thrill of returning thought; and somehow the mystery of language was revealed to me. I knew then that "w-a-t-e-r" meant the wonderful cool something that was flowing over my hand. That living word awakened my soul, gave it light, hope, joy, set it free! There were barriers still, it is true, but barriers that could in time be swept away.

I left the well-house eager to learn. Everything had a name, and each name gave birth to a new thought. As we returned to the house every object which I touched seemed to quiver with life. That was because I saw everything with the strange, new sight that had come to me. . . .

I learned a great many new words that day. I do not remember what they all were; but I do know that mother, father, sister, teacher were among them—words that were to make the world blossom for me, "like Aaron's rod, with flowers." It would have been difficult to find a happier child than I was as I lay in my crib at the close of that eventful day and lived over the joys it had brought, and for the first time longed for a new day to come. (pp. 36–37)

THE FLASH OF INSIGHT—THE BIRTH OF AN IDEA

The Dancing Atoms

The theory that matter is composed of various kinds of atoms (the elements) was proposed at the beginning of the nineteenth century. Most chemical substances are composed of the atoms of different elements joined together, and the numbers of atoms connected together to form the molecules of what we call inorganic substances (substances not containing the element carbon) had been found to be small. For example, when hydrogen combines with oxygen to form a molecule of water, there are exactly two atoms of hydrogen to one atom of oxygen. In hydrogen chloride, there is one atom of hydrogen to one atom of chlorine; in ammonia, three hydrogen atoms to one nitrogen atom. These simple proportions, observed in many substances, were discovered by the use of the atomic theory. In turn, they permitted prediction of many new chemical facts.

However, when it came to "organic" compounds—those in which carbon is combined with other elements such as hydrogen and oxygen—no such simple regularity was discerned. There were many more kinds of molecules than the simple behavior of the inorganic substances would lead one to expect. For example, many compounds of just the elements carbon and hydrogen were known, and they had such formulas (in modern notation) as CH_4, C_2H_2, C_2H_4, C_2H_6, C_3H_4, C_3H_6, C_3H_8, and so on. Even larger molecules, with as many as 10 or 20 atoms of each element, had been discovered. The carbon compounds seemed completely different from the simple 2-, 3-, or 4-atom compounds of inorganic chemistry. There was no way to

SCIENCE—THE SEARCH FOR UNDERSTANDING

predict what combinations were possible and no insight into how the atoms might be connected to one another.

The German chemist F. A. Kekulé realized that if carbon atoms, in forming molecules with other elements, could be connected to *each other* to form chains or rings of carbon atoms, the multitude of compounds could be explained, and their structures understood (Figure 10-1). Here is Kekulé's description of how the idea occurred to him:[2]

> One fine summer evening, I was returning by the last omnibus, "outside" as usual, through the deserted streets of the metropolis, which are at other times so full of life. I fell into a reverie, and lo! the atoms were gamboling before my eyes. Whenever, hitherto, these diminutive beings had appeared to me, they had always been in motion; but now up to that time, I had never been able to discern the nature of their motion. Now, however, I saw how, frequently, two smaller atoms united to form a pair; how a larger one embraced two smaller ones; how still larger ones kept hold of three or even four of the smaller; whilst the whole kept whirling in a giddy dance. I saw how the larger ones formed a chain. . . . I

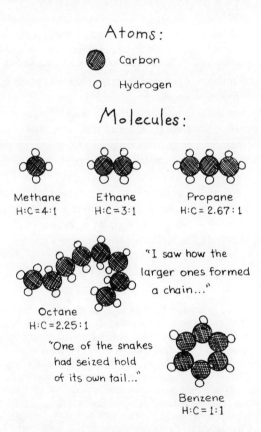

FIGURE 10-1. Kekulé's molecules.

spent part of the night putting on paper at least sketches of these dream forms. (pp. 36–37)

On another occasion, he writes:

I turned my chair to the fire and dozed. Again the atoms were gamboling before my eyes. This time the smaller groups kept modestly in the background. My mental eye, rendered more acute by repeated visions of the kind, could now distinguish larger structures, of manifold conformation; long rows, sometimes more closely fitted together; all twining and twisting in snakelike motion. But look! What was that? One of the snakes had seized hold of its own tail, and the form whirled mockingly before my eyes. As if by a flash of lightning I awoke. (p. 37) [see Figure 10-1]

We see that Kekulé's experience was not so different from Helen Keller's.

THE SENSE OF EXHILARATION

In the above quotations, the essential features of the process of understanding are shown: The person gaining it sees the world differently. Facts that previously seemed to have no special significance now stand out as part of a pattern; others that may have seemed important are no longer so. There is often a sense of exhilaration that goes with this new awareness.

It is important to see the scientific process as a search for just such experiences, as opposed to one popular but erroneous belief that science is a patient laying of one fact on top of another until an enormous and heavy structure has been achieved. While there are times when an accumulation of facts is necessary, this process is not science.

The important discoveries of science—Newton's realizing that the earth's gravity could reach out to the moon and keep it in orbit around the earth, Darwin's conceiving that species have evolved over time by a process of natural selection—were flashes of insight like Helen Keller's. They imposed an order and unity on a mass of facts.

Here as another example is Johann Kepler, the astronomer who discovered the laws of planetary motion, describing one of his insights:[3]

When I prophesied 22 years ago as soon as I found the heavenly orbits were of the same number as the five (regular) solids, what I fully believed long before I had seen Ptolemy's Harmonics, what I promised my friends in the name of this book, which I christened before I was 16 years old, what I urged as an end to be sought, that for which I joined Tycho Brahe, for which I settled in Prague, for which I spent most of my life at astronomical calculations—at last I have brought to light and seen to be true beyond my fondest hopes. It is not 18 months since I saw the first ray of light, three months since the unclouded sun-glorious sight burst upon me! . . . The book is written, the die is cast. Let it be read now or by posterity, I care not which. It may well wait a century for a reader, as God has waited 6000 years for an observer.

SCIENCE—THE SEARCH FOR UNDERSTANDING

We have noted that to gain such an insight, with the sense of excitement that goes with it, is not an experience confined solely to scientists. It is interesting to compare the emotions of scientists at the moment of discovery with those of poets at creative moments, people having religious experiences, and others at moments of intense feeling associated with a new vision of reality.

Religion, Poetry . . .

Kepler's emotions are much like the emotions associated with a religious experience. Here are the feelings reported by Jonathan Edwards, a famous clergyman and theologian of eighteenth-century New England, on his conversion:[4]

> After this my sense of divine things gradually increased, and became more and more lively, and had more of that inward sweetness. The appearance of everything was altered; there seemed to be, as it were, a calm, sweet cast, or appearance of divine glory, in almost everything. God's excellency, his wisdom, his purity and love, seemed to appear in everything; in the sun, moon, and stars; in the clouds and blue sky; in the grass, flowers, and trees; in the water and all nature; which used greatly to fix my mind. And scarce anything, among all the works of nature, was so sweet to me as thunder and lightning; formerly nothing had been so terrible to me. Before, I used to be uncommonly terrified with thunder, and to be struck with terror when I saw a thunderstorm rising; but now, on the contrary, it rejoices me. (p. 243)

It was not a farfetched metaphor Keats used when he described the emotional impact of reading Chapman's Homer for the first time:

> Then felt I like some watcher of the skies
> When a new planet swims into his ken.

We see in these descriptions the *quality* of the experience that is the goal of science, as it is the goal of much else that we do to understand the world around us. We recognize the highly subjective character of this experience—we have gained an understanding where we *feel* we have gained it. And this suggests the possibility that we might have the *feeling* of understanding and yet be wrong or misled about our perceptions.

. . . Alcohol . . .

It is common to experience some vivid sense of understanding after taking drugs or alcohol but to find on sobering up that this understanding vanishes. William James describes what this is like:[5]

> With me, as with every other person of whom I have heard, the keynote of the experience [of inhaling nitrous oxide, "laughing gas"] is the tremendously exciting sense of an intense metaphysical illumination. Truth lies open to the view in depth beneath depth of almost blinding evidence. The mind sees all the logical relations of being with an apparent subtlety and instantaneity to

which its normal consciousness offers no parallel; only as sobriety returns, the feeling of insight fades, and one is left staring vacantly at a few disjointed words and phrases, as one stares at a cadaverous-looking snow-peak from which the sunset glow has just fled, or at the black cinder left by an extinguished brand.

The immense emotional sense of reconciliation which characterizes the "maudlin" stage of alcoholic drunkenness,—a stage which seems silly to lookers-on, but the subjective rapture of which probably constitutes a chief part of the temptation to the vice,—is well known. The centre and periphery of things seem to come together. The ego and its objects, the meum and the tuum, are one. Now this, only a thousandfold enhanced, was the effect upon me of the gas: and its first result was to make peal through me with unutterable power the conviction that Hegelism was true after all, and that the deepest convictions of my intellect hitherto were wrong. (pp. 294–295)

. . . and Insanity

In Chapter 8 on the study of madness we quoted the feelings experienced by a person with a manic-depressive disorder during his manic phase (see pp. 198–199). This excerpt is worth rereading in the context of the present chapter.

Thus we conclude that while science is a search for understanding, and understanding can be recognized and identified only by the subjective feelings produced by it, that sense of understanding is not easily distinguished from the sense of understanding provided by art, religion, or insanity.

SCIENCE IS A CONSENSUS

In part, the distinction can be made on the basis that, in the long run, the subjective sense of understanding must be accepted by a scientific community: science is not concerned with "private" facts and experiences but with public ones. That this is so does not imply that questions of scientific truth are settled by majority vote. Scientific communities have wholeheartedly adhered to erroneous beliefs for long periods of time, as have other communities of human beings. There have been examples of great discoveries being initially rejected with scorn by the majority, although this happens less often than the reader may think. The key phrase in the first sentence above is "in the long run." It is not an absolute assurance of the value of the insight, but it is what we rely on. We have little choice.

The requirement that the insight of an individual must in the long run be acceptable to a community does not of course distinguish science from all other fields. The same equirement applies, and with the same qualifications, to the evaluation of works of art, to the practice of law, to theology, and so forth. What does distinguish science from these fields is the ability to subject the understanding achieved to experimental test.

REFERENCE NOTES

1. Helen Keller, *The Story of My Life* (Garden City, N.Y.: Doubleday, 1954). Copyright 1902, 1903, 1905 by Helen Keller. Reprinted by permission of Doubleday & Company, Inc.
2. Quoted in Alexander Findlay, *A Hundred Years of Chemistry*, 3rd ed., ed. Trevor I. Williams (New York: Humanities Press, 1965).
3. Johannes Kepler, *Harmonice Mundi* (Harmony of the World). Quoted in Robert K. Merton, "Behavior Patterns of Scientists," *American Scholar* 38 (1969): 197.
4. Quoted in William James, *The Varieties of Religious Experience*, Modern Library Edition, Random House, New York, Undated. From a biography of Edwards by S. E. Dwight, published in 1830.
5. William James, *The Will to Believe, and Other Essays in Popular Philosophy* (New York: Dover, 1956).

11

Science—The Goal of Generality

GENERALITY IN SCIENCE: EXAMPLES

The second important criterion of a scientific theory is its generality. To put it simply, the more it can explain, the better.

The most dramatic example of generalization in science is the laws of motion developed by Isaac Newton. The motion of the sun, stars, and planets as seen from the earth had of course been studied since prehistoric times. Fairly accurate means of predicting these motions were known to the Babylonians and Greeks. Copernicus' theory that the earth moves in a circular orbit about the sun rather than the reverse led to a *simpler* description of the motion of the planets, but it did not, at least at first, lead to *more accurate* predictions. Next, Kepler found that more accurate predictions of planetary orbits could be made by recognizing that the planets, including the earth, move around the sun in *elliptical* paths, rather than *circular* ones. What Newton showed later was that these results—the sun as the center of the planetary system, the planets moving in ellipses about it—could be understood as results of more *general* laws. These laws, which could be written in the form of mathematical equations on half a page, governed all motions in the universe, as well as on earth: falling apples, spinning tops, the vibrations of musical instruments, the tides, the motion of waves in water, the transmission of sound through air, the flow of blood in the veins, and much else. The discoveries of Copernicus and Kepler were thus reduced to special cases of much broader principles.

The general applicability of Newton's laws to so many different kinds of motion suggests the possibility of applying them to the motions of atoms and molecules as well, and thus to the kinetic theory of heat. It is not easy to do so, because even the smallest easily observable quantities of matter contain enormous numbers of atoms, but it can be done with the aid of the mathematical theory of probability. The result is that the kinetic theory of heat becomes one more special case of Newton's laws, and most of the phenomena of heat, including those that seemed earlier to contradict the kinetic theory, can be rationally explained in terms of them.

Einstein's Generalization

In this century Newton's laws were found to fail for bodies moving at very high speeds approaching the velocity of light. New laws that apply to these high speeds were formulated by Einstein. The results of Newton's laws became in turn a special case—the special case when things move slowly. Einstein's theory of relativity includes Newton's laws and explains as well things that Newton's laws could not, such as the tremendous energies that are released in nuclear explosions.

The Quantum Theory

A second failure of Newton's Laws was revealed when small objects, rather than fast-moving objects, were studied. The electrons, protons, and neutrons of which the atoms of matter are composed do not obey them either. Instead they obey the laws of quantum mechanics discovered by the physicists Niels Bohr, Werner Heisenberg, and Erwin Schrödinger. Quantum mechanics is also a generalization, which when applied to the special case of large, heavy objects like apples or planets gives exactly the same results as Newton's laws.

It is interesting to note that Einstein's theory of relativity and the laws of quantum mechanics have not yet been united into a single theory applying equally well to very fast objects, very small objects, and the heavy and slowly moving objects of our daily experience as well.

Cholera and the Germ Theory

Another example of the difference between the less general and the more general is the relation between Snow's theory of how cholera is transmitted and the germ theory of disease. Snow's theory explains specific cases of transmission—sewage seeping into the water supply, the fact that people living in close and crowded conditions with victims of cholera are likely to ingest some of their excretions in the course of handling their bedding and clothing, and so forth. But the germ theory provides a more general framework in that it explains the transmission not only of cholera but of other diseases as well: tuberculosis (through inhalation), syphilis (by sexual contact), and malaria (by the bite of mosquitoes).

THE PRICE OF GENERALITY

But there is a price we pay in science for the search for laws of great generality: it is the loss of many of the detailed qualities of the individual event.

If we look for a common pattern in a large number of different events, we must recognize all the same that no two events are ever exactly alike. We can find common patterns only at the price of selecting a few of the particulars and disregarding all the others.

SCIENCE—THE GOAL OF GENERALITY

We learn in arithmetic how to multiply any two numbers together. It requires memorizing the multiplication table up to 9×9, knowing how to add, and following a few other simple rules. Thus we can show that $11 \times 54 = 594$. But the case 11×54 is in every sense a special application of the general rule. There are no important features of 11×54 that are not already present in the rules of multiplication. The answer 594 is exact, and we can be as completely certain about that as we can of anything. Scientific knowledge is never so certain.

When we deal with the application of science to the real world, the special case will always have distinctive features that make it different from any other case. For example, we might wish to apply Newton's law of gravity to the fall of an apple from a tree. We want to ask how long the apple will take to hit the ground and how fast it will be moving just before it hits, and then see if the answers given by Newton's law agree with what we observe experimentally.

By choosing to concern ourselves with this narrow question we must be willing to ignore an enormous number of aspects of apples in general, and of this apple in particular. Most of these aspects we ignore because we have chosen to focus on one narrow kind of question, and we are sure that they are irrelevant to it. So we leave out of consideration such questions as: Who owns the apple orchard? How does the apple taste? How many other apples are on the tree? Is it red or green? Who was the queen of England at the time? and so on. Our belief that these are irrelevant to the narrow question of interest is part of our hypothesis; it is an act of faith to ignore them. Yet sometimes features that were ignored have later turned out to be crucial.

There are certain other details we omit not because we are sure that they are completely irrelevant, but because we think they are of small importance and will not change the result much. For example, is the wind blowing when we do our experiment? How fast? What is the effect of air resistance on the fall of the apple? What is its weight and size? Each of these factors will have an effect on the fall of the apple, but if we are satisfied with an accuracy of a percent or so in the quantities we want to measure, we can afford to ignore them.

We see that in choosing to be interested in the speed of the falling apple and the time of its fall, we must strip the apple of a tremendous number of the features that make *this apple* unique and interesting, and instead regard it as a drab special case of a general law of falling objects.

The Loss of Individuality

How the narrow focus of science strips most of the unique features from an object of study is well described in a passage from *Statistics*, by L. H. C. Tippet.[1] Tippet speaks of this as the act of a statistician, but it is the act of scientists in general.

> The loss of individuality results from the method of the statistician in confining his attention to only a few characteristics of the individuals and grouping them

into classes. Consider a married couple, say Mr. and Mrs. Tom Jones. As a couple their individuality consists of a unique combination of a multitude of characteristics. Mr. Jones is tall and thin, is aged 52 years, has brown hair turning grey, and is a farmer. Mrs. Jones is called Mary and at 38 years is still handsome; she is blonde and is really a little too "flighty" for a farmer's wife. The couple have been married for 16 years and have three children: two boys aged $14^{1}/_{2}$ and 11 years, and a girl aged 2. In addition to these and similar attributes the couple have a number of moral and spiritual qualities that we may or may not be able to put down on paper. It is by all these, and a host of other qualities that their relatives and neighbours know Mr. and Mrs. Jones; the uniqueness of the combination of qualities is the individuality of the couple.

The statistician who is investigating, say, the ages of husbands and wives in England and Wales is interested only in the ages, and does not wish to describe even these accurately. So he puts our couple in that class for which the age of the husband is 45–55 years and that of the wife is 35–45 years. Mr. and Mrs. Jones are now merely one of a group of some 320,000 other couples, and are indistinguishable from the others in their group. (p. 91)

Science and Maps

Scientific theories have been compared to maps.[2] Not even the most detailed maps show everything that is there, every tree or blade of grass, every rock or mud puddle. If one is interested in such details, there is no substitute for the thing itself. But for certain purposes—driving from one city to another, for example—a highway map is exactly what is needed. Such a map is highly schematic. It leaves out almost everything except roads, towns, parks, and a few other important features. Even these are presented in an idealized way: towns may be little black circles, or larger, colored areas roughly showing the shape and a few of the main streets. Not every twist and turn of the road is given, just a smoothed-out version showing its general direction. But the map serves its purposes well.

More detailed maps are possible, of course—physical maps that in addition to roads and towns also show elevation above sea level, types of vegetation, and so on. But even these give only a tiny fraction of all the details of the landscape. Thus the maps we use depend on the purpose we have in mind and the details we want described. They are necessarily always abstractions from reality, with much of reality left out.

IS HISTORY A SCIENCE?

This sacrifice of color and detail is a price not everyone is willing to pay. These qualities may well be what is most interesting to us. In an article denying the possibility of making history into a science, the historian Isaiah Berlin has stated the case for uniqueness:[3]

SCIENCE—THE GOAL OF GENERALITY

This is but another way of saying that the business of a science is to concentrate on similarities, not differences, to be general, to omit everything that is not relevant to answering the severely delimited questions that it permits itself to ask. But those historians who are concerned with a field wider than the specialized activities of men, are interested precisely in that which differentiates one thing, person, situation, age, pattern of experience, individual or collective, from another; when they attempt to account for and explain, say the French Revolution, the last thing that they seek to do is to concentrate only on those characteristics which the French Revolution has in common with other revolutions, to abstract only common recurrent characteristics, to formulate a law on the basis of them, or at any rate an hypothesis, from which something about the pattern of all revolutions as such (or, more modestly, all European revolutions) and therefore of this revolution in particular, could in principle be reliably inferred. . . . The purpose of historians . . . is to paint a portrait of a situation or a process, which, like all portraits, seeks to capture the unique pattern and peculiar characteristics of its particular subject; not to be an X-ray which eliminates all but what a great many subjects have in common. This has often been said, but its bearing on the possibility of transforming history into a natural science has not always been clearly perceived. (pp. 18–19)

The point can be made, however, that as significant and fascinating as the unique features of a historical event are, we do not usually know that they are unique until we have ruled out the possibility that they are examples of repeated patterns, by comparing this event with others.

An example of a mistake of this kind is provided by Freud's psychoanalytical study of Leonardo da Vinci.[4] One of the facts thought by Freud to be of great significance in understanding Leonardo's personality is that in his painting of the Virgin and St. Anne, St. Anne, who is Mary's mother, is portrayed as being close in age to Mary, more like an older sister than a mother. Freud relates this to certain biographical facts about Leonardo, who was the illegitimate child of a peasant woman and a notary, Piero da Vinci. Leonardo's father married another woman, and when Leonardo was about four years old he was taken into the new household. It is Freud's conjecture that Leonardo thus had two tender, loving mothers as a child—his natural mother and his father's wife—and the portrayal of the Virgin and St. Anne as two women of nearly the same age tenderly mothering the infant Jesus is an echo of his subconscious recollection of this situation. However, the art historian Meyer Shapiro has pointed out that this mode of portrayal of St. Anne was a convention of Italian painting at the time, a reflection of certain Catholic beliefs about St. Anne which were then being developed in the Church, and that examples of it had been common in paintings of this subject by other artists for a century before Leonardo's painting.[5] While this does not prove Freud's interpretation of Leonardo wrong, it makes one item of evidence for it less convincing.

Thus while science concerns itself with the general rather than the particular, it sometimes takes some scientific analysis to decide what is general and what is particular.

REFERENCE NOTES

1. Tippet, *Statistics* (Oxford: Oxford University Press, 1968).
2. Stephen Toulmin, *The Philosophy of Science: An Introduction* (New York: Harper and Row, 1977).
3. Isaiah Berlin, "History and Theory: The Concept of Scientific History," *History and Theory* 1 (1960), pp. 1–31. This essay has been reprinted in a revised form in *Concepts and Categories: Philosophical Essays by Isaiah Berlin,* Henry Hardy, ed. (New York: Viking Press, 1979).
4. Sigmund Freud, *Leonardo Da Vinci: A Study in Psychosexuality* (New York: Random House, 1966).
5. Meyer Schapiro, "Leonardo and Freud—An Art-Historical Study," *Journal of the History of Ideas* 17 (1956), pp. 147–178.

12
Science—The Experimental Test

TESTING THEORIES

The feature that distinguishes science from other ways of understanding and explaining the world is an ultimate reliance on the authority of the experimental test. There must be some agreed-on way of determining which facts are relevant to the credibility of our theories, and a willingness to place our theories at hazard in the process.

However, the reader who has followed the various experiments discussed in the case histories, and the way in which they have led to the acceptance of some theories and the rejection of others, should have realized that testing a scientific theory experimentally is not an automatic or routine procedure. Experiments that seemed decisive at one stage of understanding have been found less so at a later stage, and vice versa. Experimental results clearly in conflict with theories have not always led to the rejection of the theories. Theories that seem to have passed every possible test with flying colors are later discarded over what seemed at first a relatively minor discrepancy. Throughout, the judgment of a scientific consensus, applying criteria of truth and value that cannot easily be spelled out in objective terms, makes the decision. With the passage of time, as the criteria change, the decision can be reversed.

THE DEVELOPMENT OF THE EXPERIMENTAL METHOD

The appeal to experimental test is not an invention of modern science. Human beings—and animals as well—have the capacity to learn from experience: life would hardly be possible if they could not. The use of experimental testing in science, while not quite the same thing as learning by experience, is an outgrowth of it, and one cannot easily draw a sharp boundary between the two.

The ancient Greek philosophers quite often used observations of nature as evidence for or against theories. Aristotle, for example, who is often mistakenly thought of as an opponent of the experimental method, gave as an argument for the roundness of the earth the fact that during a lunar eclipse the earth casts a round shadow on the moon.

A striking example of a controlled experiment to test a theory is provided by St. Augustine, writing in the fifth century of the Christian era, a time not usually considered to be particularly favorable to science:[1]

> So to cure my obstinancy [in continuing to believe in and practice astrology] you [God] found me a friend who was usually ready enough to consult the astrologers. He had made no real study of their lore but, as I have said, he used to make inquiries of them out of curiosity. He did this although he was perfectly well aware of certain facts about them which he said he had heard from his father. If only he had realized it, these facts would have been quite enough to destroy his belief in astrology.
>
> This man, whose name was Firminus, had been educated in the liberal arts and had received a thorough training in rhetoric. He came to consult me, as his closest friend, about some business matters of which he had high hopes, and asked me what prospects I could see in his horoscope, as they call it. I was already beginning to change my mind in favour of Nebridius's [skeptical] opinions on astrology, but I did not refuse outright to read the stars for him and tell him what I saw, though I had little faith in it myself. Nevertheless I added that I was almost convinced that it was all absurd and quite meaningless. He then told me that his father had studied books of astrology with the greatest interest and had had a friend who shared his enthusiasm for the subject. Each was as intent upon this nonsense as the other, and by pooling their experiences they whetted their enthusiasm to the point that, even when their domestic animals had litters, they would note the exact moment of birth and record the position of the stars, intending to use these observations for their experiments in this so-called art.
>
> Firminus went on to tell me a story about his own birth. His father had told him that when his mother was pregnant, a female slave in the household of this friend was also expecting a child. Her master was of course aware of her condition, because he used to take very great care to find out even when his dogs were due to have puppies. The two men made the most minute calculations to determine the time of labour of both the women, counting the days, the hours, and even the minutes, and it so happened that both gave birth at exactly the same moment. This meant that the horoscopes which they cast for the two babies had to be exactly the same, down to the smallest particular, though one was the son of the master of the house and the other a slave. For as soon as labour began, each man informed the other of the situation in his house, and each had a messenger waiting, ready to be sent to the other as soon as the birth was announced. As the confinements took place in their own houses, they could easily arrange to be told without delay. The messengers, so Firminus told me, crossed paths at a point which was exactly half way between the two houses, so that each of the two friends inevitably made an identical observation of the stars and could not find the least difference in the time of birth. Yet Firminus, who was born of a rich family, strode along the smoother paths of life. His wealth increased and high honours came his way. But the slave continued to serve his masters. Firminus, who knew him, said that his lot had been in no way bettered. (pp. 140–141)

The above example illustrates an important but infrequently discussed aspect of experimental testing. Augustine, already turning away from astrology, was both intellectually and emotionally ready to accept such negative evidence. His friend Firminus was not ready; although he was the source of the "data," his belief in astrology was apparently unaffected by his knowledge.

In the few hundred years from Copernicus to Newton, the concept of the experimental method developed rapidly, and it was gradually accepted as one of the primary criteria of scientific truth. This development included a number of different features. In part, it represented a rejection of the authority of the Church and the Bible as a means of understanding the physical universe. But conflict with religious authority was not really an important obstacle to the acceptance of the experimental method in general. There was no conflict between church and science in such fields as anatomy and chemistry, to which the experimental method was also being applied at the time.

The End of Authority

What was more important was the rejection of authority of a different kind, the authority of past scientists and philosophers. In part, this rejection of authority required the rejection of the idea that the properties of the universe could be deduced by logical argument from philosophical principles better than from direct observation. A classic example of the philosophical approach to the universe as opposed to an experimental one is provided by the response of the Florentine astronomer Franceso Sizzi to Galileo's claims that by the use of the telescope he had discovered four new planets (actually moons) in orbit around Jupiter:[2]

> There are seven windows in the head, two nostrils, two ears, two eyes and a mouth; so in the heavens there are two favorable stars, two unpropitious, two luminaries, and Mercury alone undecided and indifferent. From which and many other similar phenomena of nature such as the seven metals, etc., which it were tedious to enumerate, we gather that the number of planets is necessarily seven. . . . Besides, the Jews and other ancient nations, as well as modern Europeans, have adopted the division of the week into seven days, and have named them from the seven planets: now if we increase the number of planets, this whole system falls to the ground. . . . Moreover, the satellites are invisible to the naked eye and therefore can have no influence on the earth and therefore would be useless and therefore do not exist. (pp. 164–165)

REPEATABILITY

Those who followed the authority of Aristotle in physics and biology, or of Galen in medicine, did not always do so through disbelief in the methods of direct observation. Rather, they often believed that the observation of nature had already been performed and did not need to be repeated. The recognition came gradually that famous, wise, and honorable men might have been wrong in their observations, and that it was always possible that new and more careful observations might contradict the old ones. This recognition was a major revolution in scientific thought.

We now take for granted that any observation, any determination of a "fact," even if made by a reputable and competent scientist, might be doubted. It may be necessary to repeat an observation to confirm or reject it. Science is thus limited to what we might call "public" facts. Anybody must be able to check them; experimental observations must be *repeatable*.

The quality of this change in attitude toward figures of authority is shown beautifully in the exuberant writing of Francisco Redi, whose book, *Experiments on the Generation of Insects,* reports his experimental demonstrations that insects do not arise spontaneously from decaying carcasses or other matter, but come from eggs laid by insects of the same kind. Redi is refuting various "fables" about the origin of insects:[3]

> Be this as it may, I shall add another fable to it, namely the origin of wasps and hornets from dead flesh, although by universal conformity of opinion it has been accepted as the truth.
>   Antigonus, Pliny, Plutarch, Nicander, Ælianus, and Archelaus, as quoted by Varro, teach that wasps originate in the dead flesh of horses. Virgil admits it to be also the origin of hornets. Ovid mentions only hornets. . . . Thomas Moufet reports that hornets are generated in the hard parts of horseflesh, and wasps in the tender parts. The Greek commentators of Nicander attribute the creative property to the horse's skin alone, adding as a necessary condition that the horse must have been bitten and torn by a wolf. But Servius, the grammarian, turned everything topsy-turvy by asserting that drones come from horses, hornets from mules, and wasps from asses. Olimpiodorus, Pliny, Cardano and Porta insist that ass-flesh gives birth to drones and beetles, but not to wasps. . . . (p. 50)
>
>   On this occasion I perceived that there was no truth in the reports of Aristotle and A. Caristio that among [scorpions] the mothers are killed by the newborn young, nor, as Pliny relates, that the young are all killed by the mother, with the exception of one more clever than the rest, who runs up on his mother's back out of reach of her sting, and afterwards avenges his brothers' death by killing his parent. . . . (p. 54)
>
>   To conclude my remarks on scorpions, I must add that the account of some of Pliny's followers, i.e., that dead scorpions come to life on being moistened with the juice of white hellebore, is an old wife's tale. As for Avicenna's assertion that a scorpion will fall dead if confronted with a crab to which a piece of sweet basil has been tied, it is likewise false, and having proved it so, I passed on to further experiments. . . . (p. 62)
>
>   Aristotle asserts that cabbages produce caterpillars daily, but I have not been able to witness this remarkable reproduction, though I have seen many eggs laid by butterflies on the cabbage-stalks and neighboring grasses; these eggs developed subsequently into caterpillars and butterflies. (p. 113)

QUANTITY RATHER THAN QUALITY—THE FAITH IN MATHEMATICS

A further feature of the development of the experimental method was the emphasis, in Galileo's work and even more so in Newton's, on mathematical formulations of the laws of nature.[4] Mathematics at that time was undergoing a very rapid development of its own, and its power was beginning to be appreciated. Galileo, for example, expressed the view that the book of the universe was written in mathematical language, in an alphabet consisting of circles, triangles, and other geometrical figures.

Now mathematical language is a precise language, and when laws of nature are expressed in mathematical form, precision is imposed upon them. The predictions that can be made from such laws are numerical predictions: some quantity that we can measure in our experiment will be found to have a certain definite value.

Thus it is much easier to tell whether the experiment confirms the theory or not. There is only one way for it to agree with the theory and too many ways for it to fail, so that when it does agree, it is convincing indeed.

Newton, using his laws of motion together with a conjecture that the force of gravity between two bodies decreases as the square of the distance between them, was able to show, among other things, that the length of a planetary year—the time it takes a planet to complete one revolution about the sun—should depend on the distance of the planet from the sun only, and not on the mass of the planet. Further, and more precisely, the length of the year should be proportional to the $3/2$ power of the distance.

Expressed as an equation,

$$T = R^{3/2}$$

T is the length of the planet's year, expressed as a multiple of the earth's year, and R is the planet's distance from the sun, expressed as a multiple of the earth's distance. The $3/2$ power may look frightening, but its calculation is simple:

1. Take the cube of R ($R \times R \times R$).
2. Then take the square root of the result.

Any hand calculator can be used for these operations.

For the earth itself, both T and R are 1, and simple arithmetic shows that $R^{3/2} = 1$, also, so the equation is satisfied for the earth. Does it work for other planets as well? Table 12-1 shows the results for the planets that were known at the time of Newton.

Notice how good the agreement between theory and experiment is; but notice also that the agreement is not perfect. Notice, finally, that the masses of the planets

Table 12-1
Test of Newton's Laws Using Data on the Planets

Planet	Planetary mass (as multiple of earth's mass)	Distance from sun R (as multiple of earth's distance)	Experiment: Length of planetary year T (as multiple of earth's year)	Theory: Predicted from Newton's Laws $R^{3/2}$
Mercury	0.055	0.39	0.241	0.243
Venus	0.815	0.72	0.615	0.611
Earth	1.000	1.000	1.000	1.000
Mars	0.107	1.52	1.88	1.87
Jupiter	317.89	5.20	11.86	11.86
Saturn	95.17	9.54	29.46	29.47

vary enormously, and yet have no effect on the period. Both the absence of an effect of mass and the $3/2$ power in the relation between T and R are direct predictions from Newton's laws. When you consider that without these laws there was *no* particular relation to be expected between T and R—T could have depended on the planet's mass, or T could have depended on the fifth power of R, or decreased as the square of R—you can begin to see why this result is such convincing evidence for their truth.

It is interesting that the relation $T = R^{3/2}$ was discovered by the astronomer Kepler before Newton was born. It was found by him purely as an experimental regularity, and he could give no reason why it worked so well.

Part of the difficulty Rumford had in convincing his contemporaries to accept the kinetic theory (see pp. 128–129) was that the experiments on the conservation of heat, which were then regarded as supporting the caloric theory, were *quantitative*—one could predict the final temperature accurately when x kilograms of hot iron was placed in a bucket containing y kilograms of cold water. In contrast, Rumford's experiments (except for the experiment on the weight of heat) were *qualitative*—friction produces heat, salt molecules move about in water. The kinetic theory became plausible when Joule, in effect, performed Rumford's experiment quantitatively, measuring *how much* work is equivalent to a certain amount of heat.

TESTING BY EXPERIMENT—PLANNED AND UNPLANNED

In this discussion of testing theories by experiment, we hope the reader has not gotten the impression that the process is always systematic, well-organized, or deliberate. The very opposite is often true.

The scientific enterprise is a complex and erratic one. It includes a large number of people with different interests, purposes, skills, and depths of understanding. At any one time, certain of them may be questioning some widely believed theory, while others take the theory for granted and are trying to apply it to some problem of special interest to them. Still others may be studying something because they need the knowledge for a purpose unrelated to any theory.

Not all experiments succeed. Sometimes they are just badly done, and the scientific consensus, recognizing this, disregards them. Sometimes, although they are carried out well, the results are confusing and raise more questions than they answer. But from time to time, out of this welter of uncoordinated activity, emerge results that are recognized as having an important bearing on the credibility of some theory.

The Experiment Must Make a Difference

When we perform an experiment, we perform it because we do not know what the result will be. If we knew in advance, we would not bother. There must be two, or several, or a large number of possibilities. We may expect one of several outcomes, or we may not know at all what to expect. In order for the experiment, whatever its purpose, to be considered a test of some theory, the *outcome must make a difference*. If the experiment has one result, we must be led to a greater degree of confidence in our theory; if it has another result, we must be led to a greater degree of doubt. If the degree of our belief remains unaffected by the result, the experiment cannot be said to have been a test, although it may have been valuable or interesting for other reasons.

Our degree of belief in a theory is not something that can be measured in any quantitative way: it is a subjective thing that each of us must judge for himself. But we can tell easily if something we have observed has changed the strength of our belief. Sometimes the change in degree is small: we conclude that a theory is less likely than we thought yesterday, before we learned of some new experimental observation. At other times it is dramatic: we change from confidence to complete disbelief, or vice-versa, all at once, in the manner of a religious conversion.

An Awareness of Alternatives

For the experiment to affect our belief, we must be emotionally prepared for the possible outcomes (as St. Augustine was and Firminus was not). We must have an awareness of alternatives, a readiness to consider that what we believe may be wrong, and that what others believe may be right. Sometimes this awareness is a simple one: the alternatives might be (1) the theory is right, or (2) the theory is wrong. In other situations we have two or several competing theories, like the caloric and kinetic theories of heat, or psychogenic and genetic theories of the causation of schizophrenia. What experiment we do will depend on the alternatives among which we are trying to choose. However, unless there is some sense of the existence of alternatives, no experimental test is possible, because no outcome can alter belief.

Women Drivers and the Lisbon Earthquake

An example of a person lacking a real awareness of alternative possibilities is given by W. V. Quine and J. S. Ullian:[6]

> Manny grumbles, "It must be a woman driver," whenever he sees an inept maneuver on the road. When, as often happens, it becomes apparent that the offending driver is not in fact a woman, Manny shifts his grumble to "He drives just like a woman." (p. 83)

One despairs of doing any experiment whatever that would change Manny's mind.

Another example is Pangloss in Voltaire's *Candide*. Pangloss is a philosopher who believes that "Everything is for the best in this best of all possible worlds." It is his practice to explain how everything that happens to Candide or to himself illustrates this principle. Unfortunately, almost everything that happens is disastrous. Pangloss is infected with syphilis and almost dies of it; a gentle and kind benefactor of Candide's is drowned at sea; the city of Lisbon is destroyed in an earthquake and thousands of people are killed; Pangloss is hanged for heresy (fortunately, he survives), whipped for making love to a pretty Moslem girl in a mosque, and chained in the galleys. All of these events are interpreted by him as proving his theory. One soon realizes that Pangloss can "explain" any event whatever: nothing can faze him. His view of the world is thus impervious to any experimental test. This is not to deny that it is of great generality in its application and provides him with a satisfying explanation of the nature of the world.

Freud, Marx, and Others

The philosopher of science Karl Popper describes in one of his essays a revelatory experience he had, as a young man in Vienna, while working with the psychologist Alfred Adler.[7] Adler, originally a disciple of Freud, had come to disagree with his master over the great importance Freud gave to sexual issues, and had developed an approach of his own, called *individual psychology*, in which the concept of the *inferiority complex* played a central role.

According to Popper the theories of Freud and Adler on psychology, and of Karl Marx on history and social change, were widely discussed among intellectual circles in post-World-War-I Vienna, and had many adherents. Popper himself worked with Adler in clinics that the latter had established for children and adults in working-class districts of the city. However, Popper was skeptical about these theories, and troubled about their scientific status.

> I found that those of my friends who were admirers of Marx, Freud, and Adler, were impressed by a number of points common to these theories, and especially

by their apparent explanatory power. These theories appeared to be able to explain practically everything that happened within the fields to which they referred. The study of any of them seemed to have the effect of an intellectual conversion or revelation, opening your eyes to a new truth hidden from those not yet initiated. Once your eyes were thus opened you saw confirming instances everywhere: the world was full of verifications of the theory. Whatever happened always confirmed it. Thus its truth appeared manifest; and unbelievers were clearly people who did not want to see the manifest truth; who refused to see it, either because it was against their class interest, or because of their repressions which were still "un-analyzed" and crying aloud for treatment.

The most characteristic element in this situation seemed to me the incessant stream of confirmations, of observations which "verified" the theories in question; and this point was constantly emphasized by their adherents. A Marxist could not open a newspaper without finding on every page confirming evidence for his interpretation of history; not only in the news, but also in its presentation—which revealed the class bias of the paper—and especially of course in what the paper did not say. The Freudian analysts emphasized that their theories were constantly verified by their "clinical observations." As for Adler, I was much impressed by a personal experience. Once, in 1919, I reported to him a case which to me did not seem particularly Adlerian, but which he found no difficulty in analysing in terms of his theory of inferiority feelings, although he had not even seen the child. Slightly shocked, I asked him how he could be so sure. "Because of my thousandfold experience," he replied; whereupon I could not help saying: "And with this new case, I suppose, your experience has become thousand-and-one-fold." (pp. 35-36)

The same point had been made three hundred years earlier by Francis Bacon:[8]

The human understanding when it has once adopted an opinion (either as being the received opinion or as being agreeable to itself) draws all things else to support and agree with it. And though there be a greater number and weight of instances to be found on the other side, yet these it either neglects and despises, or else by some distinction sets aside and rejects; in order that by this great and pernicious predetermination the authority of its former conclusions may remain inviolate. And therefore it was a good answer that was made by one who when they showed him hanging in a temple a picture of those who had paid their vows as having escaped shipwreck, and would have him say whether he did not now acknowledge the power of the gods,—"Aye," asked he again, "but where are they painted that were drowned after their vows?" (p. 50)

Bacon goes on to add:

And such is the way of all superstition, whether in astrology, dreams, omens, divine judgements, or the like: wherein men, having a delight in such vanities, mark the events where they are fulfilled, but where they fail, though this happen much oftener, neglect and pass them by. But with far more subtlety does this mischief insinuate itself into philosophy and the sciences; in which the first conclusion colours and brings into conformity with itself all that come after, though far sounder and better. Besides, independently of that delight and vanity which I have described, it is the peculiar and perpetual error of the human intellect to be more moved and excited by affirmatives than by negatives; whereas it ought properly to hold itself indifferently disposed towards both alike. Indeed in the establishment of any true axiom, the negative instance is the more forcible of the two. (p. 50–51)

Refutability

Theories like those of Manny and Pangloss, that are beyond the reach of experimental test, must be regarded as lying outside the domain of science itself. For a theory to be part of science we must be able to imagine the possibility that some kind of evidence, if it were available, would tend to make us doubt the theory. As Popper has put it, for a theory to be scientific, it must be refutable.[9]

The reason for stressing the negative property of refutability as a criterion for scientific theories is perhaps a psychological one. As noted by Francis Bacon, people, and this includes scientists, are conservative. Whatever theories they hold they tend to stick with. They are more open to evidence that confirms their theories than evidence that refutes them, and are commonly so tenacious in defending what they already believe, that it seems that no evidence, no matter how cogent, could shake them. Of course, we are more conscious of this in others than in ourselves. But nobody needs to be told that theories should be confirmable, in the sense that new experiments might be able to increase our confidence in them—we all take that for granted. We do need to be reminded from time to time that we might be wrong, and should be open to evidence that might show it.

Confirmability and refutability are two sides of a single coin. New facts should be able to change our degree of belief one way or another. Only if this is so is our belief scientific.

Note that the issue of refutability as raised by Popper is *not* "Are these theories true or false?" It is rather "Can you tell?"

As to whether the theories of Freud or Marx, or for that matter many other theories in the social and behavioral sciences, meet this test, opinions differ. Maybe the proper question to ask is not whether the theories are scientific, but rather: Are those who believe in them scientists, in the sense that they are capable of recognizing the relevance of cogent refuting evidence when it is presented to them? The answer to that question, in most cases, would probably be that some are and some are not.

You Cannot Prove a Theory Right

Theories can pass many experimental tests and still turn out to be wrong. We mentioned in an earlier chapter how Newton's laws of motion fit every experimental test they were given for almost 250 years. The more tests they passed, the more certain scientists were of their truth; that they could ever break down had become almost inconceivable. Yet at the end of the nineteenth century, experiments on fast-moving bodies and on small objects such as electrons proved that they did not apply in all cases.

We have given examples in our case histories of incorrect theories that fit at least some of the experiments used to test them. The caloric theory explained quite well the conservation of heat in Black's experiments. Another example is Farr's theory relating the rate of cholera in London to the height of residence above sea level. There was indeed a strong correlation: the higher the residence above sea level, the lower the cholera rate. Snow, however, while recognizing the strength of the relation, pointed out that in London the districts of higher elevation would tend to have less polluted water supplies, a fact that could explain the lower rate more plausibly. He noted also that some other cities with higher elevations than any part of London had more cholera.

Wide-ranging and plausible experimental evidence at one time seemed to show that the disease pellagra, characterized by a skin rash, diarrhea, nausea, and, in terminal stages, a type of psychosis followed by eventual death, came directly from eating corn, or ingesting some fungus or parasite that grew on corn. It had not been known in Europe before the introduction of corn which followed the discovery of America, it occurred solely in corn-growing districts of Europe and the United States, and it was always more common during famines, when starving people ate spoiled corn. The possibility that it might be an infectious disease was ruled out by the fact that nurses and doctors working in the hospitals where the victims were treated never got it.

In 1914–15, Joseph Goldberger showed that it was a nutritional-deficiency disease. Corn lacks vitamins present in wheat and other grains. The disease is caused not by eating corn, but rather by not eating enough other foods, such as milk, eggs, or meat.[10]

We conclude that no matter how many tests a theory has passed, it is always possible for some new kind of experiment to reveal a weakness. This is why, when we spoke earlier about testing theories, we claimed no more than that a successful test increases our confidence. "Proof" of correctness is not possible.

You Cannot Prove a Theory Wrong

The lack of certainty works both ways: a theory that fails an experimental test is often but not invariably discarded. No theory we have today fits perfectly every conceivable experiment used to test it. We are satisfied at any one time with whatever theory seems to work best, and "best" involves a considerable degree of subjective judgment. There are often reasonable alternative explanations why a good theory will fail in some particular circumstances, and even when there are not, if we think the theory better than any available alternative, we will stick with it and try to find special explanations of why it failed to work in these circumstances.

Examples have been seen in our case histories of apparent failures of good theories. One was the case described by Snow of a man who by mistake drank a glass of the evacuation of a cholera victim but did not get the disease. This case gave Snow trouble, but he stuck to his theory. Although Rumford could not plausibly explain the principle of the conservation of heat, as revealed in Black's experiments, so well as it could be explained by the caloric theory, he persisted in his belief that heat was molecular motion. Fifty years later, the law of the conservation of energy was discovered, and the principle of the conservation of heat could be shown to be a direct consequence of it.

INDIRECTNESS OF EXPERIMENTAL TESTS

Most scientific hypotheses are formulated in ways that do not permit us to test them in a direct, simple way. Instead, we are forced to rely on what might be called "circumstantial" evidence. We deduce certain consequences of the hypothesis that are directly observable, and then see if they are indeed observed. The statement "The world is round like an orange" is not testable the way we test the roundness of an orange, by visual examination and touch. The evidence for the roundness of the world is quite indirect but no less compelling. The kinds of evidence for it that led finally to its general acceptance included such data as the shape of the shadow of the earth on the moon during a lunar eclipse, the fact that as one sails from port on a ship the buildings of the port gradually seem to sink below the surface of the water, that the stars visible from any particular point on the earth surface depend on which point the observer is at, as does the height of the sun above the horizon, and, most convincing, that it is possible to sail all around the world and return to the starting point of the voyage.

Generality and Indirectness

As we seek for more general laws in science, laws designed to explain a tremendous variety of phenomena from a few simply stated principles, it is inevitable that the relation between our few general principles and the variety of phenomena we seek to explain becomes more indirect and less obvious, and, further, that those few general principles begin to have applications to phenomena other than the ones we were trying to explain when we formulated them. The result is that the variety of experimental tests a theory can be given becomes very large, and the relation of the experimental test to the theory in question becomes obscure, unless one has taken the trouble to follow the steps that show the logical connection between the two.

The conflict between the kinetic and caloric theories of heat provides an example. The theory that heat is a substance sounds simple to test directly: collect some heat in a container and study it. But the substance the supporters of the caloric theory had hypothesized had to have properties that made it impossible to collect it in a container by itself for study. It could be handled only in association with matter, and its properties studied indirectly by the changes it produced in the matter.

On the other side there were other problems that prevented a "direct" proof that heat is atomic motion: there was no way to observe atoms directly and see them moving. So that while both theories are rather simple to state, the kinds of experiments used to resolve the controversy were indirect and covered an extraordinarily wide range of phenomena. In many cases it would be hard to explain these various experiments without going into a lot of physics and some rather long and circuitous discussions often involving mathematics. We limited our detailed discussion to Black's experiments on specific and latent heat and to Rumford's experiments on the weight of heat, its production by friction, and the diffusion of salt molecules in water. We described briefly the relevance to this problem of the theories of radiation of heat through empty space, and the experiments of Joule on the conversion of mechanical and electrical energy to heat. We also mentioned Fourier's studies on the flow of heat and Carnot's studies on the efficiency of steam engines. We made no mention, however, of an enormous variety of phenomena involving heat that were being studied during that period, such as the speed of sound in gases, the heat given off in chemical reactions, the changes in volume and pressure of gases when heated, and many more. All these phenomena provided experimental information on the nature of heat. Some of them could be explained equally well by either the kinetic or the caloric theory, some seemed to support one theory over the other, and some did not at the time seem relevant to either.

The point here is the enormous range and variety of possible experiments that have a bearing on making the choice between two fairly simple-sounding and quickly stated scientific theories.

WHAT DO WE TEST, AND WHEN?

We repeat that the experimental testing of a scientific theory is not a mechanical, automatic process. There is no prescribed set of procedures we can go through, at the end of which we give the theory some stamp of approval that says it has passed its tests. The process of testing a theory, like the process of making one up in the first place, is a never-ending process, and a creative, imaginative one. We have to excercise some subjective judgment about what kind of experimental evidence will make a real difference one way or another in our degree of belief.

We can see this creative spirit in operation in our case studies: Snow seizing on the fact that two different water companies supplied different houses in a single district of London, Rumford recognizing the importance of the old and common observation that friction makes things hot, Heston and Kety seeing that cases of adoption provide a way of distinguishing the relative importance of genetic and psychogenic factors in schizophrenia.

APPENDIX 12-1: THE EXPERIMENTAL METHOD IN THE HUMANITIES

In the introduction to this book we suggested that there are parallels between the methods of scientific research and the methods of scholarly research in the humanities that should be recognized and appreciated by members of both camps. We also stated there, and again at the beginning of this chapter, that reliance on the authority of the experimental method is the distinguishing feature of science. To bring out these parallels, we would like to give one brief example of the use of the experimental method in reseach in the humanities. In the "Suggested Reading" section of this chapter, we give references to other works that deal in more detail with the similarities and differences.

The example we will use is the study *Shakespeare's Imagery and What It Tells Us* by the English scholar Caroline Spurgeon.[11] Spurgeon made a detailed analysis and classification of the images used by Shakespere in his plays and poems in the belief that two kinds of knowledge could be obtained from them: first, biographical information, such as what kind of a man Shakespeare was, what he thought, what he liked and disliked; and, second, a deeper knowledge and understanding of the themes and characters in the plays themselves.

Images as Facts

Spurgeon uses the term *image* to cover similes, metaphors, and all other figures of speech in which something is compared to something else. For example, when Hamlet speaks of "the slings and arrows of outrageous fortune," he is comparing ill fortune to a besieging army; when Portia in *The Merchant of Venice* says that "mercy . . . droppeth as the gentle rain from heaven," the image is obvious.

Imagery is one of the major components of imaginative writing, in prose no less than in poetry. Spurgeon's belief is that the images used by the poet, being more spontaneous, and arising more from the poet's unconscious mind, than the subject matter of the poem or play, may be more revealing of his innermost nature.

> The imagery he instinctively uses is thus a revelation, largely unconscious, given at a moment of heightened feeling, of the furniture of his mind, the channels of his thought, the qualities of things, the objects and incidents he observes and remembers, and perhaps most significant of all, those which he does not observe or remember. (p. 4)

It should be apparent that what is or is not an image in Shakespeare, and what kinds of things are being compared in an image, are "facts" in the sense we have used the term in this book. While there occasionally may be doubtful cases, most of the time one would expect that all informed observers—those who are sufficiently familiar with the vocabulary and culture of Elizabethan England—will agree.

Whether these "facts" will support the theories proposed about Shakespeare the man, or about the meaning of his plays, is of course another matter.

A Controlled Experiment

Now, for the claim that Shakespeare's imagery reveals Shakespeare's life and personality to be taken seriously, it is necessary to show that Shakespeare's imagery is in fact distinctively his own. If all Elizabethan writers tended to use the same images—for example, if they all compared their mistresses' eyes to the sun, misfortunes to the assaults of an army, and life to the babbling of an idiot—nothing personal could be inferred about Shakespeare from his use of these images. So Spurgeon begins her study with a comparison of the images used by different writers. This comparison deserves to be called a "controlled experiment." She makes an exhaustive survey of all images used in five representative plays by Shakespeare with the images used in comparable amounts of work by other Elizabethans: Marlowe, Bacon, Jonson, Chapman, Dekker, and Massinger. All except Bacon were dramatists; Bacon was included only because of the widespread belief that he was the real author of Shakespeare's plays, a view that Spurgeon, like most other serious scholars, does not share.

Results

The results of this comparison offer strong support to the idea that each of these writers' choice of images is unique.

Shakespeare's choice of images, as tabulated by Spurgeon, is at first surprising. It is largely based on quite humble things, on the common experiences of daily life in a small town: the changes of the weather, plants, gardening, animals and birds, food and cooking, skills and trades like carpentry and sewing, and sports and games.

Examples from one play, *Macbeth,* are as follows:

(Banquo demanding of the witches that they foretell his future if they can):

> If you can look into the seeds of time,
> And say which grain will grow and which will not,
> Then speak to me.

(Duncan promising rewards to Macbeth for his faithful services against a rebellion):

> I have begun to plant thee, and will labour
> To make thee full of growing.

(Lady Macbeth urging her husband to murder Duncan):

> But screw your courage to the sticking place,
> And we'll not fail.

(Macbeth in a soliloquy expressing his guilt at having murdered his sleeping guest Duncan):

> Sleep that knits up the ravell'd sleave of care.

(Macduff informed that his wife and children have been murdered by Macbeth):

Oh, hell-kite, all?
What, all my pretty chickens and their dam
At one fell swoop?

The images above are drawn, respectively, from gardening (the first two), carpentry, weaving, and the attack of a bird of prey on barnyard fowl. They are so apt in context that one feels no incongruity.

In contrast, Marlowe, who studied at Cambridge University, draws his images from books, mostly the classics of Greek and Latin literature, and from the heavens, the stars, sun, moon, and planets. "He seems more familiar with the starry courts of heaven than with the green fields of earth, and he loves rather to watch the movements of meteors and planets than to study the faces of men."

Representative images in Marlowe are given by Spurgeon: the loveliness of a woman is described as ". . . fairer than the evening air/Clad in the beauty of a thousand stars"; the conflicting emotions of a woman's heart are ". . . like a planet, moving several ways/At one self instant . . ."; a conqueror predicting future victories boasts that "I will persist a terror to the world./Making the meteors . . ./Run tilting round about the firmament/And break their burning lances in the air."

Although Shakespeare occasionally uses such images, they dominate in Marlowe.

Bacon uses very few images from nature. His greatest number are drawn from what Spurgeon calls "Domestic Life," but it is the life in a nobleman's mansion rather than a cottage: light and fire, furnishings, textiles, jewels, needlework, and so on. The contrast between light and darkness, and between different kinds of light, is of great interest to him, but not, apparently, to Shakespeare. Further, Bacon is much more intimately acquainted with the Bible; Shakespeare's use of Biblical images is stereotyped, being limited to "well-known characters and incidents, familiar to any grammar-school boy." Such images as Bacon uses from nature are drawn from farming rather than gardening, and seem specifically to do with farming a large estate. He uses very few images from sports or games.

Similarly, the imagery of the other Elizabethan dramatists differs from Shakespeare's in significant ways. Ben Jonson, for example, uses sports images also, but his most vivid ones are from fencing, while Shakespeare's are from bowling and archery. Unlike Shakespeare's, his images from warfare show detailed experience; he is known to have served as a soldier in the Netherlands. Chapman's war images are also vivid and suggest direct experience, but not enough is known about his life to confirm that he had been a soldier. Dekker uses a very large number of sports images, comparable to Shakespeare's, but the bulk of them are from fishing and show a close and loving acquaintance with it; Shakespeare's images from fishing are much fewer and more perfunctory.

Other differences among the choices of imagery of each of the writers considered are discussed in detail by Spurgeon; they make quite a convincing case for the distinctiveness of Shakespeare's. Had Spurgeon been unable to show this, her hypothesis that something of Shakespeare's personality could be inferred from his images would have been built upon sand. Whether, having established this much, she succeeded in her further endeavours, we cannot answer here. Readers are referred to her book and to critical comments on it by other scholars for an answer.[12,13]

APPENDIX 12-2: THE CONTROLLED EXPERIMENT REVISITED

Introduction

In the case histories given earlier we described the use of controlled experiments. While not all experiments in science are controlled in the sense that Snow's study of water supply and cholera was, the concept is of special importance. It represents the essence of the experimental method: once its basic logic is clearly grasped, the rest of science comes more easily. Fortunately the logic is easy to understand, in fact so easy that the controlled experiment is used not only by scientists in their laboratories, but also very widely in ordinary discourse and debate as well.

In this section we will give some examples of such reasoning, in which the concept of a controlled experiment is central to the argument.

Two Groups

A controlled experiment is essentially a comparison of (at least) two groups: a test and a control group. John Snow, in testing his general hypothesis that drinkers of Southwark and Vauxhall water, which was more contaminated with sewage, would have a higher death rate from cholera than drinkers of Lambeth water, was performing such an experiment.

The hypothesis tested in such a comparison is a cause-and-effect type of relationship in which both cause and effect were all-or-none propositions. A person either got cholera or not, and the home he lived in was supplied either by one or the other of the two companies. No attempt was made to relate the severity of the individual case of cholera, on a scale of say 1 to 10, to the amount of Southwark and Vauxhall water drunk daily by the victim, and none would have made much sense. There was no reason then, nor is there one today, given what we know about the properties and reproduction rate of bacteria, the variable degree of individual natural immunity, etc., to expect any quantitative relation between the severity of the disease and the amount of polluted water drunk.

The outcome of Snow's experiment was presented in a simple table (Table 4-2, p. 47). Note that Snow's hypothesis did not exclude the possibility that *some* people living in houses supplied by the Lambeth Company might get cholera—there are too many other ways to catch the infection during an epidemic—nor did it require that *all* people living in houses supplied by the Southwark and Vauxhall Company must get the disease—some individuals may have used the water only for cooking, others may have been immune, or gotten cholera in a mild or unrecognized form.

Yet the experiment, in spite of the fact that the hypothesis being tested is not formulated in quantitative terms, is quite convincing.

But if we look closely at the table alone, without taking into account some additional information that Snow gave us, it might strike us that there is a weakness in the reasoning.

We note that the Southwark Company supplied about 40,000 homes and that the Lambeth Company supplied a somewhat smaller number (26,000). The number of cholera deaths in the homes supplied by Southwark was more than 12 times the number in the homes supplied by Lambeth. But the table alone does not tell us what the total number of people living in the two categories of homes was. What if the homes supplied by Southwark were in an overcrowded slum, so that the number of people living in those homes was 12 times the number living in homes supplied by Lambeth? If this were so, the 12-fold excess in deaths would merely reflect the 12-fold excess in population, and one would have to conclude that the water supply made no difference whatever.

It is only because Snow has told us that all the homes were in a single district of London, and that it was a matter of chance which water company supplied a particular home, that we can conclude that the rates given by Snow—cholera deaths per 10,000 homes—implicated the water of the Southwark Company.

A properly prepared table, when there are two groups being compared and when there are two possible outcomes for each individual in the groups, would list the numbers of individual subjects according to two criteria: group, and outcome. This means the table should show four numbers, as indicated below.

Number of individuals (1) living in homes supplied by Southwark and (2) who died of cholera = 1263	Number of individuals (1) living in homes supplied by Southwark and (2) who did *not* die of cholera = ?
Number of individuals (1) living in homes supplied by Lambeth and (2) who died of cholera = 98	Number of individuals (1) living in homes supplied by Lambeth and (2) who did *not* die of cholera = ?

Once these four numbers are known, certain subtotals and a grand total can easily be found.

An example of a correctly prepared four-fold table was given in the appendix to Chapter 9, which represents the division of groups of male and female students, according to whether they are or are not science majors:

	Science majors	Non-science majors	
Men students	13	9	Subtotal: Men students = 22
Women students	3	7	Subtotal: Women students = 10
	Subtotal: Science majors = 16	Subtotal: Non-science majors = 16	Grand total: Students = 32

Dichotomies

The all-or-none character of the cause and effect in Snow's experiment is intrinsic—there is no meaningful way to convert these particular factors into quantitative magnitudes. The same would apply to an experiment to test a drug for a disease which is often fatal. The outcome is either life or death: there is nothing in between. The technical term for such a factor or variable is a *dichotomized variable*. There are situations where a variable may be dichotomized for convenience, even though it is capable of being assigned a numerical measure. In a study of alcohol consumption we may, for the purpose of a preliminary study, divide people up into drinkers and nondrinkers. If we find interesting differences between the two groups, we may want to make further distinctions—looking for example at an individual's average daily alcohol consumption, or dividing people up into five categories of drinkers instead of just two.

A Common Mistake

As noted earlier, in many examples of argument or reasoning outside of a scientific context, the evidence offered for some hypothesis or belief is implicitly given in the form of a controlled experiment. Yet surprisingly often a fundamental mistake is made: the control group is missing.

Such a mistake is common when people are arguing a case for some dearly held political position, but it is not by any means limited to such situations.

The Women's Christian Temperance Union, an antidrinking organization that played a leading role in the fight for the adoption of the 18th Amendment to the Constitution, and other laws prohibiting the manufacture and sale of alcoholic beverages in the United States, often used such statements in its propaganda as "In 72% of the divorces in this country, the husbands drank." The figure given may well be correct, but as evidence for the implied hypothesis that drinking causes divorce, it is useless. No information is given about what proportion of husbands drank in those marriages that did *not* end in divorce. It may also have been 72%, reflecting the possibility that drinking has nothing to do with divorce; it may have been more, or it may have been less. For all we know, it may have been 90%, implying that, other things being equal, men who drank are *less* likely to get divorced, though we doubt it.

A More Subtle Mistake

Suppose the control group has not been omitted, and suppose further that only 30% of nondivorced husbands drank, the results would then have supported the WCTU's hypothesis. But supporting the hypothesis is not the same as providing completely convincing evidence for it. Certainly if the percentage of drinking husbands who do not get divorced was the same as the percentage who do, we could reasonably draw the negative (or "null") conclusion that drinking and divorce are unrelated. The difference between 72% and 30% is however large enough to permit us to draw the conclusion that there is *some* relation between drinking and divorce. But whether the relation is a *causal* one—drinking is a cause of divorce— is a more difficult question. For example, maybe the husbands drank because their marriages were already unhappy. Or as an alternative explanation, maybe people whose views on alcohol are tolerant are also likely to believe that divorce is an acceptable solution to an unhappy marriage. But to establish these alternative hypotheses as explanations of the data would require additional experiments and analysis.

The moral: experimental evidence does not *prove* theories, it only supports them; it adds to their credibility until such time as additional experimental evidence contradicts them.

Back to the Missing Control Group

Another example of a missing control group is provided by an article in the *New York Times* by Senator Charles Mathias of Maryland, in which he expresses his concern about what he feels are burdensome tax laws:[14]

> In 1940, Frank Perdue joined his father's chicken business on Maryland's Eastern shore. They had one employee but many good ideas and a lot of drive. Today, that family operation employs 3,800 people and markets millions of tender chickens. But Frank says that if he were starting out in business now,

he'd never make it: A fledgling operation stalls before it can shift out of first gear.

For every 100 new businesses that are set up this year, it is predicted that over 50 will fail within 12 months. We must improve these odds if the free-enterprise system is to survive.

Unless we adopt a tax policy that encourages the creation of new businesses and the expansion of existing ones, we may wake up to find our economic and social structure altered beyond recognition and beyond repair.

The 50% first-year failure rate of small businesses at the time Senator Mathias was writing sounds high, but he does not tell us what it was when taxes were much lower and Frank Perdue started his chicken business. Again, it may well have been 50%, and if so, the claimed causal relation between current high taxes and business failures is contradicted, rather than supported, by the evidence.

Literary critic Alfred Kazin, in an essay on the black writer James Baldwin begins as follows:[15]

> Recently, a scholar investigating the Negro novel in America discovered that of sixty-two Negro novelists writing between 1853 and 1952, forty published only one novel; eleven published only two; only eleven published more than two. Certainly one reason for this situation is the economic difficulties that so many Negro writers have had, a lack of encouragement from publishers and a lack of audience among Negroes as among whites. But surely another reason is that too often a writer turns to the novel not because his talents lie in fiction, not even because he wants to write fiction, but because he hopes to make his experience seem as individual and artistically realized as possible. (p. 254)

Kazin assumes the failure of most black novelists to publish more than one novel is evidence for unusual difficulties faced by black novelists as opposed to white ones. But we are given no information on how many white novelists also only publish one novel. It is our understanding that most novelists of any sex, color, or political persuasion usually succeed in publishing only one novel, and our guess that Kazin would recognize this if it were pointed out to him. In criticizing Kazin's reasoning, we are not denying that black novelists have had special problems not shared by white ones; we are only concerned with whether the figures given by Kazin as evidence for this really are evidence. Ironically, if the figures are viewed from a different perspective and using a different control group they are very good evidence indeed: during the hundred years for which the data were obtained, there must have been thousands or tens of thousands of white novelists, compared to only 62 black ones. Yet the black population of the United States during this period was about 10%–12% of the total.

The last example of this kind of mistake comes from the *New York Times*, reporting the results of some studies by Japanese physicians on hearing losses among Japanese teenagers using stereo headphone sets.[16]

> In 1980, Dr. Satoshi Koizumi, an ear, eye, nose and throat specialist at Jokushin General Hospital, about 125 miles northwest of Tokyo, conducted a survey of

the hearing ability of 4,500 students at six nearby high schools. Dr. Koizumi found 29 students had hearing difficulties for "unspecified" reasons. Of those 29, Dr. Koizumi said that 21 were "headphone addicts".

These 21 students told Dr. Koizumi that they listened to stereo headphones more than 24 hours a week at volumes that averaged 87.9 decibels. In contrast, the decibel level of street noise in downtown New York or Tokyo is 40 to 60 decibels. . . .

A similar study was done by Dr. Kunishige Kambe, who is an ear, eye, nose and throat specialist in Tachikawa City, a Tokyo suburb, and serves as a school doctor at Tachikawa High School.

Dr. Kambe surveyed 700 high school students and found 150 frequent headphone listeners among them. Of those 150, eight students had hearing difficulties. These hearing-impaired students listened to their stereo headphones an average of six hours a day.

The general hypothesis in both studies is that excessive use of the headphone sets can cause hearing losses, but the specific hypotheses in the two studies were different. Dr. Koizumi found 29 out of 450 students had hearing defects. Among these 29, 21 (or 72%) were frequent users of headphone sets. The implied hypothesis here is that the 72% of students with defective hearing owed their hearing problems to the use of the sets. But no information is given as to the percentage of *all* students, or the percentage of students without hearing losses, who use the sets. The fallacy is exactly the same as that committed by the Women's Christian Temperance Union—no control group. To make the point more clearly, we represent the results in a four-fold table (Table 12-2).

Dr. Kambe's study has a different hypothesis. He found 8 students with hearing losses out of 150 students who use headphone sets. This is 5% of the headphone users. But what is the percentage of hearing losses among students who do not use headphones? Is it also 5%? Again we are given no information about a control group. The results are displayed in Table 12-3.

Ironically, each of the two studies which the *Times* reported, and which lacked control groups, provides data which could have served as a control for the other study:

Dr. Koizumi found 29 students with hearing losses, out of 4500, or about 0.6%, while Dr. Kambe found 150 students using headsets, out of 700, or 21%. One should, however, be cautious about drawing conclusions here, as the two groups of students were from different cities in Japan, and we are not told whether

Table 12-2
Dr. Koizumi's Results as Reported in the *New York Times*

	Headphone users	Nonusers
Hearing loss	21	8
No hearing loss	?	?

Table 12-3
Dr. Kambe's Results as Reported in the
New York Times

	Headphone users	Nonusers
Hearing loss	8	?
No hearing loss	142	?

the criteria the researchers used to decide whether a subject did or did not have a hearing loss were the same in the two studies.

In the *Times* article it is not clear whether the omission of these two vital pieces of information was the fault of the doctors or of the reporter, but the reporter, had he been aware of the omission, would have had a responsibility to point it out.

Conscious or Unconscious?

In giving the above examples we do not mean to imply that there was a conscious attempt to deceive in all cases, only a carelessness in reasoning.

The arguments seemed to those who proposed them to prove a point, probably because they assumed tacitly and unconsciously that in the appropriate control group the results would have been quite different from those in the test group for which the data were given. For example, the supporters of the WCTU would have expected that among stable marriages there was much less drinking by husbands. As we noted, this may well be the case, but it is dangerous to assume it without evidence. Science is concerned with the world as it is, rather than the world as we imagine it to be.

REFERENCE NOTES

1. St. Augustine, *Confessions*, trans. R. S. Pine-Coffin (Harmondsworth: Penguin, 1961). Copyright 1961 by R. S. Pine-Coffin. Reprinted by permission of Penguin Books, Ltd.
2. Gerald J. Holton, *Introduction to Concepts and Theories in Physical Science*, 2nd ed. (Reading, Mass.: Addison-Wesley, 1973).
3. Francisco Redi, *Experiments on the Generation of Insects*, trans. Mab Bigelow (Millwood, N.Y.: Kraus Reprint, no date).
4. Herbert Butterfield, *Origins of Modern Science 1300–1800* (New York: Free Press, 1965).
5. Theodore Mehlin and Charles A. Schwieghauser, *Astronomy and the Origin of the Earth*, 3rd ed. (Dubuque: William C. Brown, 1979).
6. W. V. Quine and J. S. Ullian, *The Web of Belief*, 2nd ed. (New York: Random House, 1978).
7. Karl R. Popper, *Conjectures and Refutations: The Growth of Scientific Knowledge* (New York: Harper and Row, 1968).
8. Francis Bacon, *The New Organon and Related Writings*, ed. Fulton H. Anderson (Indianapolis: Bobbs-Merrill, 1960).
9. Karl R. Popper, *The Logic of Scientific Discovery* (New York: Harper Torchbooks, Harper and Row, 1965).

10. M. Terris, ed., *Goldberger on Pellagra* (Baton Rouge: Louisiana State University Press, 1964).
11. Caroline Spurgeon, *Shakespeare's Imagery and What It Tells Us* (Cambridge: Cambridge University Press, 1935).
12. Arthur M. Eastman, *A Short History of Shakespearean Criticism* (New York: W. W. Norton, 1974).
13. Samuel Schoenbaum, *Shakespeare's Lives* (New York: Oxford University Press, 1970).
14. *New York Times,* 16 March, 1979. © 1979 by The New York Times Company. Reprinted by permission.
15. Alfred Kazin, *Contemporaries, from the Nineteenth Century to the Present* (Boston: Little, Brown, 1962).
16. *New York Times,* 17 July, 1982. © 1982 by The New York Times Company. Reprinted by permission.

SUGGESTED READING

On the question whether scholarly research in the humanities is like scientific research, the reader is referred to the article by Isaiah Berlin, "The Concept of Scientific History," in *History and Theory* 1 (1960), for a negative view. The *Modern Researcher* by Jacques Barzun and Henry F. Graff (3rd ed., New York: Harcourt Brace Jovanovich, 1977) is a guide to research in history. It does not make any claim that history is a science, but the reader is free to compare the methods of historical research described there with those of science, and come to his or her own conclusion. Another example of a work of literary scholarship which brings out the parallels between scholarly and scientific research is John Livingston Lowes's *The Road to Xanadu: A Study in the Ways of the Imagination* (Boston: Houghton Mifflin, 1927).

PROBLEMS

Collect examples of arguments from books, newspapers, and magazines, where the evidence requires a control group, but none is provided. The following are some examples:

1. An article on motorcycle safety in the December 1980 issue of a journal called *The Nation's Health* begins as follows:

    > A study done for the National Highway Safety Administration has pointed up again the need for motorcyclists to wear safety helmets. In-depth investigation into 900 motorcycle accidents in the Los Angeles area found that 78% of the riders who were killed were not wearing helmets.

2. From the *New York Times* of October 31, 1980:

    > While a borrowed band from Columbia University blared "Who Owns New York" and "Happy Days are Here Again," Miss Holtzman stood on a Manhattan street corner yesterday and denounced discrimination against women.
    >
    > "One of every four women working today can expect to be poor when they retire," she said. There is a bias against women in Social Security, in wages, in pensions and in insurance, she said.

Assume the facts given are true.

(a) State briefly and clearly the hypothesis for which the facts are offered as evidence.
(b) State what a suitable control group would be for testing the hypothesis.

13

The Experimenter and the Experiment

THE UNCERTAINTY PRINCIPLE

We stated earlier that Newton's laws failed when applied to very light objects like electrons, and that quantum mechanics was developed to replace them. One of the basic principles of quantum mechanics is called the *uncertainty principle*. It states a limitation on our ability to measure anything we want to, with any accuracy we choose.[1]

Suppose we have an object, a bullet for example, known to be moving in a certain direction at a constant speed, and we want to find out two things about it at a certain time: (1) where it is at that moment and (2) how fast it is moving at that moment. The experimental procedure for doing this is simple to understand—we might imagine using high-speed photography, photographing the bullet with two flashes of light a known short time interval apart. The first photograph is sufficient to answer the first question; the distance traveled by the bullet between the two exposures answers the second.

Electrons can also travel through empty space at a constant speed, as the tube of a television set demonstrates, and we might imagine asking the same two questions about the moving electron that we did about the moving bullet. Indeed, the experiment can be tried in much the same way, using photographic film to detect the result of two flashes of light impinging on the moving electron.

However, we find ourselves in difficulties. We find that because of the small size of the electron the accuracy with which we can locate its position from the first light flash depends on the energy of the light used: the higher the energy of the light, the better the accuracy. But the light, in the process of being reflected from the electron, gives some of its energy to the electron, and the higher the energy of the light used, the more the electron is affected. The result is that while the first flash of light can locate the position of the electron to within certain limits set by the energy of the light, the electron's speed has been so perturbed by the first flash that the position of the electron on the second flash no longer reflects its original speed. The more accurately we try to determine the position, the higher the energy of light we must use, and in turn the less we know about what the electron's speed was before the experiment. Light has a large effect on the electron, but not on the bullet, because the electron is very small and does not weigh much, while the bullet is heavy. We find we cannot know the *speed* of the electron accurately once we have measured its *position* accurately.

Thus the result of our experiment is accurate knowledge of where the electron is, combined with poor knowledge of how fast it is moving. Now there are other, different experiments we could perform to gain accurate knowledge of the electron's speed. The details of how we might do this do not matter: what matters is that the very measurement that tells us the *speed* accurately deprives us of knowledge of the *position* of the electron. The more accurately we learn how fast it is moving, the less accurately we know where it is.

So we are forced to choose; we cannot know both position and speed. The two kinds of knowledge are mutually exclusive; the better we know one, the worse we know the other. And the uncertainty principle tells us there is no way out of this dilemma. It is a fundamental law of nature. *The very act of measuring changes the system in unpredictable ways.*

A Useful Metaphor—Interaction between Experiment and Observer

The uncertainty principle is a useful metaphor (and, strictly speaking, only a metaphor) for a problem that affects all of science to a greater or lesser degree. When we do an experiment, we tend to assume that the experiment is isolated from any influences not under our control, that the instruments used to measure and the experimenter making the observations are somehow "outside" the thing being studied, and not affecting it. In fact, such an idealization is often wrong. The person doing an experiment is part of the experiment. He may be affecting the outcome by his presence, by his preconceptions, by the technique of the experiment itself.

Sometimes the difficulty lies not in the way the experimenter influences the system being studied, but rather within his own mind. What he sees is influenced by his expectations, and he reads into the situation features that are not really there. Sometimes it is a simple matter of tending to make those errors in recording data that support his preconceived idea of what the results should be.

Whatever form the problem takes, it is widespread in science. It crops up in unexpected ways, and we must be alert to it.

The Smart Mice

An assistant to the Russian psychologist Pavlov once published some remarkable results showing the inheritance of acquired characteristics. He had found that when mice were trained to solve mazes, their children could solve them faster, and their grandchildren faster still. The effect continued to the fifth generation. The first generation required approximately 300 trials to solve the maze; successive generations required approximately 100, 30, 10, and 5 trials.

Pavlov later repudiated these results; apparently the assistant, over time, had unconsciously learned to train mice how to solve mazes faster. The fifth generation of mice were not really any smarter than the first. It was the experimenter who had changed.[2,3]

Placebo Pills in Drug Trials

In several of our case histories, we have discussed the concept of a controlled experiment: the idea of taking two groups of subjects or cases identical in all relevant features except one, the one whose effect on the experimental group we wish to study.

For example, we wish to test a certain drug for its effectiveness in relieving cold symptoms. We take a large number of people with colds, divide them at random into two groups, and give one group the drug. It would seem reasonable that if the group given the drug suffers less and recovers faster we have proved the value of the drug. However, this is not so. Trials of drugs are not usually made this way anymore, because the people given a drug will usually feel better, or *think* they feel better. They have been given a medicine and they will expect it to work; this expectation is enough to make them think they have been helped.

The two groups in the experiment were supposed to differ in only one way: one group has received the drug and one group has not. But in fact they differ in two factors, not one, a situation we may represent in Table 13-1. The difference in results of drug trials may often be due to factor (2) rather than factor (1). So where possible in drug trials the control group is given some harmless and inactive pills—sugar pills, for example—so that the two groups will not differ by factor (2).

For many years, doctors have dealt with patients with whom nothing discernible is wrong, yet who complain of various illnesses. They often find honesty with the patient does not pay—telling a patient there is nothing wrong with him sometimes leaves him dissatisfied and mistrustful. It has been found easier if not more ethical to prescribe some harmless medication. Often the patient feels better—he might have anyway—and at least he feels that the doctor has taken him seriously. A "medicine" of this type is called a "placebo," from the Latin word *placēre,* "to please." The curative effect of the inactive pill in the "controlled" experiment, when it has no real curative power, is called the "placebo effect."

Obviously, since mice make no distinction between medicine and food, it is not necessary to use a placebo to avoid this particular source of error when using them in an experiment.

Table 13-1
Group Differences

Test group	Control group
1. Has received a drug	1. Has not received a drug
2. Knows it has received a drug	2. Knows it has not received a drug

BLIND AND DOUBLE-BLIND EXPERIMENTS

There are other sources of error that might arise in such an experiment even if a placebo is given to the control group. In most tests of drugs it is not sufficient to rely on the patient's subjective feelings as to whether he is better or not. More often a physician is needed to evaluate by professional criteria whether improvement has occurred in one group and not the other. When the evaluation requires some simple objective test, such as measuring the amount of sugar in the urine, there is not much of a problem. But often the evaluation of improvement in a patient's disorder requires a subjective judgment on the part of a physician—for instance, have the schizophrenic patients who received drug A improved more than the controls who received a placebo?

Such an evaluation requires paying attention to many details of a patient's behavior and balancing seemingly contradictory information to obtain a single judgment: yes, there has been improvement; no, there has not. In such a delicate procedure there is a risk that the physician's bias may cause him unconsciously to judge patients whom he knows have received a new experimental drug differently from those whom he knows have received a placebo. If he is personally biased in favor of the drug, he may lean one way; if he is biased against it—prefers some tried and tested drug he has had success with to this new and dubious one—he may lean the other.

Again, the two groups differ in more than one factor (Table 13-2). Again, a difference between the results of the tests on the two groups may be due to factor (2). Thus it is preferable that not only the patients, but also the evaluating physicians, be ignorant of who has received the drug and who has not. Such an experiment is called a "double-blind" experiment.

Unfortunately, it is not always possible to keep patients and evaluating phy-

Table 13-2
Group Differences

Test group	Control group
1. Has received a new drug	1. Has not received a new drug
2. Symptoms evaluated by a physician who knows the patients have received a new drug	2. Symptoms evaluated by a physician who knows the patients have not received a new drug

sicians ignorant of the treatment used. For example, if we wish to compare the relative effectiveness of surgery and drug therapy for some form of cancer, both patient and doctor will know. However, when the blind or double-blind design can be used, it is a better procedure. It was used in the experiments to evaluate possible genetic factors in schizophrenia. The psychiatrists who had the job of diagnosing whether the children or parents of schizophrenics were also schizophrenic were kept in ignorance of whether the individuals they examined were or were not blood relatives of schizophrenics.

The Lively Flatworms

The way in which the expectations or biases of an experimenter can influence the results he thinks he observes has been studied extensively by the psychologist Robert Rosenthal.[4] He describes one experiment, for example, where undergraduate biology students were asked to observe how many times members of a species of flatworms turned their heads or contracted their bodies. The students were led to believe that one group of worms was expected to move a lot, while a second group was expected to move only a little. In fact, the two groups of worms were identical. "[The] observers reported twice as many head turns and three times as many body contractions when their expectation was for high rates of response as when their expectation was for low rates of response."

Mental Telepathy

The results of the above experiment depended on the fact that the experimenter had to make a subjective evaluation of whether a worm did or did not, at a certain time, move its head. But the problem is present even for experimental operations that do not seem to depend on subjective judgments, such as recording the results of experimental observations. Clerical errors will always occur when results have to be written down. But, as is well known to anyone who is in the habit of checking the addition of sales clerks in stores, the errors do not occur at random. In experiments on mental telepathy, where one person, the subject, had to guess which of several symbols a second person was concentrating on and attempting to transmit telepathically, a third person recording the guesses of the subject, *and knowing which symbol the second person was "transmitting,"* tended to make errors in recording the result that depended on whether he himself believed in mental telpathy or not. If he was a believer, the errors in recording tended to increase the score of the guesser, while an unbeliever tended to make the opposite kind of error.

The Clever Horse

Another example given by Rosenthal is the following:[4]

> Probably the best-known and most instructive case of experimenter expectancy effects is that of Clever Hans [studied by the German psychologist Pfungst].

Hans, it will be remembered, was the horse of Mr. von Osten, a German mathematics teacher. By means of tapping his foot, Hans was able to add, subtract, multiply, and divide. Hans could spell, read, and solve problems of musical harmony. To be sure, there were other clever animals at the time, and Pfungst tells about them. There was "Rosa," the mare of Berlin, who performed similar feats in vaudeville, and there was the dog of Utrecht, and the reading pig of Virginia. All these other clever animals were highly trained performers who were, of course, intentionally cued by their trainers.

Mr. von Osten, however, did not profit from his animal's talent, nor did it seem at all likely that he was attempting to perpetrate a fraud. He swore he did not cue the animal, and he permitted other people to question and test the horse even without his being present. Pfungst and his famous colleague, Stumpf, undertook a program of systematic research to discover the secret of Hans' talents. Among the first discoveries made was that if the horse could not see the questioner, Hans was not clever at all. Similarly, if the questioner did not himself know the answer to the question, Hans could not answer it either. Still, Hans was able to answer Pfungst's questions as long as the investigator was present and visible. Pfungst reasoned that the questioner might in some way be signaling to Hans when to begin and when to stop tapping his hoof. A forward inclination of the head of the questioner would start Hans tapping, Pfungst observed. He tried then to incline his head forward without asking a question and discovered that this was sufficient to start Hans' tapping. As the experimenter straightened up, Hans would stop tapping. Pfungst then tried to get Hans to stop tapping by using very slight upward motions of the head. He found that even the raising of his eyebrows was sufficient. Even the dilation of the questioner's nostrils was a cue for Hans to stop tapping.

When a questioner bent forward more, the horse would tap faster. This added to the reputation of Hans as brilliant. That is, when a large number of taps was the correct response, Hans would tap very, very rapidly until he approached the region of correctness, and then he began to slow down. It was found that questioners typically bent forward more when the answer was a long one, gradually straightening up as Hans got closer to the correct number. . . .

Pfungst himself then played the part of Hans, tapping out responses to questions with his hand. Of 25 questioners, 23 unwittingly cued Pfungst as to when to stop tapping in order to give a correct response. None of the questioners (males and females of all ages and occupations) knew the intent of the experiment. When errors occurred, they were usually only a single tap from being correct. The subjects of this study, including an experienced psychologist, were unable to discover that they were unintentionally emitting cues. (pp. 129–130)

Clever Hans Reborn?

In recent years attempts have been made to teach apes the use of a language with which to communicate with human beings. Both sign languages and keyboards, on which the keys are marked with word symbols, have been used, and exciting successes with chimpanzees and gorillas reported.

Some scientists have criticized these experiments on the grounds that the experimenters have been misinterpreting what they were observing: the apes, like Clever Hans, have learned tricks rather than language, and those who have tried to teach them have been misled by their own expectations and hopes. This skeptical analysis has, however, been challenged, and the question of whether the apes have learned to "talk" remains controversial.[5,6]

INTERVIEWERS AND INTERVIEWEES

One of the obvious ways of finding out what people think about certain issues, or what they do in certain situations, is to go out and ask them. However, the process of being interviewed by a stranger is not an entirely neutral process, especially when the information sought has emotional overtones for the person being interviewed. One can imagine many questions which people are likely to answer with less than complete honesty, because of pride, fear, or other factors: how much they earn, whether they drink or take drugs, questions about sexual matters, and so on.

It has been found, for example, that people answer differently, and probably less honestly, to a live interviewer than to an impersonal questionnaire. Twice as many people answered "yes" to the question "Have you frequently suffered from constipation?" on a questionnaire than in face-to-face contact. It has been found that blacks may answer the questions of a white interviewer differently from the way they answer the same questions asked by a black interviewer. In one study in Detroit, when blacks were asked if they trust most white people, white interviewers found 35% answering yes, black interviewers found only 7%.[7]

RUMFORD'S MISTAKE

Rumford's early view of heat as a vibration of a body, with different temperatures reflecting different frequencies of vibration, led him to the idea that both heat and cold could be converted to a vibration of the ether, and hence both heat and cold radiation should occur.

He was able to demonstrate this to his own satisfaction by an experimental arrangement, originally developed by a Professor Pictet, similar to that shown in Figure 13-1. A body, B, was placed at the focus of a metal reflecting-mirror, which would concentrate any rays radiating from B on the point T, where Rumford placed a thermometer with a blackened bulb. When B was hot, he found that the temperature of the thermometer rose, demonstrating that B was radiating heat.

Similarly, if B was a cold body—specifically, he used a block of ice—the temperature recorded at T fell. He concluded reasonably that cold was radiated as well as heat.[8]

In the modern view Rumford was wrong: heat is a form of energy and energy can be radiated. Cold is only the absence of heat; it is not a separate entity and cannot be radiated. Why then does the temperature at T fall when B is a block of ice?

Rumford made what we would consider today an artificial distinction between

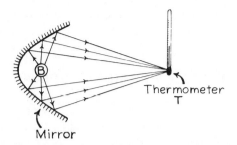

FIGURE 13-1. Rumford's experiment proving cold can be radiated as well as heat. A hot body *B,* placed close to a mirror that concentrates its radiation on the thermometer *T,* causes the temperature recorded by the thermometer to rise. A cold body placed at the same point causes the temperature to fall. In the modern view, the thermometer's temperature falls not because the cold body is radiating cold, but because the thermometer is radiating heat. The arrows are pointing the wrong way. The cold body is radiating heat also, but because it is colder than the thermometer it is radiating less.

the body *B* and the radiation from it on the one hand, which he regarded as the *objects* on which he was performing an experiment, and the curved mirror and thermometer on the other hand, which he considered *neutral instruments* that are used to study what happens to the *objects*.

But the distinction between *object* and *instrument* is unreal. The thermometer is not only a body that measures temperature, but also a body that has a temperature of its own, and is capable of radiating heat no less than body *B*. In fact, any body, no matter whether hot or cold, is always radiating *heat*. When a hot body and a cold body are exposed to each other across an empty space, the cold body warms up not because it is receiving heat rather than radiating it, but because, although it is radiating heat, it gains more from the hot body than it is losing.

So the thermometer cooled down when Rumford put the ice at *B* because it was radiating more heat to the ice than it was receiving from it. The point is not a subtle one, nor is it based on knowledge not available to Rumford. William Wells, a British physician and scientist, and an admirer of Rumford, in his *Essay on Dew,* published in 1814, interpreted the cooling of the ground at night in just such terms, rejecting Rumford's belief in "frigorific" (cold-making) radiation.[9]

THE SELF-FULFILLING PROPHECY

One test of good scientific theory is that it can predict correctly what will happen next. But this test should be applied with care. An economist might conclude, on the basis of a careful analysis of the prospects and functioning of a certain corporation, that it is in poor shape, and the price of shares of stock in it is likely to fall. If the prediction becomes common knowledge to people who speculate in the stock market, the price of shares is indeed likely to fall, but not because the economist's evaluation of the corporation was necessarily sound. It is the same thing when the rumor of a run on a bank causes a run on the bank. The fact that a prediction turns out to be correct does not always provide evidence for the correctness of the hypothesis that led to it.

The scientist who tries to understand the world is himself part of that world, and his presence in it may interfere with his understanding of it. It is a possibility one must be aware of.

REFERENCE NOTES

1. Isaac Asimov, *Asimov on Physics* (New York: Avon Books, 1978). See also Adolph Baker, *Modern Physics and Antiphysics* (Reading, Mass.: Addison-Wesley, 1970).
2. C. Zirkle, "Pavlov's Beliefs" (Letter), *Science* 128 (1958): 1476.
3. G. Razran, "Pavlov the Empiricist," *Science* 130 (1959): 916.
4. Robert Rosenthal, *Experimenter Effects in Behavioral Research* (New York: Irvington, 1982).
5. Thomas A. Sebeok, and D. J. Umiker-Sebeok, eds., *Speaking of Apes: A Critical Anthology of Two-Way Communication with Man* (New York: Plenum Press, 1980).
6. Thomas A. Sebeok, and Robert Rosenthal, eds. "The Clever Hans Phenomenon: Communication with Horses, Whales, Apes, and People," *Annals of the New York Academy of Sciences*, vol. 364 (1981).
7. Seymour Sudman, and Norman M. Bradburn, "Response Effects in Surveys: A Review and Synthesis," *National Opinion Research Center Monographs in Social Research Series*, no. 16 (1974).
8. Count Rumford, "The Nature of Heat," in *The Collected Works of Count Rumford*, ed. Sanborn Conner Brown, vol. 1 (Cambridge: Harvard University Press, 1968), pp. 360–364.
9. William Wells, *An Essay on Dew and Several Appearances Connected with It* (London: Printed for Taylor and Hessey, 1814).

SUGGESTED READING

Rosenthal, Robert. *Experimenter Effects in Behavioral Research*. New York: Irvington, 1982.

14
Measurement and Its Pitfalls

MEASUREMENT AND SCIENCE

By now we have given enough examples of measurements of quantities—such as numbers of cholera cases per 10,000 houses, temperatures, weights of hot and cold bodies, rates of mental disorders, and cancer in different places—to make it apparent that measurement plays a central role in science. While not every scientific fact is a numerical one, nor every scientific theory a predictor of numerical magnitudes, so many are that we cannot conceive of modern science without the process of measurement.

Along with a respect for the power of quantitative methods, the scientist must also have a critical sense about them: an awareness of the limitations of measurement, and of the errors to which it is subject. We have already given examples in our case histories of the operation of this critical sense. We have described the careful attention paid by Rumford to the various sources of error affecting his measurement of the weight of heat, and how he designed his experimental procedure to eliminate some sources of errors and minimize others. We made the point also that the weighing instrument used by Rumford had a certain limit of sensitivity—the smallest change of weight the instrument could detect—and that Rumford should really have concluded that the weight of the heat lost by the water when it froze was less than this limit. We have also described how the enormous differences in the measured rates of schizophrenia and depressive disorders between the United States and Great Britain did not reflect reality, and the painstaking way this was demonstrated by the Project scientists.

In what follows, we would like to give some examples of the many ways in which a measurement can give a wrong or misleading answer, beginning with a discussion of two different kinds of questions that can be asked about the results of an attempt to measure something.

Reliability and Validity

In Chapter 8 on mental disorders, in discussing the procedures used for diagnosing the various psychoses, we considered two questions:

1. Are the procedures *reliable?* Do different psychiatrists using them on the same patient tend to agree most of the time?
2. Are the procedures *valid?* Do the diagnostic procedures correspond in some sense to what we believe to be the "truth" about the patient and the disorder?

The same questions can be asked of the results of measurement generally, except that, when the result of some observation is expressed by a numerical answer, we use the terms *precision* and *accuracy* (see Figure 14-1) instead of *reliability* and *validity*. The concepts are equivalent.

Precision

In learning arithmetic at school, we work with numbers that are exactly known. When told that a certain room is 4.21 meters wide and 6.83 meters long, and asked the area, we unhesitatingly multiply the two numbers to obtain 28.7543 square meters. If we were asked whether the length of the shorter side should be written as 4.21 or 4.2100 meters, we would think the question pointless.

In the real world, when we measure something, we are aware through experience that the result is subject to some uncertainty. When we measure the length or width of a room, for example, we lay a meter stick down touching one wall, make a pencil mark on the floor corresponding to 1 meter, then slide the stick over

FIGURE 14-1. Accuracy versus precision. The two ducks on the top have been shot at with poor precision, the two at the bottom with good precision. The two on the left have been shot at with good accuracy, the two on the right with poor accuracy.

and repeat the process as often as needed. We are not surprised to find that if the measuring process is performed again later, or by someone else, the results may differ a little. The first measurement may give 4.21 meters; a second or third measurement may give 4.25 or 4.19. The best we can get from a series of measurements is some range of values rather than an exact answer. When we take the average of these values, we are making the most reasonable estimate possible of the length on the basis of the data we have. The most we can expect, however, is that the true value of the length probably lies close to the average, somewhere within the range of values found.

The range of values we have found from our series of measurements is related to the *precision* of the measurement: the smaller the range, the greater the precision.

Random Errors

In the above we have assumed our errors are *random*. Since the probability of a positive error is the same as the probability of a negative one, the result is that in a long series of measurements, positive and negative errors will appear about as often, and in calculating the average, their effects will tend to cancel. This assumption is not always justified, and when it is not the measurement is said to be *biased*. *Bias* is discussed in the next section.

To analyze the effect of random errors on a measurement we must make some further assumptions about the errors, beyond their tendency to cancel. Experience has shown that in most situations, large errors are less probable than small ones, and very large errors are extremely improbable. In Figure 7-3 (p. 155) a mathematical curve called the Gaussian curve was illustrated, which represented the probability of getting a particular outcome when 100 coins were tossed. A graph displaying the results of repeated measurements of some quantity will often resemble this curve. The tossing of 100 coins can be metaphorically regarded as a measurement of some quantity whose "true" value is 50. Sometimes the measurement gives 50 as the answer, more often it does not. The average number of heads obtained from a series of tossings of a hundred coins approaches the "true" answer of 50 the more times we "measure" the quantity.

In the appendix to this chapter we discuss precision in more quantitative terms, using these ideas about how random errors are distributed.

Accuracy

Accuracy is our estimate of how close the value we measure might be to the true value. The word *estimate* in the preceding sentence is the critical one. If we *knew* the true value, it would be pointless to do an experiment to measure it. The only reason we measure quantities is to learn their true values. Yet we recognize that our measurement may not give us this true value for many reasons.

The concept of accuracy involves much more in the way of subjective judgment than that of precision. Precision can be determined in a straightforward way by performing the measurement several times. Accuracy however is associated with independent knowledge of possible flaws in our measuring procedures. By overlooking such flaws we will mislead ourselves about our accuracy.

For example, a person choosing to take pains with the measurement of the room, making the pencil marks and lining the meter stick up with great care, making sure the meter stick is kept parallel to one wall of the room, and so on, may succeed in achieving a precision of a few millimeters—the measured values all lying within such a range about the average. But meter sticks, if made of wood, are likely to have undergone some shrinkage or expansion since manufacture, and they undergo further transitory shrinkage or expansion as the local humidity and temperature vary. The effect of such changes may be that any one wooden meter stick, if compared with the standard for the length of the meter, may prove to be a millimeter or two longer or shorter than the true meter. If such a stick is picked up and placed down four times, the cumulative error could make the apparent length 4 to 8 mm longer than the true one. The *precision* of the measurement found by the careful observer may have been 2 millimeters, but someone aware of the properties of wooden meter sticks will know that the result should not be trusted to that degree of fineness.

The measurements made with the shrunken meter stick are said to be *biased*. They tend to give a result whose difference from the true value lies consistently in one direction—in this case greater than the true value. As long as we know the bias of our measuring instrument, we can correct for it to produce an unbiased measurement, or at least a measurement unbiased by this particular source of error. But we are not always so fortunate.

Of course, steel meter sticks are not subject to the dimensional changes of wood, but their lengths do change with temperature to a small extent. They are more accurate, but not perfectly so.

The meter stick provides one example of how we estimate accuracy; having reason to consider a stainless-steel meter stick more accurate than a wooden one, we compare the results of measurements made with wooden sticks to the "true" value obtained with a steel one. This done, we know something about the limits of accuracy of the wooden sticks. The accuracy of a steel stick can be determined, in turn, if we know of some other more trustworthy method of measuring length.

However, we are more often in the situation where the exact value of what we wish to measure is not known to us. How far will the moon be from the earth at midnight tonight? What is the weight of an atom of gold? How fast does light travel in a vacuum? All of these are quantities that can be measured, but we must always be aware that the method of measurement may for one reason or another be subject to errors, and that there is no "right" answer to look up, nor an infallible measuring procedure that can be used for comparison.

When to Stop

While accuracy in measurement is a desirable goal, one should not conclude that the main concern of science is to measure all things with greater and greater accuracy. One important component of scientific judgment is to know when additional accuracy is worth the additional trouble it takes to get it.

If the purpose of a measurement, for example, is to test a theory, the question should be asked: how accurate do we expect the predictions of the theory to be? Most scientific theories are known to be simplified approximations to a complex reality; sometimes we expect them to predict very closely indeed, but other times only roughly.

If we estimate that the best to be expected of some theory applied in a given situation is that it will predict observable quantities to within 5%, it would be foolish to go to great trouble to measure them to 0.1%.

The Fall of a Leaf

Newton's laws give the following formula for the speed, s, of a body that has fallen a distance, h, in the earth's gravitational field:

$$s = 4.43\sqrt{h}$$

(s is the speed of fall in meters per second and h the distance fallen in meters). The law, as applied here, ignores various sources of error, such as air resistance, and the fact that the force of the earth's gravity depends on the height of the body above the earth. It is therefore accurate only for bodies that are falling near sea level and from not too great a height, and whose weight relative to their size is large enough to minimize the importance of air resistance.

From knowledge of the magnitudes of the errors that result from neglecting these factors, we might estimate that in one experiment the law should be obeyed to within 1%, in another, to within 10%, and in a third, say, the falling of a leaf on a windy day, it should not be obeyed at all.

The Point of Diminishing Returns

Quite generally, quantities in nature usually have some natural limit to which it is worthwhile to measure them, some point beyond which additional accuracy of measurement adds nothing to our real knowledge.

An example is the pulse, the rate at which the heart beats. It is measured by counting the number of heartbeats in a certain time interval. The pulse rate is given in beats per minute, but it is not necessary to time it for 1 minute to measure it. One can count for 20 seconds and multiply the result by 3. But, if we do this, our answer can only be a multiple of 3—we can have a pulse rate of 66, 69, 72, 75, and so on, but not 67 or 73. Obviously, if our true pulse rate were 73, we could know it only by counting for a full minute, not 20 seconds. But must the answer be a whole number? If we counted for 2 minutes and counted 145 beats in that time, the pulse rate would be 72.5 beats per minute. That is a result a one-minute count will never give us. So we conclude that the longer we count, the more accuracy we can get.

But no doctor spends a whole minute taking someone's pulse. The reason is that what a doctor cares about is not whether the pulse rate is 72 or 72.5, but rather whether it is 72 or 90. The difference between 72 and 72.5 can be measured, but is of no great significance. A person getting up from a chair to get a drink of water may cause his pulse rate to increase by 3 to 5 beats per minute or more. We can therefore measure a pulse more accurately than it is worthwhile bothering to do; the additional accuracy is meaningless. We only want to know the pulse rate to within a few beats per minute, and a 20-second count gives us that.

Counting

There is one exception to the rule that states that measured quantities are always uncertain: numbers obtained by counting something. If we find 27 cents in our pocket, we do not mean that we are not quite sure whether we have 26.8 or 27.2 cents. And we are sure, if we have counted carefully, that there is no chance we really have 26 or 28 cents.

But even counting is not always so simple. The number of people living in a large city, for example, is never known with certainty, even though it is a number which we try to measure by counting everyone. Census-takers try to visit every home, but there are always some they miss. There are always some people in the process of moving out or moving in, and others dying or being born. Whatever the number is, it is always changing, and in ways that are hard to estimate. Further, the census-takers make mistakes. So do the people who record the data gathered by the census-takers on punch cards. A count that purports to give the population of a large city as, say, 1,719,463 people cannot be as reliable as it sounds, even though it is the result of a census. We should be content with a figure of 1,720,000 with an error estimate, say, of 10,000 people, more or less, an error of $\pm 0.6\%$.

How to Fool People

Figures that are specified very precisely, such as the population of 1,719,463, carry a connotation of exactness that makes them more credible to the unsophisticated. This is a fact known to every advertising agency. The statement, "63.2% of doctors in our survey smoke Marigold cigarettes," may be true. The survey may, however, have covered 19 doctors, of whom 12 said they smoke the brand in question. The figure 63.2% gives a spurious air of scientific exactness. If one less doctor of the 19 smoked Marigolds the figure would have been 57.8%, and one more would have made it 68.4%. The statement that 58–68% smoke Marigolds would have been more honest, but less impressive, and not so honest as admitting that only 19 doctors were surveyed. One does not need an outright lie to mislead.

Outright lies in impressive-sounding statistics are not unheard of. There are often major political or economic stakes in the value of some numerical quantity. It is always worth asking: Does the individual, agency, or institution reporting a figure have a personal stake in its magnitude? If so, there is always a possibility of deliberate distortion. Examples might be the profit figures that determine the taxes of a large corporation, the production figures in a collectivized economy of an industry whose directors have been given certain production goals to meet, or the death rates from cholera in a backward country deriving a large share of its income from tourism.

How to Fool Oneself

The Speed of Light

The subjective nature of an accuracy estimate can be illustrated by an example from that most exact of the exact sciences, physics. The speed of light in a vacuum, symbolized by the letter c, is one of the most important quantities in physics. Light is only one kind of electromagnetic radiation; radio waves, X-rays, ultraviolet light, and radar are others, and all travel with the same speed. Einstein showed that this speed is the upper limit at which *anything*, not just electromagnetic radiation, can travel. Furthermore, through the familiar equation $E = mc^2$, it determines the relationship between the mass, m, of a body and its energy, E.

A. A. Michelson (1852–1931), one of the most eminent experimental physicists of his time and the first American Nobel laureate, devoted the last few years of his life to an attempt to measure the speed of light more accurately than he and others had previously done. He died before the experiment was completed, but his work was continued by his collaborators, who published the results of almost 3,000 separate determinations in 1935. The average of their determinations was 299,774 kilometers per second.

Their accuracy was evaluated by another eminent physicist, R. T. Birge, an authority on measurement and errors, who studied their techniques and data and concluded that their measurement was probably within 4 kilometers per second of the correct value.

A few decades later, the development of radar during the Second World War permitted more accurate determinations of the speed of light. Michelson's value was found to be 16 kilometers per second low. Michelson's accuracy thus proved much poorer than Birge's estimate. No satisfactory explanation for the error in measurement of Michelson and his associates has ever been found; even with hindsight it is not easy to criticize Birge's optimistic estimate of Michelson's accuracy.[1,2,3]

The Crime Problem

The error in Michelson's measurement was 50 parts per million. While achieving an accuracy greater than this is a matter of great concern to physicists, those scientists who wish to measure the activities of human beings have to be satisfied with much less. We can illustrate some of the problems of measurement in the social sciences by a consideration of crime statistics.

The primary source of data on crime is the local police. It occurred to a presidential commission on law enforcement to wonder what proportion of crimes might not be reported to the police, and hence never appear in the crime statistics.[4] A survey was carried out in several large cities in the United States by asking large numbers of individuals, chosen at random, (1) if they had ever been victims of crimes, and (2) if they had reported these crimes to the police. The results were startling: only one-third of the burglaries, less than one-third of the rapes, one-half of the assaults, and two-thirds of the robberies were reported. In certain districts of these cities, the proportion of unreported crimes was even larger. The people questioned in the survey gave various reasons for failing to report: among them were a belief that the police would not be able to do anything anyway, a fear of reprisals, and, in the case of rape, shame. Not all crimes are under-reported however: thefts of things covered by insurance, such as automobiles, do appear to be represented with fair accuracy in police figures, and might possibly be exaggerated.

It is obvious that what we call crime statistics are, at best, really statistics of *reported* crimes, with their relation to statistics of *actual* crimes problematical. It is known that people's attitudes to reporting crime may depend not only on their social class, ethnic group, and so forth, but may also change with time. There is reason to believe, for example, that people who live in the slum areas of large cities are expecting more of the police than they used to, and report crime more frequently than in the past. This alone could cause the apparent crime rates in such areas to increase.[4]

New York versus Chicago. Even if crime statistics are interpreted as *reported* crimes rather than real ones, they might still be in error. In 1935, Chicago, a city with less than half the population of New York City at that time, had eight times the number of robberies as New York, and the reputation of being a crime-ridden city. However, F.B.I. officials studying the police system in New York felt that the local precincts in New York were not reporting to the central police authority all the crimes reported to them. In 1950, New York City adopted a new system that required the immediate report of all crimes to a central office. In one year the number of robberies rose 400% and the number of burglaries rose 1,300%, surpassing Chicago's rate for these crimes.

In 1960 Chicago adopted a central reporting system similar to New York's. Its "crime rate" immediately jumped ahead of New York's, and has remained ahead ever since.[4]

Classification Again. There are other problems in the meaning and validity of crime statistics arising from the legal definitions of various crimes. The distinction

MEASUREMENT AND ITS PITFALLS

between grand larceny and petty larceny—crimes bearing different legal penalties—is arbitrarily set by the value of the goods stolen: the legal dividing line varies from state to state, but once was commonly $50. Obviously, in a time of inflation, the number of larcenies classified as "grand" committed each year will increase as the price of portable radios or leather jackets increases, with no real increase in the numbers and kinds of actual thefts. It would seem to make sense to change the legal dividing line from time to time because of inflation, but when the legal dividing line is raised from $50 to $100, one must not be surprised if the rate of grand larceny suddenly decreases.

Burglary used to be defined as the act of entering a house illegally for the purpose of committing a felony, but many states have extended the legal definition to cover business establishments also, and a few have added ships and airplanes. Obviously, burglary rates in different states cannot be compared unless they are all drawn up according to the same definition of the crime.

Rape has been legally defined to cover two quite different acts: (1) forced sexual relations; and (2) sexual relations with a girl under a statutory age. Not surprisingly, the statutory age varies from state to state.

The Teenage Widowers

Some numerical magnitudes measured with reasonable care by honest and disinterested observers may not reflect reality because of mistakes in recording or transcribing data. We have mentioned one such example in Chapter 13, where observers recording guesses in mental telepathy experiments were found to make errors in recording that supported their own preconceptions about mental telepathy.

Another example of a recording error, this time not resulting from any bias on the part of the person doing the recording, is "the case of the teenage widowers" described by A. J. Coale and F. F. Stephan,[5] who noticed that according to the 1950 United States census there were 1,670 14-year-old boys who were widowers. This struck Coale and Stephan as an unreasonably large number of widowers at that age, especially since there were only 6,195 married 14-year-old boys reported in the census whose wives were living. Boys known to be married at age 14 could not have been married long—a year, at most—and most of them would be married to quite young girls. Why should about one-fifth of the wives have died? Coale and Stephan offered a very plausible explanation of this bizarre number. Data on each individual covered in the census are recorded on a punch card. The card has some 70 columns, each of which can be punched in one of twenty different places, to classify the individual according to various characteristics of interest for census purposes—e.g., age, marital status, race, income, type of dwelling. If the keypuncher, while transferring the data gathered by the census-taker to the card, punches a hole by mistake one column to the right of the appropriate one, it converts the individual's actual classification to something completely different. Specifically, white male heads of households aged 42 years would be converted to 14-year-old widowers. Given the large number of such men, a very small number of erroneous punches (about 2 per 1,000) could have produced the 1,670 teenage widowers.

The Age of Computers

Although clerical errors have always been with us, the consequences of an error in the age of the computer can be far more overwhelming. While some safeguards against errors can be built into computer programs, one must recognize that no amount of technological sophistication can replace common sense and sound judgment.

The Bulgarian Pigs

Our last example is one discussed by the economist Oscar Morgenstern:[6]

> A particularly nice illustration of how the time element can play tricks with statistics, showing at the same time limitations of complete counts (as compared with sampling procedures), is provided by the following case discussed by the late Oskar Anderson: "According to the census of January 1, 1910, Bulgaria had a total of 527,311 pigs; 10 years later, according to the census of January 1, 1920, their number was already 1,089,699, more than double. But, he who would conclude that there had been a rapid development in the raising of pigs in Bulgaria (a conclusion that has indeed been drawn) would be greatly mistaken. The explanation is quite simply that in Bulgaria, almost half the number of pigs is slaughtered before Christmas. But after the war, the country adopted the 'new' Gregorian calendar, abandoning the 'old' Julian calendar, but it celebrates the religious holidays still according to the 'old' manner, i.e., with a delay of 13 days. Hence January 1, 1910 fell after Christmas when the pigs were already slaughtered, and January 1, 1920, before Christmas when the animals, already condemned to death, were still alive and therefore counted. A difference of 13 days was enough to invalidate completely the exhaustive figures." A time series of such counts would show a sharp kink and remain high until the celebration of Christmas was also adjusted. Incidentally we should be impressed by the power of the Bulgarian (or any other) government to count every little piglet right down to the last of seven digits for one and the same day. And how important it must have been that the last digit was a 9, not an 8 or 7 or anything else! (pp. 47–48)

APPENDIX 14-1: SIGNIFICANT FIGURES

Numbers of centimeters, widowers, and Bulgarian pigs can all be measured with limited precision, and the specification of the limits is part of the measurement.

Take length, for simplicity. Saying that some object has been found to be 15 centimeters long does not tell the whole story. We need to know if it is 15 centimeters rather than 16, or 15 rather than 15.001. The accepted way to indicate this is by means of a decimal point and zeros.

15. cm, according to this common scientific practice implies some uncertainty in the last figure given, i.e. the 5. It implies that the length is more likely to be 15 centimeters than 16 or 14.

15.0 implies an uncertainty in the first place after the decimal. The measurement has been carried out in such a way that we are sure that it is very close to 15, and we think it more likely that the length is 15.0 rather than 15.1 or 14.9. We are virtually certain that it is not 16 or 14.

15.000 cm implies that we have made the measurement with such care that we think it more likely that the length is this value rather than 15.001 or 14.999. We are virtually certain that it could not be 15.1 or even 15.01.

The term "significant figure" is used here. The first measurement is said to have 2 significant figures, the second 3, and the third 5.

When two or more measurements are added, we must not claim more accuracy than we have in the least accurately measured quantity. Suppose we found the weight of a sample of salt is 2.347 grams, placed the sample in a glass vessel, and then added 15. grams of water. It would be misleading to claim that the total weight added is 17.347 grams. Saying so would imply that we are uncertain only about one-thousandth of a gram: we would be claiming that the added weight is more likely 17.347 than 17.346 or 17.348. But the 15. grams of water is known only to the nearest gram: all we can say is that its weight is likely to lie in the range 14. to 16. That considerable uncertainty washes out the precision with which the salt was weighed. It would be more honest to forget the numbers following the decimal point and call the total weight added 17. grams.

It is clear that the uncertainty in the answer for a sum of measurements is determined by the measurement with the least number of decimal places. As far as pure arithmetic is concerned, $16.103 + 4.9 + 6.02394 = 27.02694$. But if these numbers represent measured quantities, then since 4.9 is the number known to the least number of decimal places, the sum should be rounded off to 27.0.

There is a problem with a number like 180,000. Is 8 the last significant figure? Do we mean 180,000 rather than 190,000, or 180,000 rather than 180,001? The way the number is written fails to provide a clue. There is a method of writing such numbers using what is called *scientific notation:* The number is written as a product of a number containing a decimal point after the first digit (1 in this case) and some power of ten:

$$1.8 \times 100,000$$

$$1.80 \times 100,000$$

$$1.8000 \times 100,000$$

The number 100,000 is assumed to be known exactly, and the information about the precision is contained in the first number. 1.8 is known to 2 significant figures, 1.80 to 3, and 1.8000 to 5.

100,000 is usually written 10^5, where the *exponent* 5 counts the number of zeros.

When measurements are to be multiplied or divided, the precision of the result must be obtained by a different kind of calculation. We begin by expressing the uncertainty in each of the measurements as a *percentage*. For example, suppose the length and width of a piece of lumber are found by measurement to be 10.22 meters and 0.56 meters respectively. Each measurement has been carried out with a precision of 0.01 meters. The area is the product of length and width, and multiplication gives us 5.7232 square meters for the answer.

Expressing the precision of each measurement as a percentage, we obtain the following:

	Measurement	Precision	Precision as a percentage
Length	10.22 m	0.01 m	$\frac{0.01}{10.22} \times 100 = .098\%$ or about 0.1%
Width	0.56 m	0.01 m	$\frac{0.01}{0.56} \times 100 = 1.786\%$ or about 2%

The percentage precision with which the area is known will not be better than the larger of the percentage precision of the measurements to be multiplied (2%), so we estimate as follows:

2% of 5.7232 square meters is 0.114 square meters, or about 0.1 square meters

The area should therefore be rounded off to 5.7 square meters.

APPENDIX 14-2: DESCRIBING PRECISION QUANTITATIVELY

When a scientific measurement is performed, it is always performed a number of times. One reason is to make sure that no gross error was made, such as misreading the instrument—reading a 2 instead of a 5—or making a mistake in recording the result. A second reason is to be able to estimate the precision of the measurements.

Usually the individual measurements of the series will not all agree exactly, so we take the average of the series as the best guess for the quantity measured.

The *precision* of the measurement may be described in a quantitative way using the *standard deviation* of the series of individual measurements, described in Chapter 9, p. 269, and one additional quantity called the *standard error*, which can easily be calculated from the standard deviation.

MEASUREMENT AND ITS PITFALLS

The standard deviation of a series of individual measurements tells us something about the repeatability of the measurement we are performing: how close together successive measured values of the property are expected to be.

The standard error addresses the following kind of question: If we measure a quantity ten times, we would expect to have more confidence in our result than if we measured it five times. If we measure it a hundred times we would have still greater confidence. How much more confidence would we have? Is the additional confidence worth the additional effort?

Suppose we make a series of five measurements and calculate an average. Then if we make a new series of five measurements and calculate a new average, it is not likely to agree exactly with the first one. On the other hand it is not likely to differ greatly. Both averages are estimates of the true value, and we can see intuitively that the best guess we can make about the true value is that it is probably close to both averages, and more likely to be between them than otherwise.

From our first series of five measurements we can calculate the standard deviation (SD) of our series, as well as the average. The SD is an estimate, as noted earlier, of the spread in values of a large number of individual measurements. Although it is not intuitively obvious, the SD of the second series of five measurements is likely to be close in value to the SD of the first series. The average of each series provides an estimate of the true value, and the SD an estimate of the variability of the measurement. The more times either the average or the SD is measured, the more confidence we have in the estimate. Now to repeat the question asked earlier: how much more confidence should we have in the average?

To answer this question we introduce the standard error (SE), given by the formula

$$SE = \frac{SD}{\sqrt{N}}$$

where N is the number of individual measurements in the series. To show how it is used we give the following example.

Suppose a metal casting is weighed five times on a scale which can be read to 1 gram with the following results:

Measured weight	Deviation from average	Squared deviation
15,140 grams	−7 grams	49
15,157	+10	100
15,154	+7	49
15,145	−2	4
15,139	−8	64
Average 15,147.0		Average 53.2

Square root of average = SD = 7.3 grams; $SE = \dfrac{7.3}{\sqrt{5}} = 3.3$

(Note that the average is calculated to one more decimal place than the original measurement.)

A second series of five weighings would be expected to give an average close to 15,147.0, and a standard deviation close to 7.3.

The answer to the question about our confidence in the measurement is therefore as follows: The probability that the average of a series of measurements is exactly the true value is of course very small. However the probability that the average is "close" to the true value is large. To specify what we mean by "close," we use the SE. It can be shown that if the outcomes of a series of measurements resemble the Gaussian curve, the true value has a probability of 0.682 of lying within one SE of the average for the series of measurements.

In the example of the casting, the true value has a 0.682 probability of lying in the range

$$(15{,}147.0 + 3.3) = 15{,}150.3$$
$$\text{and } (15{,}147.0 - 3.3) = 15{,}143.7$$

It can also be shown to have a 0.95 probability of lying within two SEs of the average:

$$(15{,}147.0 + 6.6) = 15{,}153.6$$
$$\text{and } (15{,}147.0 - 6.6) = 15{,}140.4$$

Had we performed 25 individual measurements instead of five, the SE (assuming the SD came out close to 7.3, as before) would now be

$$\frac{7.3}{\sqrt{25}} = \frac{7.3}{5} = 1.46$$

or about 1.5, and the range of uncertainty about the average of the 25 measurements would be reduced accordingly.

It is customary to represent the result of a series of measurements in the form: Quantity = average ± SE. For the above example this becomes: Weight of casting = 15,147.0 ± 3.3.

The above analysis assumes, of course, that our measurement is *unbiased*. As indicated earlier, this belief is often a matter of faith.

REFERENCE NOTES

1. Joseph F. Mulligan, "Some Recent Determinations of the Velocity of Light," *American Journal of Physics* 20 (1952):165; "Some Recent Determinations of the Velocity of Light, II," *American Journal of Physics* 25 (1957):180.

2. Joseph F. Mulligan, "Some Recent Determinations of the Velocity of Light, III," *American Journal of Physics* 44 (1976): 960.
3. J. H. Sanders, *The Velocity of Light* (Oxford: Pergamon Press, 1965).
4. The President's Commission on Law Enforcement and the Administration of Justice, *Task Force Report: Crime and Its Impact—An Assessment* (Washington, D.C.: U.S. Government Printing Office, 1967).
5. A. J. Coale and F. F. Stephan, "The Case of the Indians and the Teen-age Widows," *Journal of the American Statistical Association* 57 (1962):338.
6. Oscar Morgenstern, *On the Accuracy of Economic Observations* (Princeton, NJ: Princeton University Press, 1968). Reprinted by permission of Princeton University and Professor Morgenstern. Copyright 1968 by Princeton University Press.

SUGGESTED READING

Errors of measurement are discussed in more detail in Chapter 6 of Freedman, David, Robert Pisani, and Roger Purves, *Statistics*. New York: W. W. Norton, 1968.

PROBLEMS

1. A Chemistry class of 12 students takes a quiz, and all the students, strangely, get a grade of 8 out of a possible 10.
   (a) What is the average grade?
   (b) What is the r.m.s. deviation from the average?
2. A student measuring the period of a pendulum obtained the following results: 2.40 sec, 2.49 sec, 2.48 sec, 2.50 sec. The laws of physics give for this particular pendulum a period of 2.44 sec. Express both the *precision* and the *accuracy* of this measurement in quantitative terms, for example as a percentage.
3. Bring in any data of your own choosing, preferably from your field of major interest, and calculate the average, the standard deviation, and the standard error.
4. Extensive experience with a set of bathroom scales shows that the standard deviation of a series of weighings is 0.5 kilograms. Someone on a diet wants to know his weight on the starting day of the diet to within 0.2 kilograms. How many times should he weigh himself?

15
Graphs and Sketches

WHY GRAPHS?

Science is concerned, among other things, with relationships between quantities. Sometimes the relationship is only vaguely stated: Is A bigger than B? Other times it can be precisely formulated: The gravitational force between two bodies is proportional to the product of their masses, and decreases as the square of the distance between them. Statements of the latter kind can best be described in the language of mathematics, but the language of mathematics has few native speakers. There is often no choice but to use it whether we like it or not, but sometimes the relationship embodied in an equation can be understood more easily when represented by a picture. Indeed at times a picture is better than an equation, even for the mathematically minded.

So we need to know how to "read" pictures, in particular the kind of picture specifically concerned with quantitative relationships: the graph. Graphs are not unfamiliar: a good daily newspaper will always have a few, on the financial pages at least, and often elsewhere. In earlier chapters of this book we introduced a number of graphs (Figures 4-2, 4-3, 6-3, 7-1 to 7-6).

Comparison Graphs

The simplest kind of graph provides a means of comparing magnitudes.

Figures 15-1 and 15-2 show two different common ways of doing this. Figure 15-1 is called a bar graph: the heights of the bars are proportional to the quantities being compared, in this case the total electric energy production in one year of a number of European countries. Figure 15-2 is a pie graph, showing the percentage composition by weight of a graham cracker and a corn muffin: the areas of the wedges (or, equivalently, their angles) are proportional to the percentages of protein, fat, carbohydrate, and "other" in the cracker or muffin.

Note that in the bar graph, no particular sequential order for the bars is better than any other. They happen to have been placed in alphabetical order according

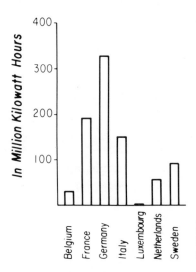

FIGURE 15-1. Electric energy production, in million kilowatt-hours, in six European Common Market countries in 1976.[1]

to the name of the country, but they could as well have been arranged in order of decreasing size.

Relationships

More often the sequential order matters: in fact it can be the whole point of the graph. Figure 15-3 shows the growth in the white population of the United States since 1790.* What is portrayed is a relationship between *population* and *time;* in each decade there has been an increase. This is a fundamentally different situation from that displayed in Figure 15-1. The difference can be brought out by asking the question: would you make a guess about what the white population is likely to be in the year 2000 from the information given about 1790–1970? The answer is *yes:* it is likely to be about 250 million. You might turn out to be wrong: some factor not present in the preceding few decades might come into play and cause a sudden jump to 280 million or an abrupt fall to 150 million, but the guess is a reasonable one. On the other hand, from the bar graph of electrical energy production would you make a guess as to a country not on the list? The answer is clearly *no*. Figure 15-3 describes a relationship between two variables that may extend or "extrapolate" to cases not given in the figure; Figure 15-1 does not. The dotted line in Figure 15-3 represents the extrapolation.

A verbal summary of the information portrayed in Figure 15-3 is: "The United States population has increased steadily with the passage of time." The order of

---

* The reason for using the white population for this graph is that in the early years of the United States neither the Indian nor the black population were counted as accurately as the white (a source of error that some claim operates even in recent census figures).

GRAPHS AND SKETCHES

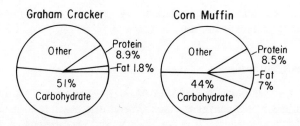

FIGURE 15-2. Nutritional compositions of a graham cracker and a corn muffin.[2]

the terms in that summary matters. We do not say, "Time passes as the United States population increases." The relationship between population and time is asymmetrical, like the relationship between a cause and its effect. Time goes on at its own pace, regardless of whether the population has increased or decreased during its passage. Not all relationships between two quantities are asymmetrical, but when they are we can usefully borrow some terms from the mathematicians for them: the independent variable, time, and the dependent variable, population.

Time is of course not the only independent variable possible. Figure 7-2 (p. 154) shows the *probability* of a particular outcome when 10 coins are tossed as the dependent variable, and the *outcome itself,* for example 6 heads, as the independent variable. The question "What is the probability of the outcome of 6 heads?" is a sensible question and has a sensible answer. The question "What has a probability of 0.2051?" is not as sensible, and has many different answers. Thus outcome is the independent variable, and is plotted on the horizontal axis.

It is customary, when one of two variables can be categorized as independent, or causal, and the other as dependent, or the effect of the cause, to use the horizontal

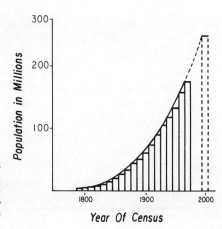

FIGURE 15-3. U.S. white population, in millions, 1790–1970. The dashed curve represents an extrapolation that permits us to estimate a population of 250 million in the year 2000, assuming the present trend continues.[3]

axis of a graph for the independent variable. When the two variables are not so related, the choice of axes is arbitrary.

Discrete versus Continuous

In Figure 15-3, populations were given in 10-year intervals because that is how often a United States census is taken. While it is possible to determine population at shorter time intervals, it is not as easy, and very expensive. Besides, no one could possibly care exactly what the United States population was between 12:18 and 12:19 P.M. on August 4, 1975. In tossing a coin 10 times, an outcome of $6^3/_4$ heads is not even possible. The two situations have a common feature in that the variable does not take on all values between its lower and upper limits, but is limited to certain definite values. The year 1963 does not appear in Figure 15-3, even though the United States population has a definite value in that year. Variables so limited are called *discrete* variables, and a bar graph or a histogram is the usual way to picture them.

There are other situations where we may wish to represent *any* value of the variable. The distance of an airplane from its destination while it is in flight has a definite value at every instant of time, which we may wish to portray for each instant that the airplane is in the air. Graphs in which the variables take on all values within their ranges are called *continuous*.

Mathematical Functions

The mathematical relationships we first studied in high-school algebra are examples of relationships between continuous variables. Let us consider a special case:

$$y = x^2 - 2x$$

For each value we assign to x, y has a definite value; we are not limited to integral values like $x = 0,1,2,3,4$, but can, if we wish, calculate y for $x = {}^1/_2$ or 1.974. All it takes is a hand calculator. But to make the graph we do not bother to calculate y for each and every possible value of x. Instead we can limit ourselves for convenience to discrete values, such as positive and negative integers.

The most common method used to represent such a relationship graphically was invented by René Descartes (1596–1650), and hence is called a Cartesian graph. We begin with two perpendicular straight lines, called the axes, subdivided into marked intervals of equal lengths. One axis is called the x axis, and a particular value of x, say 6.5, is denoted by a line perpendicular to the x axis at a point 6.5 units to the right of the point of intersection of the two axes (the origin). Values of y are represented by lines perpendicular to the y axis, above the origin if positive, below if negative.

GRAPHS AND SKETCHES

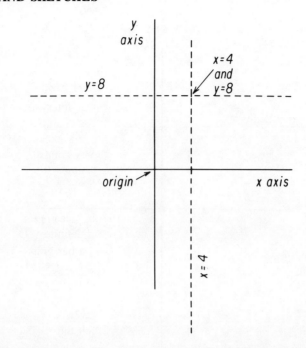

FIGURE 15-4. Constructing a Cartesian graph: locating the point $y = 8$ when $x = 4$.

The equation given above associates a y value with any given x value. For example for $x = 4$, $y = (4)^2 - 2 \times 4 = 8$. The result (the conjuncture when $x = 4$ and $y = 8$) is represented by the single point on the graph formed by the intersection of two imaginary lines, one perpendicular to the x axis at the point $x = 4$, and the other perpendicular to the y axis at the point $y = 8$ (Figure 15-4).

In Figure 15-5 we have plotted the Cartesian points which give the y values for each integral value of x from -3 to $+5$. So far this is a discrete graph. Now we make it continuous by drawing a smooth curve connecting the plotted points (Figure 15-6). If we are sufficiently skilled we can do this freehand, if not there are devices called French curves to help us. The smooth curve now permits us to determine y values for the x values we did not use in plotting, such as $x = \frac{1}{2}$ or $x = 1.974$—not that we could not also calculate them with our hand calculator just as easily, and certainly with greater accuracy.

Continuous graphs appear not only in mathematics: Figure 15-7 shows the concentration inside a home of one particular air pollutant, NO_2, plotted on a strip-chart at each moment throughout a particular day. The horizontal axis represents the time of day and the vertical axis the NO_2 concentration in parts per million at each particular time. The measure of concentration is said to be a *function* of time, just as y in the first example is said to be a function of x.

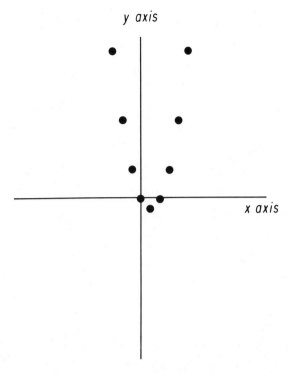

FIGURE 15-5. Points for constructing a graph of the relation $y = x^2 - 2x$. The values of y for $x = -3, -2, -1, 0, 1, 2,$ and 3 are plotted.

Graphing the Laws of Physics

The laws of physics predict a simple relation between the period of a simple pendulum and its length. A simple pendulum can be constructed by attaching a thread or string to a support, and hanging a weight on the end of it. The weight can be set swinging back and forth, and the period is the time it takes for one complete swing, forward and back to the starting point. It is interesting that the laws of physics predict that the period depends neither on how large the weight is, nor on how great an angle the pendulum swings through, as long as this is not "too" large. As a pendulum swings, the friction of the string against the support and the resistance of the air cause the pendulum to swing through a lesser and lesser angle. If the period did depend on the angle of swing, the time of a complete swing would change as the motion died down.

We can therefore test this prediction of the laws of physics by an experiment varying the weight suspended, and measuring the period for each weight, and by a second experiment repeatedly measuring the period while the angle of swing gradually decreases.

GRAPHS AND SKETCHES

But we can also test it in a more critical and quantitative way through the predicted relation between period and length. The relation is:

$$P = 0.2007 \sqrt{L}$$

The numerical factor, 0.2007, takes into account the strength of the earth's gravitational field, which varies from place to place on the earth's surface. Since the experimental results about to be given were obtained in New York City, we have used the gravitational strength appropriate to this location. On the moon, pendulums would swing much more slowly. The formula relates the period P, measured in seconds, to the length L, measured in centimeters. In Figure 15-8 we have plotted P as dependent variable against L as independent variable. We can calculate from the formula, with the aid of a hand calculator, that a pendulum of length 24.83 centimeters will have a period of one second.

To perform an experimental test of this relation we measure P and L for a particular length, then change L by shortening or lengthening the string, and measure

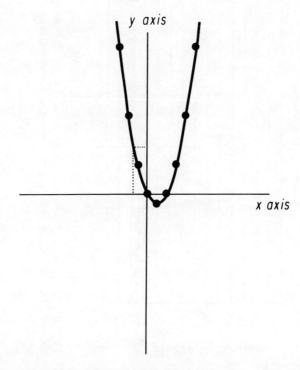

FIGURE 15-6. The points of Figure 5 have been connected by a smooth curve. From this curve the value of y when $x = -1.5$ is estimated as 5.1. The exact value given by the equation is 5.25.

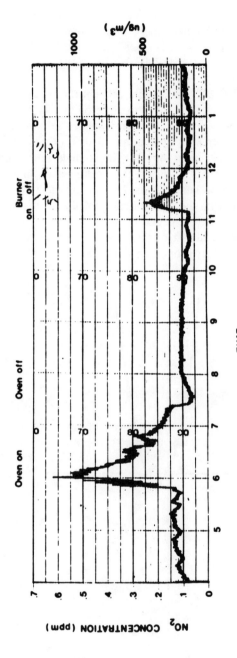

FIGURE 15-7. A continuous curve from the real world. Concentration of NO_2 in parts per million inside a home is determined at each instant of the day by a recording chemiluminescence monitor. The peaks occur whenever a gas stove is used.[4]

GRAPHS AND SKETCHES

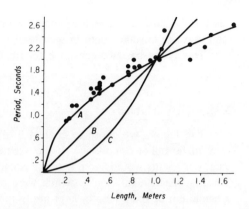

FIGURE 15-8. Data obtained by students in a science course on the period of a pendulum as a function of its length. Curve A is that predicted from Newton's laws of motion (period proportional to the square root of length). Curves B and C show, respectively, what would be observed if the period were proportional to the length and to the square of the length.

them again. The result of our experiment would then be a set of values of P corresponding to different values of L. We can then plot these points on a graph (Figure 15-8), and if they all lie on the smooth curve predicted by the theory, we have confirmed the theory. They never do.

Noise

As noted in Chapter 14, any measurement is subject to error. We make errors every time we measure the length, and errors when we measure the period. We are not likely to get the same answer for either twice in a row. The best that can be done is to measure the quantities a number of times and take an average. How many measurements must be made to get a reliable answer is a question statistics can help us answer. But statistics can help only if the errors are *random*—equally often positive and negative. There are unfortunately also systematic errors or biases, that tend to make our answers too high or too low. For example, our stopwatch may be running slowly, so that the period P is always underestimated.

Further, our "law" of physics—embodied in the equation given—neglects a number of things, partly in the interest of obtaining a simple relation between period and length, and partly because it would be too hard to take such things into account. They include various forms of friction, the weight of the string, the spatial dimensions of the weight, and a very small effect of amplitude on period which only becomes important when the amplitude is large.

The combination of random errors, systematic errors, and neglected factors together insure that our experimental data-points will never lie exactly on the curve.

It is clear from the figure that the relationship predicted from Newton's laws is pretty much what is observed, but the agreement is not exact. The observed periods are close to the curve obtained from theory, but differ, lying both above and below it. They do so not only for the various reasons mentioned above, but in addition because these measurements were made by inexperienced students who tend to make mistakes both in measurements and in calculations.

This is the best we can hope for in science, and often we are satisfied with less.

Neglected Variables

The factors we chose to neglect in the physics experiment described above were of no major importance. If the experiment were carried out with reasonable care, our results would agree with the equation within a small fraction of a percent. The first mechanical clocks, which were based on the constancy of the period of a pendulum of fixed length, kept pretty fair time.

In many fields of science—not only in the social sciences but even areas of the physical sciences—the factors we are forced to neglect are not so small. We study complex systems—winds, clouds, and rain on the surface of the earth, diseases affecting large and varied human populations, whole social systems. There are many things that influence the outcomes we are interested in, and their effects are often unpredictable. The number that we can control and measure may be small compared to the number that influences results.

This can be illustrated with an example chosen quite deliberately from the physical rather than the social sciences.

Sulfur dioxide (SO_2), an invisible but odorous gas, and smoke or haze, which consists of small particles of solids or liquids suspended in the air, are both common pollutants of urban air. Certain processes produce both simultaneously, such as the combustion of coal or high-sulfur oil for power production. Others tend to produce one or the other only: particulates are emitted from the exhaust of cars or trucks, but sulfur dioxide is not. The concentration of these pollutants in the air is determined not only by their rates of production, but their rates of dispersal or removal by wind and rainfall. It is to be expected that many situations that favor high sulfur dioxide concentrations will favor high particulate concentrations as well (coal-burning, low wind speeds) but an exact relationship is not likely. An example of an observed relationship between the two, obtained from air quality measurements in New York City, is shown in Figure 15-9.

Each point in this figure shows the monthly averages of the measured values of sulfur dioxide and particulate concentrations over a two-year period. For example, for the month of March 1967, the particulate concentration, measured by something called the *"coefficient of haze,"* was 2.2 units, as indicated by the distance of the point along the horizontal axis, and the sulfur dioxide concentration was 0.23 parts per million (weight of SO_2 relative to the weight of air itself) as indicated by the distance of the point along the vertical axis. Note that in this method of plotting, the chronological order in which the observations were made is irrelevant, and cannot be inferred from the figure.

One sees that in months in which one pollutant is high, the other tends to be

GRAPHS AND SKETCHES

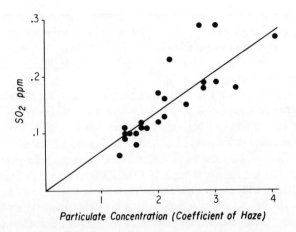

FIGURE 15-9. Relation between monthly averages of sulfur dioxide (SO_2) concentration, in parts per million, and particulate concentration (units of coefficient of haze) in an air-pollution-measuring station in Harlem, New York City, in the years 1967–1968.[5] Note that the scales in the two axes are different: the length on the vertical axis which corresponds to 0.1 unit of SO_2 concentration corresponds to 1.0 unit of concentration of particulates on the horizontal axis.[5]

high also, but this is not invariably true, as shown by the point for March 1967, when SO_2 was high and particulates about average. Thus, although we could use this figure to make a reasonable guess as to the SO_2 concentrations for a month in which the particulate concentration was measured but the SO_2 concentration was not, the guess might well be far off. A straight line is drawn on the figure to represent the approximate relationship: it shows the best estimate we could make on the basis of the data given.

Note that in this particular problem neither variable is dependent on the other: SO_2 does not *cause* haze, nor does haze *cause* SO_2. We could therefore have plotted either one on the horizontal axis: the choice of SO_2 for that axis was arbitrary.

Useful Distortions

In the graph of the mathematical relationship $y = x^2 - 2x$, given in Figure 15-5, a given distance on either the x or the y axis corresponded to the *same* change in the value of y or x.

In Figure 15-9, this was no longer true. The two quantities plotted, sulfur dioxide concentrations and particulate concentrations, were measured in different units, and those units had different numerical magnitudes, SO_2 concentrations ranging from 0.06 to 0.3 parts per million, particulate concentrations from 1 to 3.5 units of coefficient of haze.

Had we required 0.1 of a unit to have the same length along both axes we would have produced a graph much more difficult to comprehend than Figure 15-9; such a graph is shown in Figure 15-10.

The use of different scales on the two axes is an example of necessary distortion made in the interests of clarity.

Another useful distortion has to do with the location of the zero of the scale. In the plot of Figure 15-6 the origin, the point where the two axes cross, represented a value of 0 for both x and y. There are occasions where a scale is made to start not at zero, but at some other numerical value. In Figure 15-3 the horizontal axis represents time. The numbers are the numbers of the years according to the Western (Gregorian) calendar, and the years from 1790 to 2000 A.D. have been plotted. This graph does not begin with the United States population in the year 0!

Nor does a "fever chart" for a patient in a hospital (which is a plot of body-temperature against time) use a temperature of 0 degrees Fahrenheit (or Celsius) as its origin: the only temperatures ever observed lie between 96° and 106° F, or about 36° to 41° C.

Sometimes the middle portion of a scale is deleted: as when both small values and large values of the variable are interesting, but the middle values are not. An example of such a graph is shown in Figure 6-3 (p. 114), where the temperature in a container holding ice at $-10°$ C and placed in a warm room is plotted against time. The ice warms up to the melting point in a few minutes, then remains at 0° C for a long, long time, until the last bit of ice melts. The temperature then rises within a few minutes to the temperature of the room. The graph shows by a break in the horizontal scale that most of this uninteresting interval, during which the temperature remains unchanged at 0° C, is not plotted.

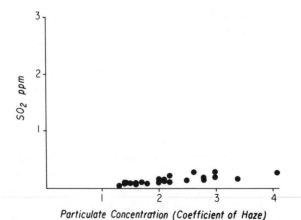

FIGURE 15-10. The same data replotted using scales in which a given numerical value of a concentration corresponds to the same length on both horizontal and vertical axes.

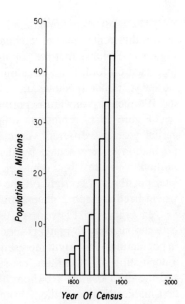

FIGURE 15-11. U.S. population since 1790 plotted using an expanded vertical scale. The length along the vertical axis that corresponded to 25 million of population in Figure 15-3 now corresponds to 2.5 million.

Exponential Growth

A more complicated kind of distortion is one in which we use a scale in which equal increases in magnitude of the plotted variable are not represented by equal increases in distance along the axis. (All the graphs we have so far given do use equal distances on the axes to represent equal magnitudes: such scales are called *linear*.)

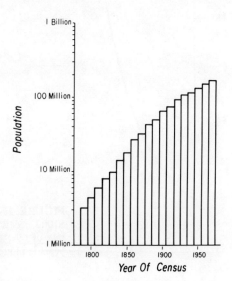

FIGURE 15-12. U.S. population since 1790 plotted using a logarithmic scale for the vertical axis.

The most common example is the so-called logarithmic scale, which can be used if different values of the quantity to be plotted in the graph differ enormously. Figure 15-3 shows that the population of the United States since the year 1790 has increased 60-fold, from three million to almost 200 million. The population in the earlier years barely shows up in the graph, but the population from the middle of the 19th century onward is portrayed clearly.

Figure 15-11 represents an attempt to portray the growth in population in the earlier years. A different vertical scale has been used, on which 2.5 millions of population is represented by the same distance along the axis as represents 25 million in Figure 15-3. While the growth from 1790 to 1840 is well portrayed, the later populations go right off the graph. To show recent years on Figure 15-11 we would need a graph 2 meters high!

In Figure 15-12 we have used a scale for population in which the distance between successive markers along the vertical axis represents a ten-fold increase in population. The origin represents a population of 1 million, the next major marker a population of 10 million, the second major marker a population of 100 million, and the third marker 1 billion. The *increase* in population from the origin to the first marker is nine million, from the first to the second 90 million, and from the second to the third 900 million, in contrast to the fixed increase of population per unit-length plotted in Figures 15-3 and 15-13. The population growth over the entire time interval now fits on this graph easily.

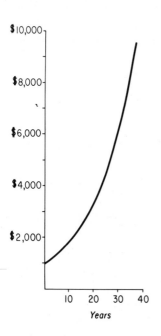

FIGURE 15-13. Balance of a bank account, initially $1000, receiving interest compounded daily, at an annual rate of 6%. The vertical scale is a linear one. Time in years is plotted horizontally.

GRAPHS AND SKETCHES

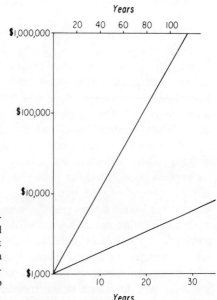

FIGURE 15-14. The same bank balance, using a logarithmic vertical scale. The horizontal scale on the top of the figure uses a shorter unit of length to correspond to one year, so that a longer period of time can be represented. The upper straight line shows the bank balance referred to the upper time scale.

Another example of the use of a logarithmic scale is shown in Figures 15-13 and 15-14, in which the balance of a bank account, initially $1,000, and receiving interest compounded daily at an annual rate of 6%, is plotted against time (number of years since the account was opened). Compound interest accounts receive interest on the accumulated interest as well as on the principal, so the more interest that has accrued, the faster the balance grows. Figure 15-13 uses linear scales both for the balance and the time elapsed. The increasing rate of growth of the balance is clearly displayed. Figure 15-14 uses a logarithmic scale for the balance. The same relation portrayed as a curve in Figure 15-13 is now a straight line. Both figures permit us to calculate that the initial balance is doubled in about $11^1/_2$ years. In 115 years the initial balance would have doubled 10 times, and be about $1,024,000. Try to plot that with a linear scale!

A quantity whose rate of growth is proportional to its size, so that the bigger it gets the faster it grows, is said by mathematicians to grow *exponentially*. The phrase *exponential growth* is sometimes used in ordinary language as a metaphor for "awfully, awfully fast growth." It doesn't mean that necessarily: 6% annual interest compounded daily does not lead to an awfully, awfully fast growth, not initially anyway.

Anyone who wishes to make such a graph should first review logarithms (as encountered in high-school algebra) and then go to a well-stocked stationery store and ask for semilogarithmic graph paper, which makes it really very easy.

SKETCHES

We put the cart before the horse in this discussion by describing in detail the quantitative aspects of drawing graphs—choices of origin, scale and distortion—before arriving at what always comes first: a sketch.

A sketch, for which we discard the fancy apparatus of graphing—carefully sharpened pencils, good rulers and meter-sticks, French curves, specially printed graph paper—for any old pencil and the back of an envelope, portrays the essence of a relationship qualitatively rather than quantitatively.

We roughly sketch a table or chair we mean to make before rendering the detailed picture, drawn to scale, with all dimensions shown precisely. Sometimes we procede to the detailed picture immediately, but often we tear up the rough sketch, dissatisfied with the general appearance it suggests. The same is true for graphs. The sketch comes first. Sometimes it alone suffices, sometimes it is simply a preamble to the detailed work of drawing a graph. The sketch is qualitative in purpose, so there is no need to worry so much about scales and rulers. But a particular mode of representation must be chosen, and if it is a matter of the relationship between two quantities we will most likely use a Cartesian diagram. We then need to sketch two perpendicular axes and label them, so that we and whoever else sees the sketch will know what it is about.

REFERENCE NOTES

1. John Paxton, ed. *The Statesman's Yearbook 1978–79* (New York: St. Martin's Press, 1978).
2. *The World Almanac and Book of Facts, 1983* (New York: Newspaper Enterprise Associates, 1983).
3. U.S. Bureau of The Census, *Statistical Abstract of the United States, 1982–83,* 103rd Edition (Washington, D.C., 1982).
4. Frank E. Speiser, Benjamin Ferris, Jr., Yvonne M. M. Bishop, and John Spengler, "Respiratory Disease Rates and Pulmonary Function in Children Associated with NO_2 Exposure." *American Review of Respiratory Disease* 121 (1980): pp. 3–10.
5. Department of Air Resources, *New York City Data Report, Aerometric Network, 1957–1968* (New York: Department of Environmental Protection, 1968).

SUGGESTED READING

Jacques Bertin, *Semiology of Graphics* (Madison: University of Wisconsin Press, 1983).

PROBLEMS

An ability to make simple sketches to illustrate a relationship is more fundamental than the ability to make neat and accurate graphs. The first ability reflects how we think, the second our manual dexterity and carefulness.

Make simple sketches to illustrate the following statements:

GRAPHS AND SKETCHES

1. The gross national product, which has been growing at a rate of 6–8% per year over the last decade, is expected to grow at a rate of only 2% per year during the next five years.
2. $y = 2x$
3. Honey catches more flies than vinegar.
4. Fuel economy, in miles per gallon, is greatest when driving at 55 miles per hour.
5. Fuel economy is usually but not invariably greater the lighter the car.
6. The patient tends to run a fever only in the evening.
7. $y = x^2 + 1$
8. Interest rates, which this month showed signs of levelling off, are rising again.

   Make graphs for the following:

9. Plot the data of Table 4-5, showing how the rates of lung cancer in men and women have changed with time in recent years. What conclusion may be drawn from the trends?
10. Toss 10 coins at once, record the number of heads, and repeat till the 10 coins have been tossed 50 times.
    Make a histogram of the outcome.
    Make a histogram of the expected number of each of the outcomes for comparison.

16
Where Do Hypotheses Come From?

POSING HYPOTHESES

We All Make Them

In this chapter we turn to the question, where do we get scientific hypotheses? The way the question is phrased may be misleading. Put this way, it gives the impression that forming hypotheses is something unique to scientific activity. However, if we understand by the word "hypothesis" the perception of some pattern in phenomena, the establishment of some expectation as to what will happen next, we realize that "forming hypotheses" is something we do all the time and have been doing since birth. We are by nature hypothesis formers. At what age our practice of interpreting the world in such structured terms begins is not known, but long before we learn to talk in sentences we have already gone far beyond the raw impressions given us by our senses. We organize things coherently into such concepts as "mother," "father," "food," and "doggy," each of which implies a whole complex set of recognitions and expectations. We are not normally aware of how much of what we "see" is seen by inference and memory rather than with just our eyes. But there are occasions when this is brought home to us, as was discussed in Chapter 2, by the study of the kinds of optical illusions favored by psychologists and the puzzle pages of newspapers, by encounters with people of different cultures, by occasions where something radically unexpected happens, when, in the graphic but hackneyed phrase, "our whole world collapses about us."

In making the statement that what we perceive as reality is actually hypothesis, the result of a culturally and personally determined interaction between ourselves and what is out there, we are not arguing that there is an unbridgeable discrepancy between "reality" and what we think is reality. As scientists we must believe in a real world which exists independently of our sense-perceptions and the interpretations we put on them, or at least act as though we do. But all we can ever know are those perceptions and interpretations. We cannot adopt a critical attitude toward everything we think we know; it would not be possible to function in the world if we did. But it is not surprising that, even in situations where some serious discordance has appeared between our expectations and what actually happens, we should cling to the ways of thinking and perceiving that we are used to. The discovery of new hypotheses in science or in daily life is difficult, because it is not so much a question of finding a new pattern where none was previously seen, but rather of replacing a pattern we are used to—so used to that we take for granted that it is really there—by a new one. The point was put very well by Josh Billings: "It aint ignorance that hurts us, it's what we know that aint so!"[1]

We are not trying to suggest that conservatism in ideas is wrong. It is impossible to subject everything we believe to doubt, and it is reasonable, when faced with new problems, to try to cope with them by the methods we have found to work with old ones. Since not all problems have easy answers, we would be foolish to assume that because our accustomed methods do not seem to work right off the bat, they must be wrong and should be discarded.

Today we "know" that the world is round. But we should be sympathetic to our ancestors who refused to believe it. It flies in the face of common sense: a wealth of experience with falling off of things like trees, rocks, and steep hills tells us that if the world were round one would fall off the other side. We know better now, because we have been taught differently, but the first men to conjecture that the world might be round did not have this advantage. They had to make an imaginative leap beyond the "facts" known by everyone, and see things in a completely new way. Even today, a child told for the first time that the world is round is startled and skeptical.

Because scientific discovery has this character, one should not be surprised to learn that it is not a routine, mechanical process but rather one in which the subconscious mind plays a part (as it does in artistic creativity, also) and that chance and circumstance contribute. There is a popular but completely misleading belief about scientific discovery, that it is an orderly process in which facts are patiently gathered and neatly arranged, at which time the scientist sits down and contemplates them until some new pattern emerges. One can contrast this tidy picture with the description quoted earlier, by the chemist Kekulé, of his dreamlike reverie during which the dancing images of the atoms—images, of course, arising from his own unconscious mind—forced the new concept into conscious awareness.

THE MOMENT OF INSIGHT

The act of discovery as a flash of insight rather than the patient assembling of the pieces of a jigsaw puzzle is shown beautifully by the following episode, described by the psychologist W. Köhler:[2]

Nueva [a young female chimpanzee] was tested 3 days after her arrival. . . . She had not yet made the acquaintance of the other animals but remained isolated in a cage. A little stick is introduced into her cage; she scrapes the ground with it, pushes the banana skins together in a heap, and then carelessly drops the stick at a distance of about three-quarters of a metre from the bars. Ten minutes later, fruit is placed outside the cage beyond her reach. She grasps at it, vainly of course, and then begins the characteristic complaint of the chimpanzee; she thrusts both lips—especially the lower—forward, for a couple of inches, gazes imploringly at the observer, utters whimpering sounds, and finally flings herself on to the ground on her back—a gesture most eloquent of despair, which may be observed on other occasions as well. Thus, between lamentations and entreaties, some time passes, until—about seven minutes after the fruit has been exhibited to her—she suddenly casts a look at the stick, ceases her moaning, seizes the stick, stretches it out of the cage, and succeeds, though somewhat clumsily, in drawing the bananas within arm's length. Moreover, Nueva at once puts the end of her stick behind and beyond her objective. The test is repeated after an hour's interval; on this second occasion, the animal has recourse to the stick much sooner, and uses it with more skill; and at a third repetition, the stick is used immediately, as on all subsequent occasions. (pp. 32–33) [see Figure 16-1]

Another example of the role of the unconscious mind in discovery is given in a description by the French mathematician Poincaré of his discoveries of some new classes of mathematical functions and their properties. One does not need to understand the mathematical terms used by Poincaré to appreciate his story:[3]

It is time to penetrate deeper and to see what goes on in the very soul of the mathematician. For this, I believe, I can do best by recalling memories of my own. But I shall limit myself to telling how I wrote my first memoir on Fuchsian functions. I beg the reader's pardon; I am about to use some technical expressions, but they need not frighten him, for he is not obliged to understand them. I shall say, for example, that I have found the demonstration of such a theorem under such circumstances. This theorem will have a barbarous name, unfamiliar to many, but that is unimportant; what is of interest for the psychologist is not the theorem but the circumstances.

For fifteen days I strove to prove that there could not be any functions like those I have since called Fuchsian functions. I was then very ignorant; every day I seated myself at my work table, stayed an hour or two, tried a great number of combinations and reached no results. One evening, contrary to my custom, I drank black coffee and could not sleep. Ideas rose in crowds; I felt them collide until pairs interlocked, so to speak, making a stable combination. By the next morning I had established the existence of a class of Fuchsian functions, those which come from the hypergeometric series; I had only to write out the results, which took but a few hours. . . .

Just at this time I left Caen, where I was then living, to go on a geologic excursion under the auspices of the school of mines. The changes of travel made me forget my mathematical work. Having reached Courtances, we entered an omnibus to go some place or other. At the moment when I put my foot on the step the idea came to me, without anything in my former thoughts seeming to have paved the way for it, that the transformations I had used to define the

FIGURE 16-1. A flash of insight.

WHERE DO HYPOTHESES COME FROM?

Fuchsian functions were identical with those of non-Euclidean geometry. I did not verify the idea; I should not have had time, as, upon taking my seat in the omnibus, I went on with a conversation already commenced, but I felt a perfect certainty. On my return to Caen, for conscience's sake I verified the result at my leisure. (pp. 52–55)

Poetry Also

The similarity of the process to other creative activities, such as the writing of poetry, is shown by this quotation from A. E. Housman:[4]

> In short I think that the production of poetry, in its first stage, is less an active than a passive and involuntary process; and if I were obliged, not to define poetry, but to name the class of things to which it belongs, I should call it a secretion; whether a natural secretion, like turpentine in the fir, or a morbid secretion, like the pearl in the oyster. I think that my own case, though I may not deal with the material so cleverly as the oyster does, is the latter; because I have seldom written poetry unless I was rather out of health, and the experience, though pleasurable, was generally agitating and exhausting. If only that you may know what to avoid, I will give some account of the process.
>
> Having drunk a pint of beer at luncheon—beer is a sedative to the brain, and my afternoons are the least intellectual portion of my life—I would go out for a walk of two or three hours. As I went along, thinking of nothing in particular, only looking at things around me and following the progress of the seasons, there would flow into my mind, with sudden and unaccountable emotion sometimes a line or two of verse, sometimes a whole stanza at once, accompanied, not preceded, by a vague notion of the poem which they were destined to form part of. Then there would usually be a lull of an hour or so, then perhaps the spring would bubble up again. I say bubble up, because so far as I could make out, the source of the suggestions thus proffered to the brain was an abyss which I have already had occasion to mention, the pit of the stomach. When I got home I wrote them down, leaving gaps, and hoping that further inspiration might be forthcoming another day. Sometimes it was, if I took my walks in a receptive and expectant frame of mind; but sometimes the poem had to be taken in hand and completed by the brain, which was apt to be a matter of trouble and anxiety, involving trial and disappointment, and sometimes ending in failure. I happen to remember distinctly the genesis of the piece which stands last in my first volume. Two of the stanzas, I do not say which, came into my head, just as they are printed, while I was crossing the corner of Hampstead Heath between Spaniard's Inn and the footpath to Temple Fortune. A third stanza came with a little coaxing after tea. One more was needed, but it did not come: I had to turn to and compose it myself, and that was a laborious business. I wrote it thirteen times, and it was more than a twelvemonth before I got it right. (pp. 47–50)

Folk Wisdom

Not all discoveries have arisen unexpectedly out of the subconscious of the discoverers. There are a number of examples where ideas were in the air, so to speak, or present in the form of folk beliefs. The achievement of the discoverer was to take them seriously and think of ways to test them experimentally. Snow was not the first to think of the water supply as a mode of transmission of cholera: he states in his book that a number of people had suggested this possibility, some of whom were professionals actively concerned with finding the cause of the disease, and others just ordinary people expressing a conviction they derived from their own experiences. Snow's genius was to think of experiments that could prove it.

Milkmaids often caught a relatively mild disease called *cowpox* from the cows they milked. When Edward Jenner (1749–1823) encountered the belief that this disease might confer immunity against smallpox, he took this "superstition" seriously enough to test it, even though it required taking the risk of deliberately infecting people with a disease in the hope of preventing a more serious one. He had to fight violent opposition to his program: there are contemporary cartoons showing people growing cow's heads (see Figure 16-2) from their shoulders at the site of the inoculation.[5]

One of the first conclusive demonstrations that the bite of an insect is capable of transmitting disease was accomplished by Theobald Smith and E. L. Kilborne in the case of Texas cattle fever in 1893. This work led to the identification of insect transmission of such diseases as malaria, bubonic plague, and yellow fever.

FIGURE 16-2. "The cow pock." Etching by James Gillray, 1802. The physician performing the vaccination is a portrait of Jenner. (From the Smith, Kline, and French Laboratories Collection of the Philadelphia Museum of Art, and reproduced with the permission of the Museum.)

Smith and Kilborne did not make the hypothesis themselves that the ticks that bite the cows transmit the disease; it was the cattle ranchers who first proposed it. They had noticed a relation between the onset of tick infestations and the appearance of the disease in their cattle.[6]

CHANCE

Sometimes, although more rarely than one might think, discoveries are made by accident. The following is an example reported by A. V. Nalbandov:[7]

> In 1940 I became interested in the effects of hypophysectomy of chickens. After I had mastered the surgical technique my birds continued to die and within a few weeks after the operation none remained alive. Neither replacement therapy nor any other precautions taken helped and I was about ready to agree with A. S. Parkes and R. T. Hill who had done similar operations in England, that hypophysectomized chickens simply cannot live. I resigned myself to doing a few short-term experiments and dropping the whole project when suddenly 98% of a group of hypophysectomized birds survived for 3 weeks and a great many lived for as long as 6 months. The only explanation I could find was that my surgical technique had improved with practice. At about this time, and when I was ready to start a long-term experiment, the birds again started dying and within a week both recently operated birds and those which had lived for several months were dead. This, of course, argued against surgical proficiency. I continued with the project since I now knew that they could live under some circumstances which, however, eluded me completely. At about this time I had a second successful period during which mortality was very low. But, despite careful analysis of records (the possibility of disease and many other factors were considered and eliminated) no explanation was apparent. You can imagine how frustrating it was to be unable to take advantage of something that was obviously having a profound effect on the ability of these animals to withstand the operation. Late one night I was driving home from a party via a road which passes the laboratory. Even though it was 2 A.M. lights were burning in the animal rooms. I thought that a careless student had left them on so I stopped to turn them off. A few nights later I noted again that lights had been left on all night. Upon enquiry it turned out that a substitute janitor, whose job it was to make sure at midnight that all the windows were closed and doors locked, preferred to leave on the lights in the animal room in order to be able to find the exit door (the light switches not being near the door). Further checking showed that the two survival periods coincided with the times when the substitute janitor was on the job. Controlled experiments soon showed that hypophysectomized chickens kept in darkness all died while chickens lighted for 2 one-hour periods nightly lived indefinitely. The explanation was that birds in the dark do not eat and develop hypoglycaemia from which they cannot recover, while birds which are lighted eat enough to prevent hypoglycaemia. Since that time we no longer experience any trouble in maintaining hypophysectomized birds for as long as we wish. (pp. 167–168)

Accident also played a part in the discovery by Ignaz Semmelweis (1818–1865) that puerperal fever—childbed fever—which killed thousands of women during childbirth in hospitals in the nineteenth century, was transmitted by the hands of the doctors. These doctors, who had previously examined women already sick with the disease or who had performed autopsies on fatal cases, in accord with the practice of the time had washed but not disinfected their hands. Semmelweis made this discovery when he recognized symptoms similar to childbed fever in a physician friend of his who died of "blood poisoning" contracted from a scalpel wound incurred while performing an autopsy.[5]

However, the role of chance in discovery is only part of the story—the discoverer usually plays an active rather than a passive role: he must recognize the significance of a chance event that most others would ignore. There is a famous quotation from Pasteur which sums this up: "Chance favors the prepared mind." Sir Alexander Fleming discovered penicillin from his observation that bacterial cultures on petri dishes were killed in the vicinity of mold colonies that formed accidentally on the nutrient medium. Prior to Fleming's work, this observation had been made thousands of times in bacteriological laboratories. The response had been to throw the mold-infected cultures out because they were no longer any good for growing bacteria.

The Lost Keys

These stories of scientific discovery may remind the reader of such common experiences as misplacing the car keys and searching the house for an hour in mounting frustration, finally giving up in disgust and going to work, where one suddenly remembers, two hours later, while absorbed in some detail of one's job, that one left them on the shelf in the kitchen while drinking a second cup of coffee. Of course, sometimes the frantic search succeeds, and sometimes the keys are never found at all.

THE COLLECTIVE UNCONSCIOUS

The above examples of the importance of the unconscious in discovery may give the impression that discovery itself is a very chancy business—a matter of having a person with the right unconscious mind in the right place at the right time. It may seem like a miracle that anything has been discovered at all.

But our unconscious minds are not that independent of our environments. Scientists share the broader culture of their society as well as the subculture of their own field, and through these are exposed to all sorts of influences and suggestions. They have been trained by senior members of their professions; they attend lectures, have private discussions with their colleagues, read papers and books, and so forth. While it is a truism that each individual is unique, we have to recognize how much each person shares with the community. It is hard to document the many ways in which one's ideas may be influenced or suggested by the ideas of others, but in science as in life it is a common experience to pick up ideas from others and honestly come to believe that they are original.

WHERE DO HYPOTHESES COME FROM?

There have been many remarkable examples in the history of science where important discoveries were made almost simultaneously and independently by several scientists. Newton and Leibniz both invented the calculus at about the same time, and disputed for the rest of their lives about who deserved credit for the discovery. In retrospect, the time must have been ripe for the calculus to be discovered, although it would require a careful historical study of the development of mathematics in the seventeenth century to show this. This does not imply that it did not require genius to make the discovery at that moment, but only that a century earlier not even a Newton or a Leibniz could have done it, and that a century later, even if they had never lived, the calculus would have been gradually developed by the efforts of many lesser mathematicians.

Something similar happened in biology. Gregor Mendel published a paper in 1865 describing his discovery of certain laws of heredity; his work was ignored until 1900, when the same laws were rediscovered simultaneously by three different groups of scientists. Again, we can conclude that in 1900 the time was ripe for the acceptance of these laws: biology had advanced in the 35 years from 1865 to 1900 in ways that made the subculture of biologists both more likely to discover them and more willing to accept them.

THE TACTICS OF SCIENCE

The examples quoted above may give a misleading impression about scientific discovery in general. Poincaré, studying the problem of Fuchsian functions, Nalbandov, trying to keep alive chicks whose pituitary glands had been removed, and Theobold Smith, working on Texas cattle fever, were scientists struggling with a problem who suddenly broke through to a solution. Now, although many scientific discoveries are made this way, many are not. This may sound paradoxical: how can you solve problems without struggling with them? But science does not always progress by deliberate and direct ways. P. B. Medawar uses an appropriate military metaphor: problems do not always yield to direct assault, sometimes they are solved by attrition, and sometimes they are outflanked.[8] Discoveries are often made and problems solved in completely unexpected ways, by achievements in other fields that seem to have no connection whatever with the problem at hand. It was not a biologist, a doctor, or an astronomer who invented the microscope or the telescope, but grinders of lenses, who as far as we know were motivated only by idle curiosity or the desire for amusement. But biology, medicine, and astronomy were revolutionized by these inventions.

The laws of Newton described very accurately the motions of all the planets in their orbits around the sun except for Mercury, which showed certain slight deviations. Astronomers struggled with the problem for years, proposing many hypotheses in an attempt to show that if Newton's laws were properly applied the discrepancies could be explained. None worked. Einstein, working on a completely different problem arising from certain peculiarities of the transmission of electromagnetic waves, was led to new laws of physics that replaced Newton's and explained the misbehavior of Mercury.

So it must be acknowledged that, to make scientific discoveries, both genius and patient hard work are useful, but neither is any guarantee of success. There is something about discovery that cannot be programmed.

It is this unpredictable character that makes it hard to know how to proceed, when faced with some deeply felt need. We want to cure or, better still, prevent cancer; how do we go about it? We can try to improve the tools at hand: better methods of surgery or radiation treatment, earlier diagnosis, new drugs, a search for possible environmental agents. But none of these may turn out to provide a real solution. More fundamental understanding of cell biology may provide an answer, or it may come unexpectedly from completely unrelated areas: research on hay fever, insecticides, or abnormal psychology.

Choosing a problem and deciding how to go about solving it are difficult. It is not enough that the problem should be important—it may not be solvable at the time, or by the tactics proposed. Medawar has put it as follows:[8]

> No scientist is admired for failing in the attempt to solve problems that lie beyond his competence. The most he can hope for is the kindly contempt earned by the Utopian politician. If politics is the art of the possible, research is surely the art of the soluble. Both are immensely practical-minded affairs. (p. 97)

REFERENCE NOTES

1. This quotation is apparently a paraphrase of Billings (the pseudonym of 19th century humorist Henry Wheeler Shaw), who actually wrote: "It is better tew know nothing than tew know what aint so." We copied it when we saw it quoted somewhere, but we have forgotten the source. It is phrased better that way for our purpose than the way Billings actually put it.
2. Wolfgang Köhler, *The Mentality of Apes* (New York: Liveright, 1976).
3. Henri Poincaré, *The Foundations of Science: Science and Hypothesis, The Value of Science, Science and Method*. Translated by George B. Halstead (Lanham, Md.: University Press of America, 1982).
4. A. E. Housman, *The Name and Nature of Poetry* (Cambridge: Cambridge University Press, 1933).
5. Harry Wain, *A History of Preventive Medicine* (Springfield, IL: Charles C Thomas, 1970).
6. Hans Zinsser, *Biographical Memoirs of Theobald Smith* (Washington, DC: National Academy of Sciences, 1936).
7. Quoted in W. I. Beveridge, *The Art of Scientific Investigation* (New York: W. W. Norton, 1957).
8. P. B. Medawar, *The Art of the Soluble* (Harmondsworth: Penguin Books, 1969).

SUGGESTED READING

Ghiselin, Brewster, ed. *The Creative Process: A Symposium*. New York: New American Library, 1952.
Koestler, Arthur. *The Act of Creation*. New York: Macmillan, 1964.
Perkins, D. N. *The Mind's Best Work*. Cambridge, MA: Harvard University Press, 1981.
Tweney, Ryan D., Michael E. Doherty, and Clifford R. Mynatt, eds. *On Scientific Thinking*. New York: Columbia University Press, 1981.

17
The Dispassionate Scientist

THE MYTHS

In Chapter 6 on the kinetic theory of heat, we referred to a common myth about scientists: that they are objective, dispassionate observers of nature, who care only for truth and are willing to discard without a qualm any theory they hold, just as soon as experimental disproof is provided. We pointed out in that chapter how little Rumford fit this myth, and how effective he was precisely because he did not.

The myth about the personality of the scientist is related to a myth about scientific method itself—that it is a set of prescribed rules or procedures for the discovery of truth, which can be applied mechanically, and is independent of the personality of the one using it.

THE REALITY

Scientists, being human beings, tend to be governed in what they do by a mixture of motives, some altruistic and some selfish. More conservative than they like to imagine, sticking to the ideas they have grown up with and not giving them up lightly, they care deeply about their own theories, and are thus likely to underestimate their weaknesses and to judge counterevidence more harshly than evidence that would tend to support them.

To have intense emotional commitments to one's own work is both natural and necessary—why work hard at something unless one has some stake in the outcome? The drawback, of course, is that a strong commitment to one side of a dispute tends to make us overlook negative evidence and overstress the importance of positive evidence. But what tends to protect science as a whole from such errors (though it does not eliminate them) is that science rests in the long run on the consensus of scientists, not on the authority of any one individual, no matter how outstanding.

Scientists do not always act, any more than creative artists do, from a disinterested love of truth or beauty alone. Both are motivated also by pride, greed, the hunger for fame, and the honors and rewards that go with it. This is a fact that has received some general recognition since the publication of the book *The Double Helix,* in which James Watson describes his role in the discovery of the structure of nucleic acid, for which he and F. H. Crick received the Nobel Prize.[1] But it is really an old story. There is a lot of competition in science: at times, in a particular field, there will be a sense of discovery in the air, a shared feeling about the best way to solve some important problem, and many individuals will be working simultaneously in the same direction. The result is that very often a major breakthrough will be made simultaneously or almost simultaneously by several different people, though each may come to it by slightly different paths. Such simultaneous discoveries have led to bitter arguments among scientists as to who was first, and even whether the original discovery may not have been "stolen" from its first discoverer.

For Example: Isaac Newton

The sociologist Robert Merton has described such conflicts in an article in *The American Scholar*:[2]

> Long after he had made incomparable contributions to mathematics and physical science, Newton was still busily engaged in ensuring the luster and fame owing him. He was not merely concerned with establishing his priority but was periodically obsessed by it. He developed a corps of young mathematicians and astronomers, such as Roger Cotes, David Gregory, William Whiston, John Keill and, above all, Edmond Halley, "for the energetic building of his fame" (as the historian Frank Manuel has put it in his recent *Portrait of Isaac Newton*). Newton's voluminous manuscripts contain at least twelve versions of a defense of his priority, as against Leibniz, in the invention of the calculus. Toward the end, Newton, then president of the Royal Society, appointed a committee to adjudicate the rival claims of Leibniz and himself, packed the committee with his adherents, directed its every activity, anonymously wrote the preface for the second published report on the controversy—the draft is in his handwriting—and included in that preface a disarming reference to the legal adage that "no one is a proper witness for himself and [that] he would be an iniquitous Judge, and would crush underfoot the laws of all the people, who would admit anyone as a witness in his own cause." We can gauge the pressures for establishing his unique priority that must have operated for Newton to adopt such means for defense of his claims. As I shall presently suggest, this was not so much because Newton was weak as because the newly institutionalized value set upon originality in science was so great that he found himself driven to these lengths.
>
> By comparison, Watson's passing account [in *The Double Helix*] of a priority-skirmish within the Cavendish [Laboratory of Cambridge University] itself can only be described as tame and evenhanded, almost magnanimous. That conflict largely testified to the ambiguous origins of ideas generated in the course of interaction between colleagues, touched, perhaps, with a bit of cryptamnesia. (pp. 205–206)

Freud Also

> But perhaps the most apt case of the myth [that scientists are indifferent to the credit for a discovery] taking precedence over an accessible reality is provided by Ernest Jones, writing in his comprehensive biography that "Although Freud was never interested in questions of priority, which he found merely boring, he was fond of exploring the source of what appeared to be original ideas, particularly his own . . ." This is an extraordinarily illuminating statement by a scholar who had devoted his own life to penetrating the depths of the human soul. For, of course, no one could have known better than Jones—"known" in the narrowly cognitive sense—how very often Freud turned to matters of priority: in his own work, in the work of his colleagues (both friends and enemies) and in the history of psychology altogether. In point of fact, Freud expressed an interest in this matter on more than one hundred and fifty recorded occasions (I make no estimate of the unrecorded ones). With characteristic self-awareness, he reports that he even dreamed about priority and the due allocation of credit for accomplishments in science. (p. 214)

Merton cites examples of how Freud's obsession with issues of priority entered into his relationships with both his associates and his rivals, and concludes his discussion of Freud as follows:

> Judging from this small sampling of cases in point, it may not be audacious to interpret as a sign of resistance to reality Jones's remarkable statement that "Freud was never interested in questions of priority, which he found merely boring . . ." That Freud was ambivalent toward matters of priority, true; that he was pained by conflicts over priority, indisputable; that he was concerned to establish the priority of others as well as himself, beyond doubt and significant; but to describe him as "never interested in the question" and as "bored" by it requires the prodigious feat of denying, as though they had never occurred, scores of occasions on which Freud exhibited profound involvement in the matter, many of these being occasions that Jones himself has detailed with the loving care of a genuine scholar. (pp. 215–216)

WHY SCIENTISTS CARE SO MUCH

Merton goes on to point out the very human reasons why scientists should be concerned to receive credit for their discoveries:

> From still another perspective we can see the fallacy of the new mythology that construes the thirst for priority as altogether self-serving. Often the drive for recognized originality is only the other side of the coin of the elation that comes from having arrived at a new and true scientific idea or result. The deeper the commitment to the discovery, the greater, presumably, the reaction to the threat of having its originality denied. Concern with priority is often the counterpart to elation in discovery—the eureka syndrome. We have only to remember what is perhaps the most ecstatic expression of joy in discovery found in the annals of science: here, in abbreviation, is Kepler on his discovery of the third planetary law:

> When I prophesied 22 years ago as soon as I found the heavenly orbits were of the same number as the five (regular) solids, what I fully believed long before I had seen Ptolemy's Harmonics, what I promised my friends in the name of this book, which I christened before I was 16 years old, what I urged as an end to be sought, that for which I joined Tycho Brahe, for which I settled in Prague, for which I spent most of my life at astronomical calculations—at last I have brought to light and seen to be true beyond my fondest hopes. It is not 18 months since I saw the first ray of light, three months since the unclouded sun-glorious sight burst upon me! . . . The book is written, the die is cast. Let it be read now or by posterity, I care not which. It may well wait a century for a reader, as God has waited 6000 years for an observer.*

We can only surmise how deep would have been Kepler's anguish had another claimed that he had long before come upon the third law, just as we know how the young Bolyai, despairing to learn that Gauss had anticipated him in part of his non-Euclidean geometry and with the further blow, years later, of coming upon Lobachevsky's parallel work, suffered a great fall from the peak of exhilaration to the slough of despond and never again published any work in mathematics. The joy in discovery expressed by the young Jim Watson does not outstrip that of [the French chemist] Gay-Lussac, seizing upon the person nearest him for a victory waltz so that he could "express his ecstasy on the occasion of a new discovery by the poetry of motion.". . .

In short, when a scientist has made a discovery that matters, he is as happy as a scientist can be. But the height of exultation may only deepen the plunge into despair should the discovery be taken from him. If the loss is occasioned by finding that it was, in truth, not a first but a later independent discovery, that he had lost the race, the blow may be severe enough, although mitigated by the sad consolation that at least the discovery had been confirmed by another. But this is nothing, of course, when compared with the traumatizing experience of having it suggested that not only was the discovery later than another of like kind but that it was really borrowed. The drive for priority is in part an effort to reassure oneself of a capacity for original thought. Thus, rather than being mutually exclusive, as the new mythology of science would have it, joy in discovery and the quest for recognition by scientific peers are stamped out of the same psychological coin. In their conjoint ways, they both express a basic commitment to the value of advancing knowledge. (pp. 222–224)

THE DEPERSONALIZATION OF DISCOVERY

In Chapter 16, we tried to convey some appreciation of the role of individual, personal, and chance factors in scientific discovery. However, once the discoverer has to write an article to announce his results, a process of depersonalization begins. In modern times, a dry, unemotional tone which uses a prescribed, rather colorless jargon has become fashionable. Papers written in the lively, engaging style of Rumford or the measured prose of Snow are no longer common. But in spite of the writing style new discoveries still bear the stamp of the discoverer and express

---

* We have used this quotation earlier (p. 282) in a different context.

something of his own view of the meaning of his work. We mentioned that the absorption of any discovery into the body of science is the act of a scientific consensus. As the other members of the scientific community begin to accept a new idea and apply it in new ways to new problems its discoverer may not have anticipated, the idea begins to change its form. It is now seen in new lights, and begins to alter in response to the various insights of those who use it. Eventually it reaches a degree of acceptance that allows it to become textbook material, by which time it has lost most features that reflect the idiosyncrasies of the person who found it in the first place and the way the discovery was made. As science develops, the idea changes still further, sometimes being incorporated into new theories, as Snow's hypothesis on cholera became one application of the more general germ theory of disease, and sometimes being expressed in different language or formulation. Although physicists still use the laws of mechanics discovered by Newton, they do not learn them from Newton's book, nor do they use Newton's mathematics. New kinds of mathematics have been developed since Newton's time that are easier to use, and that make it possible to solve problems Newton could not.

Science has been compared to a coral reef, where the living organisms at the surface produce the growth of the reef on top of tens or hundreds of feet of skeletons of organisms that have long since died. The life of the reef is only at its surface; the life of science is only at its frontier.[3]

There is truth in this rather unhappy image. In the light of it, science compares unfavorably with the arts, where the object in which the creative artist embodies his insight retains its value and interest. However, there is a role for the personality of the individual in spite of this eventual depersonalization of the discovery. There is such a thing as a scientific style that distinguishes one scientist from another. It is expressed in the field a scientist chooses, in the problems he chooses within his field, and in the way he attacks these problems. One feels this in reading Black's and Rumford's accounts of their researchers. While the dry, unemotional style used in scientific publication today may conceal it, scientists still use as wide a variety of approaches as did their predecessors of a century or two ago.

Because science, as we have indicated, works by consensus, each individual practitioner is not required to have all the contradictory qualities that characterize the scientific venture itself. There is room for the daring, speculative, inventive spirit who creates new theories or tries bold, imaginative experiments as well as for the cautious, critical spirit who examines theories searchingly or patiently designs and performs tedious but necessary experiments. There are those who like the power and conciseness of mathematics and those who prefer the nuances and color of words; those who prefer the laboratory and those who prefer the library. One person wants to deal with human beings and human problems and is willing to pay the price of vagueness and uncertainty; another prefers atoms and molecules, about which precise questions can be asked and given definitive answers. There is still another division, between those who are not satisfied unless their work is useful, and those who are satisfied with the knowledge itself.

But the joy and excitement of discovery, as described above by Merton and directly expressed by Kepler, Kekulé, and Poincaré, is a goal for all.

REFERENCE NOTES

1. James D. Watson, *The Double Helix: Being a Personal Account of the Discovery of the Structure of DNA* (New York: Atheneum, 1968).
2. Robert K. Merton, "Behavior Patterns of Scientists," *American Scholar* 38 (1969): 197–225. Reprinted by permission of the author.
3. Attributed to the physicist and Nobel laureate W. L. Bragg.

SUGGESTED READING

Scientific American. *Lives in Science: A Scientific American Book*. New York: Simon and Schuster, 1957.
Newman, James R. *Science and Sensibility*. New York: Simon and Schuster, 1961.

18
The Cultural Roots of Science

THE SUBJECTIVE ELEMENT

Throughout this book on scientific method we have repeatedly stressed how little "method" there really is in science. There is no set of prescribed rules which, when followed, will lead unerringly to the truth. Instead, progress is made by reliance on the judgment of individuals who must choose among a complex set of possible strategies that are often in conflict with each other. Progress in science depends more on intuition than on explicit procedures.

We have often emphasized the subjective element in science. This appears first in the realization that even "scientific facts" contain a more or less culturally conditioned component. It appears also in the creative processes of individual discovery, and in the role of the consensus of scientists who decide, on the basis of commonly shared but subjective criteria, which problems are important, which experiments are decisive, which theories are correct. And we have indicated that this consensus is not a democratic consensus of everybody, but rather a narrow consensus of interested specialists who have taken the time and effort to master the methods and problems of some particular discipline. We have described the risks of such a procedure—the frequent occasions when the informed consensus shared misconceptions that hindered understanding, when the few creative thinkers therefore had to struggle to change those misconceptions.

It follows, then, in view of the overriding importance of this subjective element, that the science of any particular time is rooted in the intellectual climate of that time, and can escape its limitations only with difficulty. One can go further: the concepts of what we call science are inextricably bound up with the particular cultures that have produced it, and that have been shaped by it in turn.

We cannot offer the reader any magical procedure for rising above the limitations of the intellectual climate to which he is exposed, or, more broadly, above the limitations of his own culture, so that he might be able to think about scientific questions in other than "culture-bound" terms. But here in the final chapter of this book we do want to discuss the role of a specific culture in determining the nature of the science that can be carried on in that culture.

THE TACIT COMPONENT

We may use as an example a fact about this book to which the reader has almost certainly given no thought, but which he or she must concede as true the moment it is pointed out. *This book is written in English.* We have assumed that the reader shares with us a common language, with its vocabulary, grammar, and nuances of expression. We did not begin this book with a course in the English language to make sure the reader understood the sentences in the way we meant them. We took the common language for granted.

In the same way, we assumed that the reader has the common experiences and common sense of members of an industrialized, scientifically oriented society, the concepts, beliefs, logic, and familiarity with things and facts shared by most people living in such a society. We took for granted what we and the reader agree on, and spent time explaining only those aspects of science that we supposed were unfamiliar. We did this because we had no alternative in writing a book of this kind, any more than we had an alternative to writing it in *some* language or other.

We do not believe that it would be possible, even if we wanted to, to explore and discuss every assumption, every concept, used by ourselves and by the reader in developing our picture of scientific method. But in this chapter we have a smaller ambition: to help the reader realize that there are shared components of our culture that we accept without conscious awareness, and that these tacit and unanalyzed components are as essential in providing the conditions for scientific activity as are the methods, procedures, and practices with which this book has been largely concerned.

The discussion so far has been abstract. To make it more concrete, we will provide an example of how a culture determines what kind of science is possible. Historians of science have studied in detail the relation between the scientific thought of a period and its cultural and intellectual climate. We have discussed one example from the history of physics in Chapter 6, where we dealt with the reasons why Rumford's experiments on heat, which in retrospect seem conclusive to us, were not accepted as such by most scientists of his time; and why half a century had to elapse, during which discoveries would have to be made in seemingly unrelated fields of physics, before the ideas of the kinetic theory prevailed. Other examples could have been chosen from similar historical studies, but we feel that the points we wish to make can be made more sharply by considering a culture as different as possible from our own. For this we have chosen a "primitive" culture—that of the Azande in eastern sub-Saharan Africa, as studied by the British anthropologist Edward Evans-Pritchard in the late 1920s.

THE CULTURAL ROOTS OF SCIENCE

THE BELIEF IN WITCHCRAFT

The Azande believe that some people of their own tribe are witches by biological inheritance, and that this can be experimentally demonstrated by the discovery of a distinct substance in their bodies after death. Witches have the power, out of spite or envy, to injure others and even cause their deaths. As we will see, the belief in witchcraft fulfills many of the criteria of a scientific system: it provides

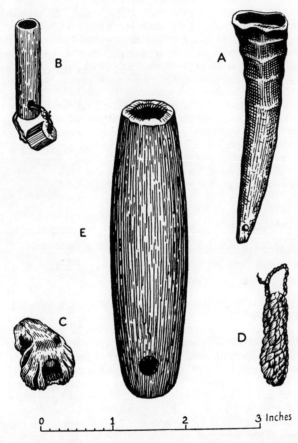

FIGURE 18-1. Zande magic whistles. (A) Gazelle's horn for preventing rain. (B) Whistle of the *Mani* association with the blue bead badge of the association attached to it. (C) Whistle to protect a man against *adandara* cats. The mouth has been scraped away to provide dust for eating. (D) *Gbau*, a charm of invisibility. (E) A whistle to give protection against witchcraft. From Evans-Pritchard, *Witchcraft, Oracles and Magic among the Azande*. (Reproduced with permission of Oxford University Press.)

understanding, it is generally applicable to a wide range of phenomena, and it is based on experimental evidence.

The pervasive role played by witchcraft (see Figure 18-1) is described by Evans-Pritchard as follows:[1]

> Witchcraft is ubiquitous. It plays its part in every activity of Zande life; in agricultural, fishing, and hunting pursuits; in domestic life of homesteads as well as in communal life of district and court; it is an important theme of mental life in which it forms the background of a vast panorama of oracles and magic; its influence is plainly stamped on law and morals, etiquette and religion; it is prominent in technology and language; there is no niche or corner of Zande culture into which it does not twist itself. If blight seizes the ground-nut crop it is witchcraft; if the bush is vainly scoured for game it is witchcraft; if women laboriously bale water out of a pool and are rewarded by but a few small fish it is witchcraft; if termites do not rise when their swarming is due and a cold useless night is spent in waiting for their flight it is witchcraft; if a wife is sulky and unresponsive to her husband it is witchcraft; if a prince is cold and distant with his subject it is witchcraft; if a magical rite fails to achieve its purpose it is witchcraft; if, in fact, any failure or misfortune falls upon any one at any time and in relation to any of the manifold activities of his life it may be due to witchcraft. Those acquainted either at first hand or through reading with the life of an African people will realize that there is no end to possible misfortunes, in routine tasks and leisure hours alike, arising not only from miscalculation, incompetence, and laziness, but also from causes over which the African, with his meagre scientific knowledge, has no control. (pp. 63–64)

Arguing with the Azande

Evans-Pritchard at times attempted to convince the Azande that witchcraft was not causing these misfortunes, using the sort of arguments that would naturally occur to a person from a Western scientific culture:[1]

> I found it strange at first to live among Azande and listen to naïve explanations of misfortunes which, to our minds, have apparent causes, but after a while I learnt the idiom of their thought and applied notions of witchcraft as spontaneously as themselves in situations where the concept was relevant. A boy knocked his foot against a small stump of wood in the centre of a bush path, a frequent happening in Africa, and suffered pain and inconvenience in consequence. Owing to its position on his toe it was impossible to keep the cut free from dirt and it began to fester. He declared that witchcraft had made him knock his foot against the stump. I always argued with Azande and criticized their statements, and I did so on this occasion. I told the boy that he had knocked his foot against the stump of wood because he had been careless, and that witchcraft had not placed it in the path, for it had grown there naturally. He agreed that witchcraft had nothing to do with the stump of wood being in his path but added that he had kept his eyes open for stumps, as indeed every Zande does most carefully, and that if he had not been bewitched he would have seen the stump. As a conclusive argument for his view he remarked that all cuts do not take days to heal but, on the contrary, close quickly, for that is the nature of cuts. Why, then, had his sore festered and remained open if there were no

witchcraft behind it? This, as I discovered before long, was to be regarded as the Zande explanation of sickness. . . . (pp. 65–66)

One of my chief informants, Kisanga, was a skilled woodcarver, one of the finest carvers in the whole kingdom of Gbudwe. Occasionally the bowls and stools which he carved split during the work, as one may well imagine in such a climate. Though the hardest woods be selected they sometimes split in process of carving or on completion of the utensil even if the craftsman is careful and well acquainted with the technical rules of his craft. When this happened to the bowls and stools of this particular craftsman he attributed the misfortune to witchcraft and used to harangue me about the spite and jealousy of his neighbours. When I used to reply that I thought he was mistaken and that people were well disposed towards him he used to hold the split bowl or stool towards me as concrete evidence of his assertions. If people were not bewitching his work, how would I account for that? (pp. 66–67)

Carelessness and Witchcraft

The Azande do not explain all misfortunes as the result of witchcraft. They recognize fully that misfortunes can be caused by one's own carelessness, inexperience, or improper behavior. A girl who breaks her water pot and a boy who forgets to close the door of the henhouse at night will be admonished by their parents, but the harmful consequences of their negligence are not attributed to witchcraft; if a clay pot cracks during firing, and it is found on examination that a pebble was left by accident in the clay, this will be blamed on the carelessness of the potter. It is only those misfortunes that occur to people who have taken normal and reasonable care in their work or their lives that need such an explanation:[1]

In speaking to Azande about witchcraft and in observing their reactions to situations of misfortune it was obvious that they did not attempt to account for the existence of phenomena, or even the action of phenomena, by mystical causation alone. What they explained by witchcraft were the particular conditions in a chain of causation which related an individual to natural happenings in such a way that he sustained injury. The boy who knocked his foot against a stump of wood did not account for the stump by reference to witchcraft, nor did he suggest that whenever anybody knocks his foot against a stump it is necessarily due to witchcraft, nor yet again did he account for the cut by saying that it was caused by witchcraft, for he knew quite well that it was caused by the stump of wood. What he attributed to witchcraft was that on this particular occasion, when exercising his usual care, he struck his foot against a stump of wood, whereas on a hundred other occasions he did not do so, and that on this particular occasion the cut, which he expected to result from the knock, festered whereas he had had dozens of cuts which had not festered. Surely these peculiar conditions demand an explanation. . . . (pp. 67–68)

In Zandeland sometimes an old granary collapses. There is nothing remarkable in this. Every Zande knows that termites eat the supports in course of time and that even the hardest woods decay after years of service. Now a granary is the summerhouse of a Zande homestead and people sit beneath it in the heat of the day and chat or play the African hole-game or work at some craft. Consequently it may happen that there are people sitting beneath the granary when it collapses and they are injured, for it is a heavy structure made of beams and clay and may be stored with eleusine as well. Now why should these particular people have been sitting under this particular granary at the particular moment when it collapsed? That it should collapse is easily intelligible, but why should it have collapsed at the particular moment when these particular people were sitting beneath it? Through years it might have collapsed, so why should it fall just when certain people sought its kindly shelter? We say that the granary collapsed because its supports were eaten away by termites. That is the cause that explains the collapse of the granary. We also say that people were sitting under it at the time because it was in the heat of the day and they thought that it would be a comfortable place to talk and work. This is the cause of people being under the granary at the time it collapsed. To our minds the only relationship between these two independently caused facts is their coincidence in time and space. We have no explanation of why the two chains of causation intersected at a certain time and in a certain place, for there is no interdependence between them.

Zande philosophy can supply the missing link. The Zande knows that the supports were undermined by termites and that people were sitting beneath the granary in order to escape the heat and glare of the sun. But he knows besides why these two events occurred at a precisely similar moment in time and space. It was due to the action of witchcraft. If there had been no witchcraft people would have been sitting under the granary and it would not have fallen on them, or it would have collapsed but the people would not have been sheltering under it at the time. Witchcraft explains the coincidence of these two happenings. (pp. 69–70)

The Poison Oracle

The Azande have a procedure for ascertaining if someone is threatened or being made to suffer by witchcraft, and who the witch is. It is by the use of what Evans-Pritchard calls the "poison oracle."

The procedure for appeal to the poison oracle requires the preparation of *benge*, a presumably poisonous extract of a particular plant (which does not grow in the country the Azande live in, but must be prepared elsewhere and brought in).

An operator forces *benge* down the throat of a young fowl (see Figure 18-2) while addressing questions to the *benge* inside the fowl. Sometimes the fowl dies during the procedure and sometimes it lives. The life or death of the fowl provides the answer to the question.

The primary question to be answered may be, for example, "Is Namarusu's health threatened by Nabani and her relatives?" That is, is Nabani using witchcraft or some other form of magic to harm Namarusu?

The operator speaks as follows, having administered the poison:

"Poison oracle, if Namarusu's health is threatened by Nabani, kill the fowl. If Namarusu's health is not threatened by Nabani, spare the fowl." If the fowl dies, the answer to the primary question is "yes"; if it does not die, the answer is "no."

One may wonder, as Evans-Pritchard did, if the Azande realize they are administering a poison that might kill the fowl regardless of the question asked:[1]

FIGURE 18-2. Operating the poison oracle. The operator contemplates the chicken during the address to the oracle (above). The chicken, held in the operator's hand, is at its last gasp (below). From Evans-Pritchard, *Witchcraft, Oracles and Magic among the Azande*. (Reproduced with permission of Oxford University Press.)

Therefore, to ask Azande, as I have often asked them, what would happen if they were to administer oracle poison to a fowl without delivering an address or, if they were to administer an extra portion of poison to a fowl which has recovered from the usual doses, or, if they were to place some of the poison in a man's food, is to ask silly questions. The Zande does not know what would happen, he is not interested in what would happen, and no one has ever been fool enough to waste good oracle poison in making such pointless experiments, experiments which only a European could imagine. Proper *benge* is endowed with potency by man's abstinence and his knowledge of tradition and will only function in the conditions of a seance.

When I asked a Zande what would happen if you went on administering dose after dose of poison to a fowl during a consultation in which the oracle ought to spare the fowl to give the right answer to the question placed before it, he replied that he did not know exactly what would happen, but that he supposed sooner or later it would burst. He would not countenance the suggestion that the extra poison would otherwise kill the fowl unless the question were suddenly reversed so that the oracle ought to kill the fowl to give a correct answer when, of course, it would at once die. When I asked a Zande whether you might not put a handful of the poison into a man's beer and rid yourself of an enemy expeditiously he replied that if you did not utter an address to the poison it would not kill him. I am sure that no Zande would ever be convinced that you could kill a fowl or person with *benge* unless it had been gathered, administered, and addressed in the traditional manner. Were a European to make a test which proved Zande opinion wrong they would stand amazed at the credulity of the European who attempted such an experiment. If the fowl died they would simply say that it was not good *benge*. The very fact of the fowl dying proves to them its badness. . . . (pp. 314–15)

If you ask a Zande what would happen if a man were to administer three or four doses to a tiny chicken instead of the usual one or two doses he does not perceive that there is any subtlety in your inquiry. He will reply to you that if a man were to do such a thing he would not be operating the oracle properly. He does not see the relevance of your question, for you are asking him what would happen if a man were to do what no one ever does and he has no interest in hypothetical actions. During the early part of my residence among them they used to say to me, "You do not understand these matters. However many doses you administer to a chicken it does not alter the verdict of the oracle." You say to the oracle, "So-and-so is ill. If he will live, poison oracle kill the fowl. If he will die, poison oracle spare the fowl." "If he is going to live, however many doses of the poison you administer to the fowl it will still survive." They responded to my questions without signs of distress. Clearly they were not defending a position which they felt to be insecure. (p. 324)

One should recognize that the Azande, in refusing to consider the "experiments" suggested by Evans-Pritchard, are not being unreasonable or unscientific by the standards of their own culture. They are responding much as an astronomer might if he were asked if he would still see the stars through his telescope if the outside of it were painted blue. He would regard the question as pointless, based on a complete misconception of how a telescope works, and he would not be in the least tempted to paint it blue to convince the questioner or reassure himself.

The Confirmatory Test

In using the *benge*, or poison oracle, a confirmatory test is required: to be sure of the answer, two tests on two different fowls must be made. In the second test the same basic question is being asked, but in order for the answer to be confirmed, the wording of the question must be such that if in the first test the fowl died then in the second test it must live, and vice versa. Specifically, in the second test (if the fowl died in the first one) the operator must address the *benge* in a form equivalent to the following: "If the poison oracle told the truth on the previous test, spare the fowl. If it lied, kill the fowl."

If the results of the second test confirm the results of the first (the fowl in the second test lives), the result is trusted. If, however, the second test does not confirm the first (the fowl dies), the result is invalid. Assuming for simplicity that about half the time the poison kills a fowl and half the time it does not, we can apply probability theory to predict that about half the time the two tests will lead to a consistent answer and about half the time they will lead to an invalid answer. How do the Azande deal with the invalid results? What happens if the answer given even by a valid result turns out to be wrong?

Dealing with Contradictory Results[1]

> What explanation do Azande offer when the oracle contradicts itself? Since Azande do not understand the natural properties of the poison they cannot explain the contradiction scientifically; since they do not attribute personality to the oracle they cannot account for its contradictions by volition; and since they do not cheat they cannot manipulate the oracle to avoid contradictions. The oracle seems so ordered to provide a maximum number of evident contradictions for, as we have seen, in important issues a single test is inacceptable and the oracle must slay one fowl and spare another if it is to deliver a valid verdict. As we may well imagine, the oracle frequently kills both fowls or spares both fowls, and this would prove to us the futility of the whole proceeding. But it proves the opposite to Azande. They are not surprised at contradictions; they expect them. Paradox though it be, the errors as well as the valid judgements of the oracle prove to them its infallibility. The fact that the oracle is wrong when it is interfered with by some mystical power shows how accurate are its judgements when these powers are excluded.
>
> A Zande is seated opposite his oracle and asks it questions. In answer to a particular question it first says "Yes" and then says "No." He is not bewildered. His culture provides him with a number of ready-made explanations of the oracle's self-contradictions and he chooses the one that seems to fit the circumstances best. He is often aided in his selection by the peculiar behaviour of the fowls when under the influence of the poison. The secondary elaborations of belief that explain the failure of the oracle attribute its failure to (1) the wrong variety of poison having been gathered, (2) breach of a taboo, (3) witchcraft, (4) anger of the owners of the forest where the creeper grows, (5) age of the poison, (6) anger of the ghosts, (7) sorcery, (8) use. . . . (pp. 329–330)

Witchcraft . . . is often cited as a cause for wrong verdicts. It also may render the oracle impotent, though impotency is usually attributed to breach of taboo. Generally speaking, the presence of witchcraft is shown by the oracle killing two fowls in answer to the same question, or in sparing two fowls in answer to the same question when it has killed a fowl at the same seance. In such cases the poison is evidently potent and its failure to give correct judgements may be due to a passing influence of witchcraft. For the time being the seance may be stopped and resumed on another day when it is hoped that witchcraft will no longer be operative. Out of spite a witch may seek to corrupt the oracle, or he may act to protect himself when the oracle is being consulted about his responsibility for some misdeed. . . . (p. 332)

But when faith directs behaviour it must not be in glaring contradiction to experience of the objective world, or must offer explanations that demonstrate to the satisfaction of the intellect that the contradiction is only apparent or is due to peculiar conditions. The reader will naturally wonder what Azande say when subsequent events prove the prophecies of the poison oracle to be wrong. The oracle says one thing will happen and another and quite different thing happens. Here again Azande are not surprised at such an outcome, but it does not prove to them that the oracle is futile. It rather proves how well founded are their beliefs in witchcraft and sorcery and taboos. On this particular occasion the oracle was bad because it was corrupted by some evil influence. Subsequent events prove the presence of witchcraft on the earlier occasion. The contradiction between what the oracle said would happen and what actually has happened is just as glaring to Zande eyes as it is to ours, but they never for a moment question the virtue of the oracle in general but seek only to account for the inaccuracy of this particular poison, for every packet of *benge* is an independent oracle and if it is corrupt its corruption does not affect other packets of the poison. (p. 338)

SCIENCE VERSUS WITCHCRAFT

As we pointed out, the Azande belief in witchcraft has many of the features of a scientific system. First, it has great explanatory power and is of great generality in application: it explains more of the events and misfortunes of daily life than any Western scientific system does. Also, it is supported by experimental evidence: the stubbed toe that becomes infected, the wooden bowl that splits, the granary that collapses. Further experimental evidence is provided by the poison oracle.

One may ask, but what about the truth or falsity of the belief in witchcraft itself? Can that not be subject to an experimental test that would convince the Azande of its falsity?

Here we must remember a point made many times in this book: testing a theory is not a routine procedure. What kind of experimental evidence will be considered relevant is always a subjective judgment of the scientific community involved. When the Azande disregard or brush aside the types of experiments we would propose in order to refute their belief in witchcraft, they are not acting so differently from ourselves. We too stick with a theory that we have found useful, in spite of awkward contradictory facts, unless we have a better one to replace it with. We may also fail to recognize contradictions and absurdities in our strongly held beliefs, although they may be apparent to others who do not share them.

All the arguments and evidence we could muster to refute the Azande beliefs may be appropriate and convincing to fellow members of our own culture, but not to members of theirs, as Evans-Pritchard points out frequently. We note that up to a few centuries ago a belief in witchcraft was widespread in Western societies also. It no longer is, but not because solid scientific evidence was provided to refute it.

There is no appeal to "common sense," either. The Azande have survived as a people for a long period of time in a hostile and difficult environment, and have developed a complex society, based on both hunting and agriculture, which includes different occupations: princes, witch-doctors, and woodcarvers, among others. Whatever we mean by common sense, we cannot say that they lack it just because ours is different.

In making these points we are raising difficult philosophical questions that are beyond our competence to answer: Is what we have called the scientific method, as developed mainly in Western societies, a better, surer road to truth and understanding than any other? Or is truth relative, is truth in one culture falsehood in another, with no objective way to decide? We are not trying to answer these questions. Our purpose is a much less ambitious one: to make the reader aware of them.

However solid and universal scientific knowledge may seem, it should be recognized as a culturally determined kind of knowledge, expressed in the language—in both the literal and conceptual senses of the word *language*—of a particular culture, and it depends in complex ways on the unspoken assumptions of that culture.

CULTURES AND SUBCULTURES

We hope that this description of Azande beliefs has helped the reader recognize the relation between science and the presuppositions of the culture of which science is a part. It should also reveal something about the relationships within a society among the various subcultures that are distinguished by their adherence to particular sets of beliefs. This includes the different disciplines which constitute science, as well as the competing schools of thought within each discipline. It also includes other subgroups within Western society: for example, those with distinct sets of religious, political, or esthetic beliefs. Members of such subgroups share not only formally expressed beliefs of which they are consciously aware, and which they recognize as setting them off from others, but also sets of tacit and unconscious concepts which provide the foundations on which the formal beliefs are based. This unacknowledged foundation of beliefs is something we dimly sense when we argue with members of subgroups other than our own, and it is the existence of such tacit components that often makes such arguments difficult, frustrating, and pointless.

SCIENTIFIC SUBCULTURES

The tacit part of knowledge creates problems for communication between adherents of different belief systems, and—from a narrower standpoint, more relevant to this book—it creates problems within science. For controversies between different scientific schools of thought are not only about what is conscious and acknowledged.

In the case histories we discussed examples of such controversies: between the kinetic and caloric theories of heat, between psychogenic and biological theories of the origin of schizophrenia. In both, the members of opposing schools of thought formed their own subcultures, each with its own distinctive philosophy, style, type of training, way of asking questions, and so forth. The differences were about much more difficult things than facts alone. In truth, the differences between scientific subcultures can seem at times as great as those between Western society and the Azande. Using the term proposed by T. S. Kuhn, they operate according to different "paradigms."[2]

The problems that arise in the course of a scientific revolution, when scientists belonging to a particular subculture are faced with the necessity of rising above its limitations and seeing things from an entirely different viewpoint, were described in a pessimistic and cynical way by Max Planck, one of the discoverers of the quantum theory of the atom:[3]

> An important scientific innovation rarely makes its way by gradually winning over and converting its opponents: it rarely happens that Saul becomes Paul. What does happen is that its opponents gradually die out and that the growing generation is familiarized with the idea from the beginning. (p. 97)

BREAKING THROUGH

One should avoid the temptation to feel superior to the conservative and culture-bound members of a scientific discipline faced with revolutionary new ideas. Most dazzling and startling new ideas are not revolutionary advances at all—they are false starts and deserve rejection. There is a lot to be said for judgment, discrimination, a measure of caution, even if once in a thousand or ten thousand times they lead one to disregard a new and valuable insight.

Fortunately, we do break through the limitations of our culture from time to time. Scientific revolutions have occurred and will occur again, with dramatic consequences for our previous conceptions of ourselves and of our universe, as happened with Darwin's statement that we are descended from animals, with Einstein's discovery that matter and energy are different manifestations of a single entity, with Freud's revelations of the sources of our deepest feelings.

Of course, most scientific discoveries are not really breathtaking new insights which change our most fundamental conceptions of what is true and what is not.

They are often humble affairs, having no dramatic consequences outside the narrow discipline in which they take place. Yet all of them do involve to some extent a process of making explicit what had previously been tacitly accepted, and challenging it. Something that had been buried in the collective unconscious of the members of some culture or subculture is forced to the surface and looked at for the first time. Much of scientific discovery, particularly that which changes us, the discoverers, is a discovery of a part of our tacit heritage, a recognition that something that has been so taken for granted that we have never dreamed of doubting it is, first, doubtable and, finally, wrong.

Fortunately, there are simple enough examples from our own history, which to some extent we have to relive in our own individual intellectual development, to convey the idea of what a scientific revolution is like. To think of the world as round rather than flat is one. To make such a leap in thought, it is necessary to overcome one basic concept we all have developed from our earliest experiences: the concept of *up* and *down*. It is easy for adults, having been indoctrinated with the idea, both in school and out of it, that *up* and *down* are determined by the earth's gravitational field, to know that up and down in China or Australia are different from up and down in New York or London. But a young child, seeing things fall down, falling himself as he begins to walk, feeling with every step the sense of the downward force of his own weight, takes as given the idea that *down* is the unique direction in which things fall and *up* its opposite. The basic experiences that define these directions do not hint to us that they are relative, and that under other circumstances—on a rocket traveling to the moon, for example—*up* and *down* are not even there. One can see what a leap of the imagination, what a denial of the obvious and unquestioned, was involved in thinking of a round earth.

A second such revolution, again repeated in our individual intellectual growth, is to see the earth as moving, rather than the sun and the stars. Again, we had to discard what was obvious from our earliest experiences. Almost all our experiences of motion are experiences of rough motion. Although we no longer ride much on horseback or in ox carts but rather use cars and planes, we associate motion with jerks, bumps, the pulls of acceleration and deceleration. How could the earth be moving if we do not feel it? To imagine a moving earth was not simply a matter of stating glibly that motion is relative. It was necessary to imagine an idealized kind of motion that lay outside ordinary experience—a smooth motion at a constant speed, without those bumps and jerks that we have learned to associate with motion as we commonly experience it. Once this was done, it became possible to speculate that our impression that we are standing still, and that the sun and stars are moving about us, might be only an impression, and that reality might be something else.

Living through such revolutions—being forced to drop our old viewpoint, which we may have relied on for most of a lifetime, and which had seemed to offer us such a solid, true picture of reality, and adopt a completely new one—can be an unnerving experience. The history of science is filled with examples of scientists, even great ones, who were unable to face it. But it can also be an exhilarating and liberating experience, and the highest reward of the scientific life.

REFERENCE NOTES

1. Edward Evans-Pritchard, *Witchcraft, Oracles, and Magic among the Azande* (Oxford: Clarendon Press, 1937). Reprinted by permission of Clarendon Press.
2. Thomas S. Kuhn, *The Structure of Scientific Revolutions*, 2nd ed. (Chicago: University of Chicago Press, 1970).
3. Max Planck, *The Philosophy of Physics*, trans. W. Johnston (New York: W. W. Norton, 1936).

SUGGESTED READING

In this chapter we quote extensively from Edward Evans-Pritchard's book, *Witchcraft, Oracles and Magic among the Azande* (Oxford: Clarendon Press, 1937) with the permission of the publishers. An abridged paperback version has also been published by Oxford University Press. More detailed discussions of the issues dealt with in this chapter have appeared in *Modes of Thought: Essays on Thinking in Western and Non-Western Societies,* edited by Robin Horton and Ruth Finnegan (London: Faber and Faber, 1973), and in a two-part article by Robin Horton, "African traditional thought and Western science," in vol. 37 of the journal *Africa* (1967). The tacit component in scientific knowledge has been discussed in detail by Michael Polanyi in *Personal Knowledge* (London: Routledge and Kegan Paul, 1958).

Index

Accuracy, 120, 328–336
   limits on value of, 331–332
Adler, Alfred, 300–301
Alice in Wonderland, 24
American Psychiatric Association, 218
Anecdotal evidence, 235
Anthropology, 237, 249, 378–387
Apes, use of language, 322–323
Arieti, Silvano, 195, 238
Aristotle, 17, 21, 294, 296
Astrology, 16, 29
Astronomy, 2, 15, 287, 369–370
   planetary motion, 90, 287–288, 369–370
Atomic theory of matter. *See also* Chemistry
   atoms and molecules, distinction, 130, 280–282
   heat, kinetic theory of, xi, 94, 280–282
   Newton's laws applied to, 287
Augustine, Saint, 294–295, 299
Azande (African tribe), xii, 378–388. *See also* Witchcraft, Azande belief in

Babylonians, 16
Bacon, Francis, 16, 95, 98, 120, 301, 307–308
Bacon–Shakespeare controversy, 307
Balance, analytical, 120, 121
Baldwin, James, 313
Barton, Russel, 219–228, 231
Bedlam (Bethlehem Hospital), 186–187, 188
Behan, Brendan, 233
Belief, degree of, 299
Bell-shaped curve. *See* Gaussian curve
Benge. *See* Witchcraft, Azande belief in

Berlin, Isaiah, 290
Bias, 321–322, 329, 330
Bible, The, 29, 155, 295
Billings, Josh, 362
Biology. *See* Evolution, theory of; Genetics
Birge, R. T., 333
Black, Joseph, 96, 103, 122, 303, 304, 305
   caloric theory, preference for, 112
   capacity for heat, 108–112
   distinction between heat and temperature, 106–107
   equilibrium of heat, 102–103, 106
   latent heat, 113–117
Bleuler, E., 193, 250, 251–252
Blind and double-blind experiments, 245, 320
Boerhaave, H., 109, 120
Bohr, Niels, 288
Boisen, Anton, 200
Bolyai, J., 374
Boyle, Robert, 95
Brahe, Tycho, 1, 282
Bulgarian pigs, statistics of, 336
Büttner (Gauss's teacher), 89–90

Caloric theory. *See* Heat, caloric theory of
Calorique, 94
Cancer, x, 56–69
   causes or contributing factors
      aflatoxin, 62
      age, 57, 58n
      age of child-bearing, 62
      asbestos, 63
      biological and biochemical study of, 57
      diet, 69
      genetics, 60–62

Cancer (*cont.*)
  causes or contributing factors (*cont.*)
    industrial chemicals, 60, 62, 63, 68
    life-style, 57
    malarial infection, 62–63
    radiation, high-energy, 62
    smoking, 57, 62, 64–67, 68
    soot from chimneys, 63
    sunlight, 59, 60, 63
    uranium, 63
    Vitamin A, insufficient, 69
    zeolite minerals, 63
  geographical distribution, 57–61, 63, 68
  migration studies, 61
  rates
    age, 57, 58n
    blood type, 60
    child-bearing age, 62
    environment, 61–69
      industrial, 62–63
      life-style, 62–68
      natural, 62–68
    ethnicity, 57, 60, 62
    occupation, 57, 60, 62, 63
    smokers, 64–67
    urban-rural, 66
  types
    bladder, 60, 65, 68
    breast, 58, 59, 60, 61, 62, 190
    Burkitt's lymphoma, 62–63
    cervix, 59, 61, 63
    esophagus, 58, 61, 65, 68
    leukemia, 190
    liver, 58, 63, 68
    lung, 16, 57, 61, 64–67, 68
    melanoma, 59
    mesothelioma, 64, 68
    scrotum, 63
    skin, 59
    stomach, 60, 61
    uterus, 59, 61, 63
Candide, 300
Carnot, Sadi, 118–119, 305
Cataract, vision after removal of, 11–14
Catholic church, 30, 291, 295
Causation, criteria for, 312
Ceres (planet), 90
Chadwick, E., 35
Chapman, George, 283, 307–308
Chemical transmission of nerve impulses, 206

Chemistry, 5, 280–282. *See also* Atomic theory of matter
Cholera, xi, 29–56, 66, 189, 207, 375
  Broad St. Pump outbreak, 39–43, 50
  contagiousness of, 33
  germ theory and, 35, 49, 288
  history of, 30, 32–33
  preventive measures, 54–55
  rates, 327
    coal miners, 45
    doctors, 36, 45
    and elevation, 51–52, 272, 303
    in England and Scotland, 49–50
    and housing conditions, 37
    social class differences, 36–37, 45
    water company customers, 45–48, 305, 309–311
    water from sandstone and limestone, 52
  symptoms, 31
  theories of transmission
    effluvia, 34, 35, 48, 50, 51, 95
    ingestion of excretions, 31, 36–52, 235
    water from sandstone and limestone, 52
    water supply, pollution by excretions, 28, 37–52, 243, 303, 309–311, 366
  treatment by fluid replacement, 55–56
Classification in science, xii, 183, 188–190, 218
Clever Hans, 321–322
Codes and cryptography, 84n
Cohen, Morris R., 25, 26, 27
Cold, radiation of, 323–324
Common sense, 8, 387
Composite numbers, 83–84
Computers, 90, 336
Consensus, role of in science, 139, 284, 371, 377
Conservation
  of atoms, 106
  of energy. *See* Energy, conservation of
  of heat. *See* Heat
  of matter, 94, 106
Contagion, 29
Controlled experiment, 43–45, 51, 244–245, 252–253, 309, 315, 319
  control group in, 245
  missing control group, 311–315
Copernicus, 287
Crick, F. H., 372
Crime statistics, errors in, 334

INDEX

393

Cultural basis of scientific belief, 377–389
Custance, John, 198–199

Danish Folkeregister, 244–245
Darwin, Charles, 2, 7, 189, 388
Davy, Humphrey, 136
Death of Ivan Illyich, The, 185
Dekker, Thomas, 307–308
DeMorgan, Augustus, 35
Deoxyribonucleic acid (DNA), 259, 372
Depersonalization of scientific discovery, 374–375
Depressive disorders. *See also* Mental disorders; Schizophrenia
  affective disorders as synonym, 196
  causes, 202
    psychosocial, 207
  classification, 190–191, 192–193
  description of, 195–196
  diagnosis, 196–197
    comparative, United States and United Kingdom, 210–216
    by "Project", 212–213
    reliability and validity, 197–198, 209–218
  drug therapies, 206, 209
  electroshock treatment, 209
  epidemiological approach to, 207
  experience of, 184–185, 198–199
  manic-depressive psychosis, 195, 198–199, 209
  rates
    change with time in New York State, 216–217
    of hospitalization, 208–209
    in United States and United Kingdom, 207–218, 236, 237
  schizophrenia, distinction from, 196
  suicide in, 195
  symptoms, 195
Descartes, René, 346
Description of the Retreat, 187
Diabetes, genetics of, 242
Diagnostic and Statistical Manual of Mental Disorders (DSM-III), 218, 250, 251n
Dichotomization, 311
Diffusion, 130–133, 134–135, 171–180, 305.
  See also Heat, kinetic theory of
Dispassionate scientist, myth of, 371–376
Dohrenwend, Bruce, 205, 253
Dohrenwend, Barbara S., 205
Dopamine, 206

Double Helix, The, 372
Drug trials, 264–265, 271, 319, 320
DSM-III, 218, 250, 251n
Dunham, H., 204

Edwards, Jonathan, 283
Effluvia. *See* Cholera, theories of transmission
Einstein, Albert, 8, 104, 132, 180, 288, 333, 369–370, 388
Electricity, nature of, 105, 171
Electromagnetic radiation, forms of, 333. *See also* Light
Electrons, speed and position of, 317–318
Elizabethan writers, 306–309
Elliptical orbits of planets. *See* Astronomy, planetary motion
Empirical, pertaining to both experiments and observations, 2
Energy
  Calories and British thermal units (BTUs) as measure of, 139–141
  conservation of, 139, 140–143, 304
  forms of, 139–141
  kilowatt-hour as measure of, 139, 141
  kinetic, 139–141
  laws of, 139–143
  mechanical, 129
  nuclear, 18, 140, 141
  potential, 118, 140, 141
  storage of, 112
  thermal, 140–141
  thermodynamics, laws of, 119, 141–142
    and perpetual motion, 141–142, 143
Epidemiology, 56, 67. *See also* Cancer; Cholera; Mental Disorders
Epstein-Barr virus, 63
Errors of measurement, 327–340. *See also* Accuracy; Precision; Statistics
Eskimos, 237
Essay on Dew, 324
Ether, as medium for wave motion, 137–138
Euclid, 74, 75
Euler, L., 85–87, 89
Evans-Pritchard, E., 378–387
Evolution, theory of, 2, 7, 190, 259, 388
Exact figures, misleading nature of, 333, 336
Exact sciences, 1, 104–105
Experimental test, as distinguishing characteristic of science, 3, 4–6, 293–315. *See also* Controlled experiment
  history of, 293–297

Experimental test (*cont.*)
  in the humanities, 306–309
  proof and disproof in science, 302–304
  quantification and, 297–298
  refutability and, 302
  repeatability of experiments, 295–296
  untestable theories, 300–302
Experimenter and experiment, interaction between, 317–325
Experiments on the Generation of Insects, 296
Exponential growth, 355–357

Facts in science, 4, 8–19
  public nature of, 296
  relation to theories, xi, 8–19, 41, 128, 189, 228, 362
  verification of, 208, 236
Fahrenheit, G. D., 109
Fahrenheit temperature scale. *See* Heat
Falling bodies, law of, 79, 81–82, 331
Faris, R., 204
Farr, William, 51, 272, 303
Flatworms, movements of, 321
Fleming, Alexander, 368
Focus in science, 228
Fordyce, George, 122–123
Fourier, J., 117, 118, 305
Freud, Sigmund, 2, 202, 291, 300–302, 373, 388
Friction. *See* Heat

Gage, T. (British General), 119
Galen, 185, 295
Galileo, 8, 17, 82, 95, 99, 295, 297
Gamma rays, 139
Gauss, K. F., 89–90, 153, 374
Gaussian curve (bell-shaped curve), 153, 155, 175, 176, 329
Genetics, 240–251
  carrier of a trait, 241
  chromosomes, 259
  of diabetes, 242
  dominant and recessive characteristics, 241–242, 257–258
  genes, 259
  inheritance of acquired characteristics, 318–319
  Mendelian laws, 240–242, 254–259, 309
  mutations, 242, 259
  natural selection, 259
  polygenic inheritance, 242
  probability applied to, 170

Geology, 1, 2, 189, 235
German measles, consequences during pregnancy, 190
Germ theory of disease, 31, 34–35, 49, 56, 66, 95, 190, 288, 375. *See also* Cholera
Goffman, Erving, 231–235
Goldberger, Joseph, 303
Graphs, xii, 343–357
  axes, choice of, 345, 346
  bar graph, 43, 343, 344
  Cartesian graphs, 346–347
  comparing magnitudes, 343–344, 345
  construction of, 346, 347, 348, 349
  continuous, 346
  different horizontal and vertical scales, 353–354
  discrete, 346
  extrapolation, 344–345
  Gaussian curve, 155
  histogram, 346, 153, 154, 155, 159, 160
  interrupted scale, 114, 354
  logarithmic scale, 355–357
  origin not at zero, 354
  "pie" graph, 343, 345
  random errors, 351–353
  relation between variables, 344–351
Gruenberg, E. N., 223, 226, 227, 229–230

Hamlet, 5, 306
Hearing loss and stereo headphones, 313–315
Heat
  British thermal unit (BTU), 109, 116–117, 122
  caloric theory of, 93–145
    analogy to water, 107, 108, 117
    capacity for heat and, 112
    and conservation of heat, 106, 110, 112, 117, 127, 303, 304
    explanations of heat produced by friction, 125–127
    flow, 95, 117–118
    heat as an element, 93–94
    and heat capacity of mercury, 112
    latent heat, 96, 116–117
    quantitative character, 97
    statement of, 94
    survival after Rumford, 97, 136–138
    thermal energy as equivalent to caloric, 140–141
    weight of heat, 96, 120
  Calorie, definition, 108
    in foods, 108n

INDEX

Heat (*cont.*)
  capacity for heat, 96, 108–117, 305
    brass, bulk metal vs. shavings, 125–126
    copper, 111
    water, 112
    water and mercury, 108–110, 123
  cold, as distinct entity, 105, 138
  conductivity of, 118
  conservation of. *See* Heat, caloric theory of
  convection, 122, 132
  energy, heat as form of, 129, 139–141, 323. *See also* Energy
  engines, 93, 118–119, 305
  equilibrium of, 102, 105
  flow of, 95, 117–118, 305
  heat capacity. *See* Heat, capacity for heat
  insulation, 98, 118
  kinetic theory of, 93–145, 287, 298, 299, 304–305, 378, 388
    atomic and molecular motion, 95, 97, 120, 129–136
    cannon-boring experiments of Rumford, 97, 120–129
    capacity for heat of mercury and, 112
    conservation of, refuted by Rumford, 127
    diffusion and, 170–180
    diffusion of salt as evidence for, 129–133
    friction producing heat, 97, 135–136, 305
    Joule's measurements confirming, 97, 129
    Newton's laws applied to, 287
    qualitative character in Rumford's time, 97, 129
    statement of, 94–95
    vibration as heat, 94–95, 120, 129, 138
    weight, absence of as evidence for kinetic theory, 96–97, 120–124, 136
  latent heat, 96, 113–117, 120, 305
    of melting of ice, 113–116
  measurement. *See* Heat, Calorie; Heat, British thermal unit (BTU)
  radiation of, 97
    in conflict between caloric and kinetic theories, 136–137
    as form of light, 138
  specific heat. *See* Heat, capacity for heat
  storage of, 112
  temperature, distinction from, 106–108
  weather and, 93–94
  weight of, 120–124
    use of latent heat, 122–124, 327
Heisenberg, Werner, 288
Heliocentric theory, 287, 389

Helmholtz, H., 139
Heston, L., 244, 245, 305
History, status as science, 290–291
Hogarth, William, 188
Holmes, Sherlock, 17
Homer, 283
Hotness, sensation of, 96, 98, 101–102, 103
Housman, A. E., 365
Huff, Darrel, 273
Humanities, relation to science, 5
  experimental method in, 306–309
Huxley, Thomas, 7
Hypotheses
  and facts, 49
  formation of, 361–370
  as natural human activity, 361
  perception and, 361–362
  sources of
    chance, 367–368
    folk wisdom, 365–367
    the unconscious, 368–369

Images, in writing, 306
Inferiority complex, 300
International Classification of Diseases, 218
IQ test, 103–104, 214, 249

James, William, 283
Jargon in scientific writing, 374–375
Jarvis, Edward, 203
Jenner, Edward, 366
Jones, Ernest, 373
Jonson, Ben, 307, 308
Joule, James, 97, 129, 138, 139, 298

Kambe, Kunishige, Dr., 314–315
Kazin, Alfred, 313
Keats, John, 283
Kekulé, F. A., 281–282, 362, 376
Keller, Helen, 279–280, 282
Kelvin, *Lord*, 139
Kepler, Johannes, 282–283, 287, 298, 373–374, 376
Kety, Seymour S., 244, 245, 250–251, 305
Kilborne, E. L., 366–367
Kinetic theory of heat. *See* Heat, kinetic theory of
Koch, Robert, 31
Koenigsberg, bridges of, 85–88
Köhler, W., 362–363
Koizumi, Satoshi, Dr., 314

Kraepelin, E., 192, 193, 195, 218, 227, 251–252
Kuhn, Thomas S., 388

Labelling. *See* Schizophrenia
Lancisi, G. M., 53–54
Latent heat. *See* Heat, latent heat
Lavoisier, A., 94
Lea, John, 52
Leibniz, G. W. von, 369, 372
Length, measurement of, 328–330
Leonardo da Vinci, 2, 291
Light
  infrared, 138, 139
  nature of, 97–105, 121, 137–138
  ultraviolet, 139
  velocity of, 18, 333–334
Lilienthal, A. M., 242
Literary Digest, 266
Locke, John, 95, 120
Logic, xii, 21–28, 73, 171
  and experimental testing, 23
  fallacies, 23–25
  as valid inference, 21

Macbeth, imagery in, 306–307
Malaria, 53, 288
  Burkitt's lymphoma, relation to, 62, 63
  mosquito as vector, 53–54, 62, 288
  sickle-cell anemia, relation to, 258
Maps, as metaphor for science, 290
Marlowe, Christopher, 307–308
Marx, Karl, 300–302
Mathematics, xi, xii, 21, 73–91, 163, 171.
  See also Probability; Statistics
  as abstraction, 73, 74
  branches
    algebra, 77, 80
    arithmetic series, 89
    calculus, 369, 372
    geometry, 74–76
    group theory, 164
    number theory, 82–85
    topology, 85–88
  discovery in, 363, 364
  fear of, 74
  games, similarity to, 75
  graphing functions, 346–348
  reasons for, 77–79, 82
  in science, 76, 297
  truth of, 75
  truth vs. validity, 74

Mathias, Charles, 312–313
Mayer, J. R. von, 139
Mead, Margaret, 249
Measurement, errors of, 327–340
Medawar, P. B., 369, 370
Mendel, Gregor, 240, 241, 258, 369
Mendelian laws. *See* Genetics, Mendelian laws
Mental disorders. *See also* Depressive disorders; Schizophrenia
  Alzheimer's disease, 192
  biochemical and biological factors, 202, 205, 206
  causes, 262
  cerebro-arteriosclerosis, 192
  classification, 183, 185, 188–191
  community care of patients, 227
  cretinism, 191
  dementia, 191, 192, 193
  dementia praecox. *See* Schizophrenia
  diagnosis of, 196–198
    reliability and validity, 197–198
  Down's syndrome, 191
  drugs in, 187, 225–227
  epidemiological approach to, 207–210
  genetic factors in. *See* Schizophrenia
  history of, 185–188
  hospitals for, 186–187, 188, 228–229
    as "total institution," 231, 234. *See also* Mental disorders, institutional neurosis
  hydrocephalus, 191
  idiocy, 203
  institutional neurosis, 219–230
    causes, 224–228
    distinction from other disorders, 223
    patients, behavior and appearance, 219, 221, 222
  mania, 185, 187, 191, 192, 193
  manic-depressive psychosis. *See* Depressive disorders
  melancholy (melancholia), 185, 191, 192, 193
  mental retardation, 190, 193
  mongolism. *See* Mental disorders, Down's syndrome
  organic vs. functional, 192
  paresis, 191, 192, 223
  pellagra, psychosis of, 303
  "prison psychosis," 223, 235
  psychoanalysis and, 202–203
  psychogenic approach, 202, 203
  psychosis, definition, 193n

INDEX

Mental disorders (*cont.*)
  rates, United States and United Kingdom, 207–218, 235–237, 273, 327
  scientific study of, possibility, 183
  social breakdown syndrome, 229–230
Mental telepathy, bias influencing recording errors, 321
Mercury, orbit of, 369–370. *See also* Astronomy, planetary motion
Merton, Robert K., 372–374, 376
"Methods of experimental inquiry" (John Stuart Mill), 16
Meyerson, A., 223
Mice, maze-solving, 318–319
Michelson, A. A., 333
Microscope, invention of, 369
Mill, John Stuart, 16, 17
Models in science, 172, 173
Morgenstern, Oscar, 336
Mosquito as vector in malaria, 53–54, 62, 288
Murphy, J. M., 237, 252

Nabakov, Vladimir, 194, 201
Nagel, Ernest, 2, 25–27
Nalbandov, A. V., 367
National Cancer Institute, 59, 68
"Natural" experiment, 44, 243
Nature vs. nurture controversies, 249, 253
Nerve synapses, 206
Newton, Isaac, 1, 95, 128, 282, 287, 297, 369, 372, 375
Newton's Laws, 1, 2, 8, 17, 90, 104–105, 139, 287–288, 302, 317, 369–370, 375
  for falling bodies, 79, 81–82, 331
  for pendulum, 348–352
Nobel Prize, 333, 372
Null hypothesis. *See* Statistics

Objectivity of scientists, 138–139, 371, 376
Observational sciences, 2, 235
Occam's Razor, 137
"On the Characteristics of Total Institutions" (Erving Goffman), 231–234
Optical Illusions, 9–14, 361

Pangloss, 300
Paradigms, 388
Parkinson's disease, relation to dopamine, 206
Pasteur, Louis, 1, 31, 368
Pattern recognition, 197
Pavlov, Ivan P., 1, 318
Pellagra, 262
  relation to corn, 303

Penicillin, discovery of, 368
Perception, 9, 11
Perdue, Frank, 312–313
Perpetual motion machines, 141, 142, 143
Pfungst, Oskar, 321
Physics, x, xi, 1, 5, 295. *See also* Astronomy; Heat; Newton's Laws; Quantum Mechanics; Relativity, theory of
Pictet, Professor, 323
Pietro Leopoldo, *Grand Duke*, 186
Pinel, Phillipe, 186, 202, 223
Placebos, 319
Planck, Max, 388
Plato, 95
Poetry writing, similarity to scientific discovery, 365
Poincaré, Henri, 363–364, 369, 376
Popper, Karl, 300–302
Pott, Percival, 63
Precision, 328, 330
  quantification of, 337
  standard deviation as measure of, 338–340
  standard error, 338–340
  of sums and products of measured quantities, 337–338
Preconceptions, 230
Prime numbers, 82–85
Priority in discovery, 372–373
Probability, xii, 147–181, 263–265
  atomic motion, application to, 287
  averages, use in calculation of, 161–169
    of deviations, 164–169
    root-mean-square, 165
  coin-tossing, 148–158, 263–265, 329, 346
  colored balls, 158–160, 263–265
  in gambling, 147
  Gaussian curve, 153, 155
  genetics, use in, 254–259
  independent and dependent events, 149–150
  "law of averages" (law of large numbers), 156–157, 161
  mutually exclusive events, 150–151
  rules of, 150
  simple and compound events, 150–151
  skewed curve, 160
  unique events, 160–161
Probable inference, 27
"Project" (United States–United Kingdom Diagnostic Project), 211–217
Proof and disproof in science, 66, 147, 302–304
Proportionality, 80–82

Psychoanalysis, 202, 291
Psychology, x, xi
   abnormal. *See* Depressive disorders; Mental disorders; Schizophrenia
Puerperal fever (child-bed fever), 367–368
Pulse, accuracy of determination, 331

Quantification, xi, 97, 98, 103, 129, 212, 297–298
Quantitative thinking, xii
Quantum mechanics, 288, 317–318
Quine, W. V., 300

Radiation, 305. *See also* Heat
Radio waves, 139
Ramazzini, Bernardo, 62
Randomness, 329
Random walk, 171–180
   by coin-tossing, 173–174
Reality and perception, 361–362
Redi, Francisco, 296
Reil, J. C., 186
Relativity, theory of, 8, 18, 104, 288, 328, 333, 369–370, 388
Reliability, 328. *See also* Mental disorders, diagnosis of
Robison, John, 106
Roosevelt–Landon election race, 266
Root-mean-square, 165–169
Rosenthal, David, 244, 245
Rosenthal, Robert, 321–322
Rumford, Count (Benjamin Thomson), 96, 97, 119–139, 170, 175, 179, 298, 303, 305, 323–324, 327, 371, 374, 378
   biographical, 109–110
   cannon-boring experiments, 124–129
   heat from friction, 124–129
   lack of objectivity, 138–139
   molecular motion revealed by diffusion, 129–135
   radiation of cold, 138
   radiation of heat, 136–138
   weight of heat, 120–124

Saul, *King*, madness of, 185
Schizophrenia
   adoption studies. *See* Schizophrenia, genetic studies
   causes, theories of, 388
      biological, 205–206. *See also* Schizophrenia, genetic studies; Schizophrenia, labelling theory

Schizophrenia (*cont.*)
   causes, theories of (*cont.*)
      psychogenic, 203, 238–239, 251, 252–253, 305
      psychosocial, 204–205, 207–208, 236, 253
   classification, 190, 191, 192–193, 194–195, 218, 249–252
   dementia praecox, 192–193, 251
   depressive disorders, distinction from, 196
   description of, 194–195, 196
   diagnosis of, 196–197, 249
      "blind," in adoption studies, 245
      comparative, United States and United Kingdom, 211–216
      reliability and validity, 197–198
   dopamine and, 206
   drugs, producing schizophrenialike reactions, 206
   drug therapies, 206, 320
   experience of, 199–201
   family disturbance in, 203, 238–239, 252, 253
   genetic studies, 205, 240–251, 299, 305
      adoption studies, 243–247
      critiques of, 249–251
      relative importance of psychogenic factors and, 243
      twin studies, 247–249
   hospital treatment of, 218, 227–228, 236
   labelling theory, 235, 236–238, 252
   latent schizophrenia, 250
   myth of. *See* Schizophrenia, labelling theory
   nerve synapses, 206
   rates
      changes over time in New York State, 216–217
      in Denmark, 244–246, 250
      in different social classes, 204–205
      in families of schizophrenics, 244, 247
      in general population, 244
      of hospitalization for, 208–209
      in primitive societies, 237–238
      social mobility hypothesis, 204
   schizotypal personality disorder, 250
   symptoms, 194, 223, 237
   thought disorder in, 194, 196
   twin studies. *See* Schizophrenia, genetic studies
   types
      catatonic, 194, 251
      hebephrenic, 194, 251

Schizophrenia (*cont.*)
  types (*cont.*)
    paranoid, 194, 199–200, 251
    process, 251, 252
    simple, 194, 251
  urinary phenolic acids in, 205, 218
  uterine environment as factor, 246
Schrödinger, Erwin, 288
Schulsinger, F., 244
Science
  consensus in, 18–19, 284
  cultural roots of, 377–389
  definition, 1–5
  experimental test as criterion, 3, 4–5, 279, 293–315. *See also* Controlled experiment
  in the humanities, 306–309
  generality as goal of, 1, 4, 287–291, 304
  understanding as a goal, 3–4
Scientific method, lack of well-defined, ix–xi, 371
Scientists
  dispassionate, myth of, 371–376
Self-criticism in science, 139
Self-fulfilling prophecies, 324–325
Selikoff, Irving, 63
Semmelweis, Ignaz, 367–368
Shakespeare, William, 5, 306–309
Shakespeare's Imagery and What It Tells Us (Caroline Spurgeon), 306
Shapiro, Meyer, 291
Sickle-cell anemia, 254–258
Simultaneous discovery, 369, 372–374
Sizzi, Francesco, 295
Sketches, 358
Smallpox, vaccination for, 366
Smith, Theobald, 366–367, 369
Snow, John, xi, 2, 29–56, 66, 207, 235, 243, 288, 303, 305, 309–310, 366, 374, 375
  biographical, 29
  Broad Street pump epidemic, 39–43
  hypothesis on transmission of cholera, 31–32, 36
    alternative hypotheses, criticisms of, 48–52
    anecdotal evidence for, 33–34
    applications to other diseases, 53–54
    preventive measures for cholera, 54–55
    water companies and cholera rates, 43–48
Sociology, status as science, 235
Soranus of Ephesus, 185
Spurgeon, Caroline, 306–309

States of matter, 133
Statistical mechanics, 170
Statistics, 56, 147, 263–278
  average, value of, 267–268
  chi-squared test, 273–278
  "null" hypothesis in, 265, 270, 272, 275–277
  as plural noun, 267
  probability, relation to, 263–265
  random selection, 265–266
  root-mean-square in, 269
  sampling, 263–265, 274
    bias in, 266–267
  standard deviation in, 269–270
  statistical significance, 271–272, 277
  as summary of data, 267
  surveys, 266–267
  two-by-two table, 273–278, 310–311
Steam engines. *See* Heat, engines
Style in research, 375–376
  in scientific papers, 374–375
"Subtle fluids," 105
Sullivan, Anne, 279
Surveys, errors, in, 323
Syllogism, 73–74
Symmetry, 164–169
Syphilis, as cause of paresis, 191

Tacit component in science, 378, 387–389
Teen-age widowers, 335–336
Telescope
  Galileo's discoveries with, 295
  invention of, 369
Temperature. *See also* Heat
  Celsius (Centigrade) scale, 101
  equilibrium of, 103
  Fahrenheit scale, 101
  molecular speed, relation to, 135
Texas cattle fever, 366–367
Theory and hypothesis, relation between, 18
Thermodynamics. *See* Energy
Thermometer, 96, 99–103
  calibration of, 100–101
  standards for, 100–101
Thomson, Benjamin. *See* Rumford, *Count*
Thomson, J. J., 1
Tippet, L. H. C., 157, 289
Tolstoi, L., 183–184, 199
"Total institutions," 223, 231–235
Tuke, Samuel, 187, 202, 223, 230
Tuke, William, 186

Two-by-two table, 310–311
Typhoid fever, 30, 53, 55, 190

Ullian, J. S., 300
Uncertainty principle, 317–318
United States Surgeon General, 57

Validity, 197–198, 328
  vs. truth, distinction, 22, 74

Watson, James, 372
Wave nature of light, 137–138
Weight of heat. *See* Heat
Wells, William, 324
Wender, Paul, 240, 243, 244
White, R. W., 199
William of Ockham, 137
Witchcraft, Azande belief in, 377–387
  benge, 382–386
  experimental evidence for, 380–381, 382–386

Witchcraft *(cont.)*
  explanatory power, 381–382
  generality, 381–382
  "poison oracle," use of, 382–386
Women's Christian Temperance Union, 312, 315
World, roundness of, 25, 76, 304, 362, 389
World Health Organization, 218
Wrong theories, usefulness of, 35, 95

X-rays, 139

Yellow fever, Snow's hypothesis for, 53–54
York Retreat, 187
Yoruba (Nigerian tribe), 237
Young, John Z., 11

Zande. *See* Azande; Witchcraft, Azande belief in

Vienna.

Having flouri[shed under] the Habsburg Dynasty, Vienna o[ffers all the] attractions of an imperial city. B[ut economic] stagnation and the fading of its fin-de-siècle glory forced the Austrian capital to re-invent itself.

Even towards the end of the 19th century, a new movement was brewing. Artists such as Gustav Klimt and Oskar Kokoschka forged new approaches by breaking out of prevailing conventions. They and their peers became known as the Vienna Secessionists: they laid foundations for how the city would evolve throughout the next century.

Vienna is as delicate as its pastries, as rude as the waiters of the famous Kaffeehäuser and as diverse as the former empire. Despite its rich history, the city manages to stay utterly contemporary.

And it continues to be the people who shape the city's atmosphere. In Vienna we spoke with gifted personalities: a form-busting sculptor, a fashion designer and DJ couple, a vinegar king and a multi-media artist with multiple personalities… Our picks are complemented by a photo showcase, flash fiction and a feature story by local talents. It's all about original minds and the creative vibe. Get lost in the sights, sounds and flavours of the city.
Get lost in Vienna.

3
Editorial

8
Waltzing Vienna

Neighbourhood

20
2nd District
The Island

44
3rd & 4th District
Inside Out

Testimonial

12
Erwin Wurm
Art Moves

24
Roshi Porkar & Wolfram Eckert
Export Champions

38
Carola Schmidt
Traditional Way

52
Erwin Gegenbauer
The Insider

Story

48
Singing in the Boys' Choir
Espionnage in Vienna
by Werner Sturmberger

64
Out of the Dark
by Jörg Henning

Showcase

30
Wiener Blutwurst
by Lukas Gansterer

56
Editors' Picks
Wiener Melange

58
Hotels

67
Index

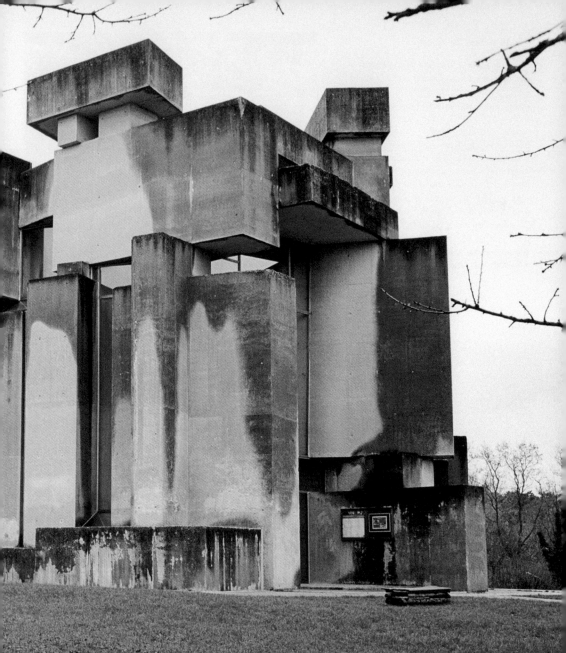

Entering the Wotruba Church is rather like immersing yourself in a work of art. That's because it was designed by a sculptor. Inspired by the Chartres Cathedral, Viennese Fritz Wotruba used 152 blocks of concrete in its construction. The monument embodies the experimental side of Vienna—rough and asymmetrical, yet entrenched in tradition.
• Wotruba Kirche, Ottillingerplatz 1, 23rd District, georgenberg.at

| Food | Kiss the Cook

... is the legend stitched onto Mario Bernatovic's chef's jacket. After several years in some of the city's finest restaurants, the Croatia-born chef opened his own place in 2014. *Kussmaul* divides neatly into three parts: restaurant, café-pâtisserie and bar—completed by a beautiful terrace. For the full experience, head to the fine-dining section. Here, Mario serves regionally inspired, creative cuisine, like his standout shitake-raviolo and pig's trotter ragout. Recipient of 16 points from the Gault Millau guide, this is a star of Vienna's contemporary dining scene. Surely worth a kiss.
• Kussmaul, 7th District, kussmaul.at

From Pharmacy-Restaurants to Hipster Heurigen

Waltzing Vienna

| Outdoors | Great Escape

Every city has a spot locals go to for peace and quiet. In Vienna, it's Kahlenberg, a 480-metre-high hill in the Vienna Woods. Perfect panoramic views of the city are visible from lush patches of green grass, making it a romantic choice for summer picnics. Plus the hill is laced with hiking trails and bike paths. If you want proof of its credentials, even Albert Einstein used to come here for inspiration. Reachable on tram line D direct to the last stop, Nussdorf, it's only a few minutes from the city centre... Yet it feels so far.
• Kahlenberg, 19th District, kahlenberg-wien.at

| Culture | Contemporary Kunst 101

Sitting proudly on the northern edge of Stadtpark is the red-brick grandeur of the *Museum for Applied Arts*. Not only is it one of the city's most impressive buildings but it's also home to world-class collections of design and contemporary art. Past the illustrious columned hall a wealth of design wonders await, ranging from original Biedermeier furniture to the mosaics of Gustav Klimt. There is no better place to see how Viennese heritage interacts with 21st century creativity. Case in point: the traditional comfort food and futuristic interior design of their *Österreicher im MAK* restaurant.
• MAK, 1st District, mak.at

| Food · Night | Imperial Wining and Dining

The *Palmenhaus* is a first-rate botanical garden in its own right as well as a lively, three-pronged establishment: café, brasserie and bar. First opened in 1882, its grand cast-iron arches and jugendstil architecture is evocative of imperial Austria, making it one of the finer settings in which to while away an evening. The food is perfectly decent, but the atmosphere is second to none. Seating arranged beneath exotic plants and palms provides a wonderfully incongruous setting for dark winter nights—until warmer months arrive and the terrace is suddenly the place to be. Choose a nice bottle of local Donau Riesling and enjoy the music and conversation as it flutters across the palace gardens.
• Palmenhaus, 1st District, palmenhaus.at

| Night | Under the Pearly Light

Working at the legendary Loos Bar for 15 years, Roberto Pavlovic went from dishwasher to head barman to business partner. Then in 2014, he ventured out alone to open his own place, just a couple of blocks away. And what a crackerjack of a bar it is. Shunning the current obsession with mixology, it's all about straight-up classics—served to absolute perfection. The dark, sumptuous surroundings feature a showstopper of a chandelier as a centrepiece. Featuring around 80,000 freshwater pearls, the soft glow of the light draws patrons back to the bar, time and time again. Top tip: Roberto's favourite drink is an Old Fashioned. The finest this side of the Atlantic, some say...
• Roberto American Bar, 1st District, robertosbar.com

| Food | Disco Pizza

This is no ordinary Italian pizza joint. A giant sparkling disco ball is the most noticeable thing in the restaurant. Maybe that's not so unusual—until you see your pizza will be cooked inside it! Frequented by many of the creative cats of Vienna, it's a place for a quick lunch or a spicy start to the evening. The name refers to the yacht of evil Emilio Largo in the James Bond movie "Thunderball". But it's also a play on words: The translation means "flying saucer" and that's what this place is famous for.
• Disco Volante, 6th District, disco-volante.at

Food · Night — Heuringer Remixed

Try blending 1990s digital nostalgia with an Austrian-style bar-restaurant—and the result would probably resemble *Zum Gschupftn Ferdl*. Schnitzel and Gschupfte Krapfn dishes are served in this eclectic, Super Mario-flavoured space, accompanied by numerous live gigs and DJ sets that go on into the night. Yes, it's as mad as it sounds. Their lunch menus are also worthy of attention, unsurprising given their products are 100% organic and delivered fresh by farmers from the region.
• Zum Gschupftn Ferdl, 6th District, zumgschupftenferdl.at

Shop — Appetising Apothecary

This is not your average pharmacy. Firstly, it's occupied the same spot since the turn of the century. Natural, traditional products have always been the name of the game here—an ethos apparent in their deluxe line of organic cosmetics. But setting *Saint Charles Apothecary* apart is its dual character as a restaurant. Raising the bar in the world of regional and organic cuisine, the cooks handpick the herbs and vegetables themselves. The result is a spellbinding, ten-course romp through the Austrian forest.
• Saint Charles Apotheke, 6th District, saint.info

Erwin Wurm
He is an artist who uses and yet questions the medium of sculpture. His motto is "All is sculpture" and he acts on those words. Wurm initially applied to study painting at Mozarteum University in Salzburg but found himself majoring in sculpture instead when professors assigned him to the discipline. So Wurm decided to mix the medium with his passion for painting. The result? Sculptures that follow Wurm's unique rules rather than the prevailing norms of the art form. For his ongoing series, "One Minute Sculptures", the pieces become visible only when the viewer interacts with the material. Such are the thought processes behind Wurm's works

Erwin Wurm, Artist

Art Moves

Vienna's art scene is still thriving, shaping the new instead of following the established. Maria Lassnig, for example, has sparked conversations about females in society and the art world. Erwin Wurm is another of the contemporary artists defining the city's artscape. He questions his medium of sculpture, while challenging our perceptions of the ordinary. Here, he shares his everyday life in Vienna

You are originally from Bruck an der Mur—you lived in Salzburg and Graz during your studies and then moved to Vienna, where you graduated from the University of Applied Sciences. Was it there where you fell in love with the city and decided to stay?

I was in Berlin for half a year and in New York for one and a half years, but yes, apart from that, I have always lived in Vienna. The city has shaped me throughout my life and still does today. Vienna is like my mother's milk. It attracts me, repels me and then attracts me again. I can't imagine living anywhere else, especially since my family and friends are here.

What is it about Vienna that shapes you?

The cultural background; the intellectual horizon; the openness, and respectively, the closeness of society here; the architecture; the system of the city and the political capability that Vienna offers, or more precisely doesn't offer. All this embeds you into a society and yes, that's what connects me to Vienna.

You say you feel both attracted and repelled by Vienna. What causes these feelings?

I am fascinated by the history of the city, but at the same time, I feel repelled by the characteristic features that are a result of the history. Austria went through tough times. It was once an important country, but then it all collapsed and it kind of lost its identity for many years. During the Habsburg Monarchy, hierarchies were unalterably determined. When the monarchy disappeared, the notion of how society has to work remained. After a great misery, it's mostly the case that you try to experience different directions. And the Austrians did that too, with fascism, communism... and they're still trying different things today without finding what could serve as a central point and ensure calm. And this is what repels me. When I grew up in the 1950-60s, we were still a very restrictive society. With the 1968 movement, of course, people tried to rebel against restrictions, but the liberalisation only became established very slowly. And all these together—the new, the old, the contemporary, the conservative, the reactionary, the revolutionary—cause a very special atmosphere here.

Where in Vienna do you live?

In the 2nd District, but actually very close to the 1st. The 2nd District is the former Leopoldstadt, which was once home to a great Jewish community and is again now, too. Also a lot of young people are there—it's a very mixed and enjoyable district, especially because it's not overtaken by tourists. It's more a presentation than a form of life. The first district is kind of cemented in its lifestyle. The second is more free.

Is your studio located in the second district, too?

It was there for a long time but now it's in the countryside as I just have more space there—vehicle access for trucks and so on. It's nearly impossible to find this in Vienna.

Vienna has experienced some exciting art movements. Was there something brewing when you finished your studies?

Back then, it was the time of the "Neue Wilde"—an art movement that I, by the way, just coincidentally became aligned with.

How did this happen?

I studied in the 1970s and was taught all prevailing doctrines of minimal and conceptual art. Once

Architecture as a testimony of the imperial past: the Marble Hall in the Upper Belvedere

I read that you need to overcome the established to find yourself. As a reaction, I made classical sculptures out of not yet established material.

Such as?

Waste. And as I actually wanted to become a painter, I painted them. This was against the thinking at the time, but then suddenly, it was accepted and even gained appreciation through this new movement. And that's how I suddenly became an active part of it.

The Vienna Secession is one of those famous movements with big names such as Gustav Klimt. Which other artists also represent Vienna as an art city?

Richard Gerstl, for example. He's from Klimt's times. Or Arnold Schönberg for his paintings. And there's also Walter Pichler and Maria Lassnig, Arnulf Rainer, Hermann Nitsch and Valie Export. I would also mention names of pioneers from other disciplines who are worth a mention. Architects like Hans Hollein or Wolf Prix, for example.

Do you still go to galleries and museums to see exhibitions?

At the moment, less so, as I

The palatial building of the Kunsthistorisches Museum is topped with a 60-metre-high octagonal dome

Kunsthistorisches Museum
1st District

Albertina
1st District

Belvedere
3rd District

Büro Weltausstellung
2nd District

Mochi
2nd District

Motto am Fluss
1st District

Café Ansari
2nd District

Karmelitermarkt
2nd District

Naschmarkt
6th District

Stephansdom
1st District

Burgtor
1st District

Plachutta: a family-run mini chain whose signature dish is tafelspitz, or boiled meat in broth

travel a lot, but I normally regularly go to the *Kunsthistorisches Museum* and also to other houses like the *Albertina* or *Belvedere*. There are also some young, nice galleries and independent venues. *Büro Weltausstellung*, for example, is great.

How else do you spend your free time in Vienna?
Back in the days when I was a student and a young artist, I still had time to stroll around. Today, I mostly work and if there is free time, I meet friends, or indeed, go to a museum. And what I always do is take time for good food.

Where would you go for lunch, for example?
At *Mochi*. The kitchen is very contemporary, healthy and Japanese-inspired.

And dinner?
Motto am Fluss. Or *Ansari*—it's next door to *Mochi* and serves Georgian cuisine—very nice, too.

Would you recommend these two spots to those visiting for a weekend?
Definitely. I would also show them the area around *Praterstrasse*, the street where *Mochi* and *Ansari* are located. It's not just a good destination if you're peckish—the neighborhood itself is very nice.

Where else would you take sightseers?
Karmelitermarkt. A nice scene of very good restaurants and cool bars evolved around this market. Then, of course, I would also show the *Naschmarkt*, which is said to be Vienna's most popular market. Next stop would be the *Stephansdom*. It's really beautiful. And then I would walk with them through the *Burgtor* towards *Heldenplatz*. If you are standing there, the city kind of opens up in front of you—you will see the former Glacis where the city walls ended and the ground was left empty in order to have a clear view of who might be approaching. But during Gründerzeit in the 19th century, all this free space got blocked. Buildings, such as the big museums—the Kunsthistorisches Museum, the town hall, the Burgtheater, the Austrian parliament, are there now.

Motto am Fluss, an upmarket spot for a meal overlooking the Danube

Do you also have kind of a secret spot?
The small alleys with all the little depressing shops are very attractive. You will see shop windows that make you look at them in disbelief. There might lay three dusty umbrellas and that's it. Unbelievable that something like this is still possible today. Not in the 1st District, but in the 2nd and many others where the city hasn't been caught by modernisation yet, where it is still antiquated. It has something attractively depressing—it's in fact difficult to verbalise. That's why I would show it.

Is there actually such a thing as "Vienna cosiness"?
The so-called Vienna cosiness went through a change with a culture that still had the Heuriger as the elixir and centre of life. Heurigen are venues where you go for a wine. You could even get some food there and take it home. I don't experience it any more at all, as Vienna is shaped by a hectic and fast pace like other big cities. Starbucks is also here in Vienna, which, in my opinion, is an imposition on the city. We really don't need Starbucks with such an amazing selection of great coffee. And in the traditional coffee houses, you will also experience one remarkable feature that makes up Vienna: grumpy waiters.

People say you get the best coffee in Vienna. Do you agree?
What's fact is that there is a great variation of coffee—the so-called Einspänner, Kleiner Brauner, Großer Brauner, Großer Schwarzer, Kleiner Schwarzer, Melange, Verlängerter and so on. It's not only the variation, but allegedly the water in Vienna is very soft—perfect for

Büro Weltausstellung is the open platform of the Vienna Art Foundation

<u>Café Prückel</u>
1st District

<u>Café Museum</u>
1st District

<u>Plachutta</u>
1st District

coffee-making. And then there are all the cooked desserts that highlight the coffee indulgence. All this happens in an environment where you are mostly still allowed to smoke, as there is no general smoking ban in venues in Vienna. And then the grumpy waiters—what a set!

What coffee house would you suggest? Café Prückel or Café Museum?
Café Prückel still has its original interior from the 1950s. It looks super. You have to go there. And Café Museum still has an interior created by Adolf Loos. It's, so to speak, the still-existing old Vienna.

Which pastry should you order to go with your coffee? The typical sachertorte?
Yes, sachertorte (chocolate cake) is always great. But there's more. Cremeschnitte (vanilla and custard cream cake)—delicious schaumrolle (cream pastry) and gorgeous punschkrapfen (rum cake).

Which traditional meal is your favourite?
Prime boiled beef, for sure. Plachutta is by far the best restaurant to eat it. And kaiserschmarrn, another pastry which is also a lunch meal, goes well with stewed plums. And all these other messes—wiener schnitzel, for example, or fleischknödel (meat dumpling), krautrouladen (cabbage rolls) and schinkenfleckerl (diced ham pasta).

That all sounds very heavy.
Yes, you shouldn't eat too much of it; it's life-endangering. Everything is made with butter, traditionally. Even rice will be tossed in butter here.

Art Moves

2nd District
The Island

| Shop | **Start Making Sense**

These days when everything's virtual and cloud-based, it's good to take a break in the real, analogue world—and Vienna might be the perfect place for it. Especially at unique concept store *Supersense*. Situated in a mock-Venetian palace from the late 19th century, *Supersense* is a store with pre-digital delights for all the senses—a time machine transporting you to analogue heaven. Part workshop, part stylish retail space, part café, *Supersense* was co-founded by Florian Kaps, who—with his "Impossible Project"—saved instant photography by buying the last existing Polaroid factory in the Netherlands. Consequently instant photography is a major theme here—among other wonderful stuff, *Supersense* sports the world's largest Polaroid camera (which takes pictures in an incredible 20×24 format!). Wish you could turn your smartphone pictures into Polaroids? *Supersense* has a machine for that too. If you fancy producing a vinyl record, just enter the "flabbergasting record elevator" and record your own 90 seconds of vinyl fame. A completely analogue, living-room studio can be booked for longer sessions. And don't forget your sense of taste by having a cup of wonderful coffee (from a machine handmade in Seattle by Italian immigrants) or trying one of the store's locally produced delicacies.

• Supersense, Praterstr. 70, 2nd, supersense.com

| Night | Sunset Tower

Das Loft, the top-floor restaurant at the Sofitel Vienna designed by Jean Nouvel, not only provides a drop-dead gorgeous view of historic sights like the gothic St. Stephen's Cathedral or the Riesenrad, but it's become a landmark in itself. This is mainly due to the ceiling of the restaurant on the 18th floor, which was designed by Swiss video artist Pipilotti Rist. It looks like a kaleidoscope of autumn foliage and sky, twisted in gold, amber and blue. Because of the slight curvature of the ceiling, the reflections in the glass appear to extend out to the horizon. Perfect for a sundowner at the bar—no matter what season.
• Das Loft, Praterstraße 1, 2nd, sofitel.com

| Food | Georgian Style

Café Ansari is located at the quiet end of Praterstraße. In these magical surroundings, former artist Nana Ansari and her husband Nasser have created a casual yet stylish restaurant offering fusion food that might be described as Georgian-Oriental. Designed by renowned Austrian architect Gregor Eichinger, *Café Ansari* emphasises the small details. Black tea is served in real British bone china, Moroccan mint tea in small silver pots from Casablanca, and guests can help themselves to a drinking fountain—a statement on the ongoing discussion about the traditional free tap water policy in Viennese venues.
• Café Ansari, Praterstr. 15, 2nd, cafeansari.at

| Shop | Brandmania

This former fur coat wholesaler turned concept store on Praterstraße offers upscale fashion, furniture and art. *Song*'s minimalist aesthetic with metal elements, wooden racks and concrete slabs also creates a perfect setting for rotating exhibitions and performances in its gallery (SONG SONG). Owner Myung il Song is a tasteful curator of apparel and she rejoices in mixing younger designers with established icons. Brands include: Dries van Noten, Paul Harnden Shoemakers, AF Vandevorst, Marc Le Bihan. Items from classy Viennese glassware manufacturer J. & L. Lobmeyr can also be bought here.
• Song, Praterstr. 11–13, 2nd, song.at

| Food | Casual Avant-garde

If you're looking for the perfect wiener schnitzel, accompanied by great wine and followed by a creme brulée, *Skopik & Lohn* is the destination. It has the look of a modern French brasserie with an avant-garde edge—the ceiling's painted with a spidery scrawl by Austrian artist Otto Zitko. Waiters wear white suits, and the food is modern European and Austrian with a focus on French cuisine. *Skopik & Lohn*'s selection of Austrian and French wine is very impressive, and it's a favourite among local bohemians—so be sure to book in advance.
• Skopik und Lohn, Leopoldsgasse 17, 2nd, skopikundlohn.at

| Culture | Art in the Park

Hidden on the far side of beautiful baroque park the Augarten, is to be found *TBA 21*—a progressive art space from Francesca Habsburg's foundation, Thyssen-Bornemisza Art Contemporary. The foundation is dedicated primarily to the commissioning and exhibiting of ambitious projects that defy traditional categorisations, and its collection, featuring names like Sharon Lockhart or Heimo Zobernig, has gained a reputation among art lovers. *TBA 21*'s restaurant—*die AU*—has a bar designed by Hans Schabus. Its wonderful garden is perfect for relaxing with coffee or enjoying traditional cuisine with a contemporary twist.
• TBA 21, Scherzergasse 1A, 2nd, tba21.org

| Food | Alpine Food

Restaurant *Ü-Lokal* is not for salad and calorie counting. The eatery was created by three men from Vorarlberg. And Austria's most western and mountainous region is famous for its elaborate cheese production. All this means that within its walls you'll be sure to find hearty "jause"—an afternoon snack of bread, ham and cheese—dishes loaded with "speck" (bacon), and fine, traditional "käsknöpfle" (cheese dumplings). When you are well and truly stuffed there's just one thing left to do. Call for some original Vorarlberg schnaps—the selection at Ü is superb. Prost!
• Ü-Lokal, Obere Augartenstr. 46, 2nd, ue-lokal.at

| Food | Grätzel Style |

Vienna's former Jewish Quarter of Karmeliterviertel has seen dark times. Almost all its population was deported during World War II. Yet it later managed to revive its ethnic diversity somewhat—mainly due to the immigration of Eastern European Jews. Some fifteen years ago the area was also discovered by the creative and art scenes. And although this "grätzel" (neighbourhood) has seen considerable gentrification, it still offers some of the most interesting venues in town. A good range of authentic ethnic places jostle alongside exemplars of Viennese cuisine. A good place to start is a stroll through the market, in existence since 1671. Set in a square with architecturally picturesque surroundings, it is not overly crowded like Naschmarkt, Brunnenmarkt, or other major markets. Vendors sell vegetables next to butchers selling kosher and halal meats. On Saturday the square features a "bauernmarkt" (farmers' market). Foodwise, check out *Zimmer 37*, featuring "5 Elements" cuisine, *Kaas am Markt*, a deli and café with excellent organic and slow food dishes, or *Café Einfahrt*. Just around the corner from the market in Große Pfarrgasse, restaurant *Das Engel* welcomes guests with a charming marble-wooden reception that was once a pharmacy counter. Smokers—yes, it is still a topic in Vienna—can come here for a cigarette break. The rest of the place is non-smoking.

• Karmelitermarkt, 2nd, various locations, see Index p.67

Roshi Porkar & Wolfram Eckert
Roshi was born in Vienna and studied fashion under Véronique Branquinho and Bernhard Willhelm at the University of Applied Arts. She has worked for Lanvin Paris and Karl Templer in New York and in 2014 won the Chloé Prize for her collection at the International Festival of Fashion and Photography in Hyères. Her boyfriend Wolfram is also a globe-trotter. He performs as a DJ and music producer in clubs and venues from Bogotá to Venice under the pseudonym Wolfram Amadeus. His collaborations with Hercules and Love Affair, Patrick Pulsinger and Euro dance-legend Haddaway are some of Austria's hottest musical exports

Roshi Porkar & Wolfram Eckert, Fashion Designer & DJ-Producer

Export Champions

The creative couple sum up Vienna's vibe: a mix of high-culture and underground, tradition and innovation... Visit vintage shops and flea markets, local designers and schnitzel restaurants, tiny streets and public pools with Roshi and Wolfram's recommendations

Vienna nights: lounging at Loos Bar (below), clubbing at Grelle Forelle (above)

Albertina 1st District	
Mumok 1st District	
Leopold 1st District	
MAK 2nd District	
Wien Museum 1st District	
Kunsthistorisches Museum 1st District	
Park 7th District	
Song 2nd District	
Firis 1st District	
Flo Vintage 4nd District	
Prater Sauna 2nd District	
Grelle Forelle 9th District	
Roberto American Bar 1st District	
Schönbrunn Palace 13th District	
Burggarten 1st District	
Wurstelprater 2nd District	

Roshi and Wolfram, you both work internationally and travel a lot… What makes Vienna special to you?
Roshi: Our families, the quality of life.
Wolfram: And the great tasting tap water.

You both have a creative background. In that field, what's not to be missed when travelling to Vienna?
Roshi: There are many museums like the *Albertina, Mumok, Leopold, MAK, Wien Museum* and the *Kunsthistorisches Museum* that shouldn't be missed.
Wolfram: If you are interested in classical music or theatre there are always concerts, plays and operas that you can get last-minute tickets for.

Roshi, you're a designer. What's to look out for when going shopping? What fashion is unique to Vienna?
Roshi: *Park* has a very nice selection of fashion labels and a few hand picked local designers as well. *Song* also carries beautiful garments though mainly international brands—and the same goes for *Firis*. My favourite local designers are Femme Maison, Gon and Petar Petrov. These are friends and people I studied with. They started their own brands early on and have been creating beautiful collections ever since. If you like good vintage, the only place I can recommend is *Flo Vintage*. It's owned by a couple who claim designers all over the world come to visit them to gather inspiration for their collections. Whether this is true or not, they definitely have a huge secret warehouse somewhere in Vienna and regularly bring in new things.

Wolfram, what's the best place to have a solid party? Is there a bar you recommend before clubbing?
Wolfram: The clubs *Pratersauna* and *Grelle Forelle*. For a drink, I like *Roberto American Bar*!

Roshi—if your best friend were to visit, what would the perfect day in Vienna look like for you two?
Roshi: I would probably just walk around the city with her. Take her to exhibitions, sit in old coffee houses for hours: *Café Sperl, Prückl, Griensteidl, Havelka*, and *Central* are all worth a visit. And depending on her interest in opera or theatre, I'd take her to a play. Then I'd take her for fancy drinks at night and get her a late-night sausage at one of the many legendary "Würstelstand" (sausage stands) in Vienna—definitely a must!

And Wolfram—what would be your plan to entertain a visiting best friend?
Wolfram: I would just take him through all the tiny roads in the 1st District where only horses and carriages are allowed. Just walking through a city is sometimes better than going to a museum because the city itself is basically a museum. During summertime *Schloss Schönbrunn* (Schönbrunn Palace) is a nice spot to hang out, or in the garden of the *Burggarten* where you can grab a great coffee.

Where do you go in Vienna if you want to get inspired?
Wolfram: If you've lived in a city long enough it's hard to find places that still inspire.
Roshi: Plus, we don't really like the word "inspiration". But in terms of going to places in Vienna that are unique and memorable we would suggest the *Wurstelprater*, an old amusement park that dates back to the 18th century. Or the *Friedhof der Namenlosen* (Cemetery of the Nameless). The Saturday flea market at *Naschmarkt* can be full of interesting and strange discoveries.

Export Champions

The original streetfood: it's never too late for a würstl at one of the city's many sausage stands

Friedhof der
Namenlosen
11th District

Naschmarkt
6th District

Neuwaldegger Bad
17th District

Kuishimbo
6th District

Skopik und Lohn
2nd District

Finkh
6th District

And where do you relax?
Roshi: In the summer we love going to the *Neuwaldegger Bad*, a very old, beautiful outdoor pool surrounded by a forest. Because it's a little more expensive than other public pools it's less frequented and is the perfect spot to relax. It's run by a small family, they also serve lunch, drinks, homemade cakes and ice cream there.

What's your favourite place for dining out?
Wolfram: There is a tiny Japanese Restaurant called *Kuishimbo* in the Naschmarkt area where we've been eating for years. It's run by a lovely Japanese family and serves delicious Kyoto-style dishes. We love their udon soups, especially during the winter.
Roshi: *Skopik und Lohn* is another one of our favourites. Wolfram especially loves their schnitzel. *Finkh* is also great; it's mainly Austrian cuisine, but very refined.

Is there a place on the outskirts of Vienna worth travelling out of the city to see?
Wolfram: The *Wienerwald* and the *Kahlenberg* are definitely worth a visit. If you like walks through the forest that is. You can also just go to one of the old coffeehouses and enjoy the scenery and view of the city.

In your opinion, what could is missing in Vienna, that would make the city even more perfect?
Roshi: 24-hour supermarkets.

Fairytale garden: the grounds of Schönbrunn Palace have been a public park for more than 230 years

Wiener Blutwurst

A photo showcase by Lukas Gansterer

Following in the footsteps of photographers such as Teller and Richardson, Gansterer uses a minimalist approach to capture his subjects in their most natural state. In this photo essay, he manages to capture the fading glory of the Austrian capital through the juxtaposition of Viennese nightlife characters and half-consumed local dishes

Carola Schmidt
She is an artist of many talents. Spanning media, she works with film, video, performance art and photography and often combines techniques. After finishing her studies at the Berlin University of the Arts until 2007, she quickly scooped up numerous awards for her music and video art. In 2012 Schmidt brought her alter ego Lulu Schmidt into being. Lulu is a performative, multi-talented, erotic and romantic character, acting out a journey of self-discovery through live music performances. Lately she has collaborated as singer with Berlin band Tangowerk

Carola Schmidt, Artist-Performer

Traditional Way

Vienna is Carola Schmidt's haven. Here she reconnects to her roots, recharges her batteries and finds the antidote to her busy everyday life which brings her to stages and studios all over the world. Traditional food, stunning nature and the relaxed atmosphere of Vienna give her the feeling of going back to her youth; calming her with traditions from the old imperial times and its coffee houses and swimming spots in the River Danube

Austrian designer Lena Hoschek is famous for her "Dirndl" collection—but she is also a big fan of the 1950s

Café Bräunerhof
1st District

Café Korb
1st District

Everyone has their own way of finding the right spots in a city—how do you usually do that?
When I travel, I go to the street. I try to find people who look like me, are my age and have an aura that appeals to me, and ask them for advice. In my experience you should never follow the first tip, because that's the one that's obvious. The second one is their second guess and the third is the most private one—the one they really love. That's the one I go for. That's how I move when I am a foreigner in a place.

Now we're curious about your third piece of advice for Vienna… But before that: what makes Vienna so special?
Vienna is a city where people are still sitting in the old coffee houses, reading and talking. The place I go to is the *Café Bräunerhof*. I really love the old people you see there. The old world is still alive here. The Bräunerhof is the place my Grandma used to go to and it's still the same. A visit is like time travel. They have live music. There's this old man coming with his violin and sometimes a piano player. They play the the traditional music called "Schrammelmusik"—which is never perfect. It's a sad sound, always a bit off tune and a bit squeaky. You immediately feel the vibe.

Vienna and its coffee houses… are there any others you would recommend?
Yes, I also love the *Café Korb*. What I like about those old places is that most of them have a "Patronin", an old woman who has

Leupold Restaurant
1st District

Zum Roten Bären
9th District

Alt Wiener Gastwirtschaft Schilling
7th District

Loos American Bar
1st District

Roberto American Bar
1st District

owned the place for ages. They have pictures of her and the team on the wall from back in the day. *Korb* is also special because they have a "Kegelbahn" (the German/Austrian version of bowling) in the basement. A lot of people I know go there, because it is in the 1st District. It's easy to just pass by and accidentally end up there.

What's on the menu at Bräunerhof or Korb? Which local dishes are to try?

There are all those nice dishes that only exist in Vienna. You definitely should try one of these: "eiernockerl mit frischer Petersilie" (Austrian dumplings made from semolina and flour), "griesnockerln-suppe" (dumpling soup), "mohn-nudeln" (noodles with melted butter and poppy seeds), "selch-fleischknödel" (meat dumplings), "käsespätzle" (cheese noodles), "marillenknödel" (stuffed dumplings), "powidltascherl" (sweet dumplings) or "mohr im hemd" (chocolate cake). I personally also love "gebackene champignons mit sauce tartare" (breaded mushrooms with tatar sauce). They are best at *Leupold* at Schottentor.

And of course Wiener Schnitzel... Where can I get a perfect Wiener Schnitzel in Vienna?

I would go to *Zum Roten Bären*. Thats my favourite spot right now, because those guys go out of town to the markets or even to the mountain regions like the Steiermark to get the freshest seasonal products available. Their menu changes weekly and you have to reserve your dinner seats early, because it's booked out almost every night. Still they have a lunch menu offering a perfect schnitzel daily or all the other hausmannskost (traditional food). Vienna has good beers too. You have to try an unfiltrated zwickelbier, which is an early stage beer. *Zum Roten Bären* serves them very fresh. You also get a nice schnitzel at *Gastwirtschaft Schilling*.

If there's no time for a big lunch, do you have a recommendation for a quick snack on the road?

You have to go to the würschtelbude (sausage stand) and try an "eitrige" (cheese sausage). I like the stand next to the Albertina. It looks like a UFO. It's very modern and at least you know the sausage hasn't stayed there for more than five days. Of course you are supposed to order it with hot mustard and fresh horseradish. Those places are open all night long, so they're perfect late night food spots, when you've had a drink or two. In the daytime there is also the "extrawurstsemmel" (sausage sandwich)—try it without cucumber and order an extra portion of "extrawurst". You can order it in any regular supermarket at the meat counter.

Vienna is best when it's very traditional. My advice is: go with the traditional food. One thing that's also very basic and you run into everywhere in the city is Aida, a bakery brand. This is where old women go to buy cakes and baked goods. My recommendation is to try an "original cremeschnitte" —that's much cheaper than an original "sachertorte", which is more of a tourist snack. Vienna is the perfect spot for gaining weight!

Where do you get a decent drink in Vienna?

For drinks only, I'd go to the *Loos Bar*. Although it is featured in some tourist guides, this is a real classic. Many artists come here and it's a perfect nighttime location. *Roberto American Bar* in the 1st District also offers excellent cocktails. For good white wine there is also *Zum Roten Bären*—one

Traditional Way

Elmayer
1st District

Motto
5th District

Strandbad Gänsehäufel
22nd District

of my absolute favourite spots at the moment. It's a superb example of a "wirtshauskultur", which is basically the more traditional kind of place, offering local wines and dishes, and all the specialities mentioned above.

Speaking of wine... Can you tell us what a "heuriger" is?

These are small taverns in the wine-growing regions in Austria. Like pop-up restaurants that the winegrowers run when they have their wines made. They're very much places for everyone. Everybody in those communities helps out, even children. You can find a heurige in virtually every small village around Vienna. I especially like the whole region around Krems in the Wachau, villages like Stein. You can't go wrong with that. It's just 45 minutes by car to the Wachau and you're surrounded by winefields —it's hot and it feels like you're suddenly in the south of Italy. Just google the "heurigenkalender" (wine bar calendar). But you don't even have to do that. Just go to the countryside and watch for a little ring made of hay with a little bulb inside—when the bulb is lit that's the sign that there's a heurigen going on. So just stop and look for the next garage, sit down, and let them cater for you.

What about the nightlife—where do Austrians go on a night out?

Austrians really love going to a ball. Number one with the young people is the "lifeball". A smaller variation is the "rosenball". From April to November there is a different ball almost every day: one for the "zuckerbäcker" (confectioner), one for the "jäger" (hunter)—the list is endless. There's a small booklet which tells you where and when. Austrians learn the dances and the etiquette in school. If you plan to attend a ball you should visit *Elmayer* in the 1st District to either book a course or get their book about how to behave there.

And for the easy nightlife... In terms of live music, clubs and parties, what are your recommendation?

Vienna people always go somewhere else. I would start a night at *Motto*. That's the only spot in Vienna where they have food and club music at the same time. It's a place where you don't get tired before heading into the night because the sound is just superb. The subculture scene is mostly posted on social networks like Facebook. It's a much smaller scene than, for example, Berlin. It's word-of-mouth propaganda here. You can also grab an issue of "Falter" magazine to see what's going on. But be aware, people smoke in the clubs...

Back to daytime... What do you do during the daytime on a sunny Vienna weekend?

Actually, in the summer I love to go to the countryside. You can go there easily—even with the tram. For example, you can go to the Wienerwald (Vienna Woods) and collect mushrooms. After that you head over to the Amtsstation to get your mushrooms tested—they will sort out the poisonous ones and then you go home with only the ones that are edible and taste good. For a short trip on the weekend I also recommend going to the region of Wachau to take a swim in the Danube.

In Vienna there's the *Strandbad Gänsehäufel*—a lido on the banks of the old arm of Danube. It's a very cool spot, because you're out in nature and in the distance you see the skyscrapers of the business district. Another nice thing is to take a boat trip, which boards at Schwedenplatz. You can make it to Budapest within three hours.

Steinhofgründe
14th District

Villa Aurora
16th District

Song
2nd District

Mühlbauer
1st District

Lena Hoschek
7th & 12th Districts

We Bandits
6th District

Feinkoch
6th District

Habari
6th District

At the inn Zum Roten Bären a young team serves traditional cuisine with local ingredients

Where else would you go to relax, in autumn and winter?
When I want to rest outside, I go to the *Steinhofgründe*. That's a park where you can spot wild deer if you're lucky. There's also the beautiful Otto-Wagner-Kirche and an old hospital for the mentally challenged—the whole place is magic. And you have a great view over the city. There's a restaurant close by: *Villa Aurora*. Their speciality is "kaiserschmarrn" (shredded pancake). But it's not a sweet one; it's served with little pieces of schnitzel. That's very special. And in winter there is an open-air ice-skating rink next to it.

If I were in a shopping mood, where would you send me?
For fashion: *Song* on Praterstraße is fantastic. They offer inspiring high-end fashion plus a small selection of furniture. Or you could buy yourself a hat at *Mühlbauer*, a traditional hatmaker with innovative designs. For a "dirndl" visit *Lena Hoschek*'s shop—she's famous for spicing up traditional Austrian dresses. Another of my favourite places is *We Bandits* in Theobaldgasse. Also, *Feinkoch*—they have ready-to-cook meals with recipes, so you can make your own classic Viennese dish at home. Next is *Habari*, which offers furniture from all over the globe. And on Schleifmühlgasse street you can get all the vintage stuff.

Do you have any other special tips for a visitors not to miss?
Haha, yes. So, I told you I love simple things... You shouldn't miss our tap water. Vienna probably has the best tap water in the world. It's spring water coming directly from the mountains. It's much better than any bottled water you can get anywhere in the world. It's perfect for your skin and hair—and it tastes just unbelievably good.

Traditional Way

3rd District & 4th District
Inside Out

| Culture | Emerging Art |

The *21er Haus*, Vienna's newest museum and exhibition space for contemporary art, has seen quite a few transformations during its short history. Originally designed for the Brussels World's Fair in 1958, this modernist gem and masterpiece of minimalist, lightweight architecture was later moved to Vienna. It housed the Museum for 20th Century Art, coming to be known as the 20er Haus. After the arrival of the new century, in 2011, it was carefully modernised and remodeled by Adolf Krischanitz and consequently renamed 21er Haus. It's now a branch of the nearby Belvedere museums, and focuses on Austrian art of the 20th and 21th centuries. You will also find the Blickle Kino with its stunning 1950s design and spacious auditorium, based on the cinema of the original Expo pavilion. And the museum's surroundings are completely new, too—the 21er Haus offers a perfect starting point to explore one of Europe's largest urban redevelopment projects. Around the new Vienna Hauptbahnhof, an entirely new district is emerging, with residential areas, office blocks, shops and cultural institutions.

• 21er Haus, Arsenalweg 1, 3rd, 21erhaus.at

| Food | **Daily Bread**

This spin-off of the Joseph Brot boutique (1st District, Naglergasse) quickly became a popular breakfast, lunch and dinner place after it opened in late 2013. Owner Josef Weghaupt (pictured) only processes ingredients from local producers he personally knows—and even the terrazzo stone stables in this minimalist venue have their roots in the region. There is also a upscale take-out bakery attached where you can buy bread, fantastic patisserie items, and Viennese homemade delicacies. Be sure to try the freshly made "belegte brote" sandwiches.
• Joseph Brot, Landstraße Hauptstraße 4, 3rd, joseph.co.at

| Night | **Rock around the Malipop**

"And even if the world should end, I'm sure that inside here nothing can happen to me," croons Viennese singer-songwriter Ernst Molden, in his eulogy to *Café Malipop*. Indeed, once you're inside this little alternative café-bar, time seems to slow down. The place was opened in 1979 and since then has stayed pretty much the same. It has a French, intellectual flair and the walls are coated with a smoky patina of old and new concert posters. Frau Margit, the bar's owner, plays records from her impressive vinyl collection, mostly indie or classic rock, and serves beer and toast. A wonderful place for late-night conversations.
• Café Malipop, Ungargasse 10, 3rd

| Shop | **Silver Man**

Welcome to Vienna's flamboyant side, where extravagance meets artisan handcraft. At *And_I* jewellery, goldsmith and sculptor Andreas Eberharter has made his mark with high-end accessories and jewellery made from precious metals and leather, as well as unusual materials such as aluminium. His fanciful pieces are in international demand—the metal eye patch became a cult object after it was used in Lady Gaga's "Paparazzi" video, and he has supplied other luminaries like Beyoncé. *The Workshop* in the 3rd District, which is only open on Thursdays, also offers ready-to-wear pieces.
• And_I, Hohlweggasse 11/2, 3rd, and-i.com

Shop | **Pure Porcelain**

The basic material for all objects produced by *Feinedinge* is Limoges porcelain. The shelves in the 340-square-metre shop are stacked with vases, lamps and tableware, most in white or pastel colours. Each item is handmade in the attached workshop, and can therefore differ slightly in form, colour and décor—adding to the label's individual flair. Look out for the illuminated "Moonstruck glow sphere"—key to the development of a label that's been in existence for more than ten years. Despite its fragile appearance, *Feinedinge*'s tableware is dishwasher proof and fit for everyday use.
• Feinedinge, Margaretenstrasse 35, 4th, feinedinge.at

Food | **Granny Time**

The name of this café-bar is a tribute to the owner's grandmother Erna. And that's not the only reference to Elisabeth Bader's roots. Papa Bader renovated the wooden tables, the chairs once belonged to her hometown's fire station, and the huge industrial lamps belonged to an ice rink. As far as the food's concerned, *Erna B.* is a family business as well. Mouthwatering homemade pies and snacks are on offer and Granny Erna also pops in once in a while to help out. The perfect cosy hangout after a stroll through the city—and a charming place to have a sip in the evening. Try a pint of Czech lager Velkopopovický Kozel.
• Erna B., Große Neugasse 31, 4th, ernab.at

Food | **Sweet 'n' Sauerkraut**

If during your stay you should tire of wiener schnitzel, tafelspitz or gulasch, then *ON* might be a welcome alternative. This restaurant has the look of a modern Viennese "gasthaus", but the cuisine is in fact Chinese, mixed with international and Austrian influences. Think Oriental sea bass meets Sauerkraut. Beside being a wonderful cook, ON's chef Simon Xie Hong is also a qualified doctor of traditional Chinese medicine, so he knows about the importance of food for our health—a fact reflected in the choice of his ingredients. In the summer there are also tables in the wonderful backyard garden.
• ON, Wehrgasse 8, 4th, restaurant-on.at

| Shop | Walk of Fashion |

Luxurious high-end brands might be found in the Vienna inner city but the more diverse, interesting shops and local designers are to be found in the 7th and especially in the 4th District. Start your stroll at *Jutta Pregenzer* at Schleifmühlgasse which has been around more than 20 years. Jutta Pregenzer sells her own purist fashion as well as international labels. Turn left at Margaretenstraße where you'll find the charming *Elfenkleid* (pictured), specialised in elegant evening and bridal creations (no.39). Just a couple of houses further along, the designer duo Superated operates a store called *Samstag* (no.46), including brands like Chalayan, La Casita de Wendy, Grosser Heinrich or the Austrian labels House of the Very Island's, Andy Wolf Eyewear and of course their own collections. Diagonally across find concept store *Unikatessen* (no.45), which is worth a visit, offering unique pieces and selective collections from aspiring young designers. If you are hunting for funky vintage items, *Little Joe's Gang* Vintage Shop might be just the ticket. It has some fantastic treasures such as shoes and sunglasses, from classic brands like Chanel. There's also a Little Joe's line of bags and t-shirts.

• 4th, various locations, see Index p.67

Singing in the Boys' Choir
Espionnage in Vienna

Viennese hospitality is not just a magnet for tourist culture vultures, conference participants and international diplomats. Even the city's clandestine workers feel comfortable in the capital on the Danube. Vienna is the European capital of spies, agents and intelligence services

On a hazy summer day at the Vienna airport, a Russian and an American plane land almost simultaneously. At the aircraft parking area, the two aircraft stand next to each other. Stairs are brought down; the doors open. People leave the American plane and get into a black van that takes them to the Russian one. There, another group gets in and is driven to the American vessel. People disappear into the aircraft, the doors are closed, the stairs are moved away. The Russian machine takes off around 12:30pm; the American, a quarter of an hour later. It takes less than two hours on 9th July, 2010 to complete the biggest spy swap since the Cold War.

The fact that the Vienna airport served as the backdrop for this is no coincidence. As a spy capital, Vienna stands next to Washington and New York in stature. Although Vienna itself has not played a significant role in world history since the end of the Habsburg monarchy, it has become a stage for those who dominate world political events. There are suspected to be about 7,000 spies currently in Vienna. With liberation by the Red Army from the Nazi regime and the division of the city into four zones of occupation, this was one of the frontline sites in the Cold War. Caroll Reed's spy thriller "The Third Man" gives an unmatched portrayal of this phase of contemporary history in a dark, cynical and debt-laden Vienna. But the withdrawal of the occupying powers did not change the significance of the metropolis on the Danube. Although the international patrol of the occupation forces—aka the "Four in a Jeep"—left Vienna in the summer of 1955, their intelligence system remains. Austria's neutrality freed the country from the occupying forces and sparked the ascent of Vienna to becoming the third city of the UN. The labelling of the federal capital as a centre of espionage and the prime meridian between East and West remains as relevant as ever.

The local environment of Vienna offers intelligence services almost ideal conditions. Because the city is home to a variety of corporations, diplomatic missions and international institutions—the UN, IAEA, OPEC and OECD—the central resources of the information business can be found here in abundance. And the trade in information thrives amid a wide legal grey zone. In Austria, espionage is prohibited only if it is directed against the Republic itself. As long as intelligence services do not infringe the basic rules of hospitality and behave in a somewhat discreet manner—staying in line with the proven "Do not ask, do not tell" method—generally speaking, people usually know nothing and do not even have to ask.

The high quality of life and the safety of the metropolis enhances its standing as a spy nexus. Just like everybody else, spies also enjoy the ordinary amenities of the city: its culture, the quality of life, good food, and water you can drink from the tap. Vienna is a

"Stolen Identity" is one of many Vienna-set movies about underhand dealings

popular posting in intelligence circles and in the CIA it was so highly regarded that there was a separate expression for the placement. Agents who went to Vienna were "singing with the Boys Choir".

According to journalist and intelligence expert Emil Bobi there are also cultural characteristics that distinguish Vienna as a city of espionage. He sees the Viennese as natural spies: "It's life here. It's part of Vienna's culture." The Viennese were always eager to meet strangers and to judge their agenda. They would therefore have an advanced level of human knowledge when dealing with other cultures. They are experts in keeping their own circumstances secret; offering the other the benefit of their experience and the opportunity to make use of it. The more whispers you've heard, the more important you are—because secrets are at the core of Viennese society. "Viennese society is built on secrets and people live to have secrets," says Bobi. "If you have a secret in Vienna you are somebody."

There seems to be no shortage of Viennese with secrets—not just in literary worlds like that of Arthur Schnitzler's "Traumnovelle" (Dream Novel)—but also in real life. After the fall of the Iron Curtain it was announced that about 12,000 city residents had spied for the USSR. They included Helmut Zilk, who was responsible for the fate of the city from 1984 to 1994 and is still considered one of its most popular mayors. At the same time Zilk lived in a residential house where the CIA maintained apartments for agents to work. It is believed that Zilk was active as a double agent. That would explain why his activities have never been made public, although they were known to the State Police. The existence of a double agent is nothing more than a specific form of international understanding and inter-cultural dialogue. It's thus considered a profoundly Viennese property. Isn't it fitting that the mayor of the capital of espionage be a spy himself?

Bobi goes on to speak of the "Wiener truth"—only a part of the "whole truth". And the Viennese have a special relationship with that cardinal virtue. The residents of this city don't find the truth uncomfortable, per se. But it can be considered impractical, or worse—rude. Bobi explains that when things are clear and known—then they are no longer negotiable. Avoiding setting things in stone is typical of Vienna. It allows you to have done something differently in retrospect. For Bobi, a lot has to do with "Viennese artfulness"—usually rephrased as "Viennese charm". What this translation deliberately omits is that this charming and joyful form of behaviour is maintained even when dealing with the truth. "Austrians tend to make their lives easier. So, first of all, they are very polite and second, they do not mean it." revealed Oscar-winner Christoph Waltz when appearing on Conan O'Brien. This attitude protects the other from the truth, thereby placing them in a losing position.

While this typical Viennese behaviour can be observed in the everyday life of the city quite easily, there are of course only very limited opportunities to observe the espionage activities themselves. The places where they occur can, however, be visited—and not just the historical locations, such as the Vienna sewers depicted in "The Third Man". The documents leaked by Edward Snowden list several Viennese NSA listening posts, some of which are now well known. One is a villa in the middle-class district of Währing. If you believe the Austrian news magazine "Format", about 70 percent of telecommunications traffic in the city is monitored. Another monitoring system is located on the roof of the US Embassy and a third is suspected to be on the roof of the IDZ Towers, where the US mission is quartered at the United Nations. From the roof of the building you can look down on the United Nations headquarters just 50 metres away. Another spy hotspot will inevitably be visited by many guests of the city: the transit area at Schwechat Airport is considered one of the busiest hubs for information of all kinds. If you're lucky you might even find yourself in a handover or even an exchange of prisoners during your visit. Though you might not even know about it.

Werner Sturmberger is a Vienna-based social scientist and freelance journalist. He is a frequent contributor to Heureka, the science and humanities supplement of Viennese city magazine "Falter". For magazines like "Biorama" or "The Gap" he covers topics related to ecology and urbanism as well as music, film and theatre.

Erwin Gegenbauer, Food Expert
The Insider

Erwin Gegenbauer
In Vienna's Favoriten district, his family's company was originally founded by Ignaz Gegenbauer in 1929. In 1992 Erwin Gegenbauer developed the business from mass production of canned goods to high quality products like vinegar, juice and oil—manufactured in a traditional way with local ingredients. His vinegars are now used by top chefs all over the world. You can find Gegenbauer's products in Naschmarkt. Or take a tour and stay for the night at the Wiener Gäste Zimmer guesthouse set at his vinegar distillery in the heart of Vienna

With chefs and gourmets as your clients, you have to be a foodie. But to capture the real essence of Vienna, Erwin Gegenbauer recommends sampling more simple pleasures. A traditional meal in a "wirtshaus", enjoying the fruits of the vineyard at a "heuriger", or dining at the first-ever vegetarian restaurant with a Michelin star

Restaurant Konstantin Filippou: Michelin-starred cuisine served in a minimalist interior

Meixner's Gastwirtschaft
10th District

Gasthaus Wolf
4th District

Heuriger Spaetroth
Gumpoldskirchen

Your products are very successful in the high-class gastronomy sector, globally—and you've lived in many different cities. What keeps you in Vienna?

Yes, that's true. I spent time in Paris, London and New York when I was young. That was simply to see what other cultures are like, and what the pots and pans look like in other places. That fascinates me. I can recognise the culture of a people in its cookware. But back to the point: why I stayed in Vienna, I can rest here. Vienna is a comfortable city. I'm able to withdraw and concentrate on my work. The privacy I have here is unprecedented.

Where in Vienna should can I hope to see the city's nature in its cookware?

Go to a "wirtshaus" (tavern). You can taste traditional Viennese food and see how it developed. Viennese cuisine is mixed, with diverse influences from many regions of Europe. I would say that it's the first cosmopolitan cuisine. With sweet dishes from Bohemia and Moravia, goulash from Hungary, spicier dishes from the Balkan countries—the mix is what makes it so unique. It also still involves plenty of handicraft.

Which taverns do you recommend?

Meixner's Gastwirtschaft is great and traditional. I also find the *Gasthaus Wolf* very good. That's actually quite a hot tip, in the 4th District—first-rate.

What about "heuriger"—the traditional wine houses. Do you have a recommendation?

The best winery for me is in Gumpoldskirchen, which is quite a well-known wine town—it's called *Heuriger Spaetroth*. A Michelin-star cook serves traditional regional food there. Of course, they also have wonderful wines, especially from local grape varieties such as Zierfandler and Rotgipfler. Those

The Insider

The Naschmarkt remains the heart of Vienna's food scene—and its surrounding area is worth exploring

Tian
1st District

Konstantin Filippou
1st District

Steirereck
1st District

Naschmarkt
6th District

Karmelitermarkt
2nd District

Victor-Adler-Markt
10th District

are fantastic old wines—different from the mainstream.

Can you recommend a specific vintner?
Yes, I'd suggest two: the Johanneshof Reinsch and the Schaflerhof. Both of those are really good estates with interesting wines.

So, let's move away from tradition... A vegetarian restaurant in Vienna was the first of its kind to receive a Michelin star. Where can I find modern and innovative cuisine in the city?
That takes us into the top ranks. The vegetarian restaurant you mentioned, *Tian*—that's really high-class. It's a bit too opulent for my taste, but the effort invested in the dishes is incredible—not to mention the ingredients used. It's just excellent. Another very interesting top-class restaurant is the *Restaurant Konstantin Filippou*. It has a Michelin star too. The *Steirereck*, of course, is still wonderful—it will certainly get its third star sometime in the next

few years. The cuisine there is truly innovative; each course takes you to new depths with great clarity.

Besides taverns and star-crowned restaurants, are there any other tips to discover Vienna?
The markets—naturally, the area around the *Naschmarkt*. This is not just Vienna's culinary heart. Galleries, small bars, taverns, coffee houses... The area really has a lot to offer. Also, the *Karmelitermarkt* in the old Jewish Quarter, where one can sense the old classical Viennese culture and lifestyle. For something completely untouched, then try the *Victor-Adler-Markt* in the 10th District. There, one can have a very pure, non-touristic experience, and brush shoulders with people from anywhere you can imagine. The quarter has just been refreshed because of the new central station. The 10th District has a bad image and is considered a classic

Excellent food meets innovative architecture in a park setting at Steirereck

Café Prückel
1st District

Café Korb
1st District

Loos American Bar
1st District

Pearl Bar,
Park Hyatt Hotel
1st District

working-class area. However, it's currently undergoing rapid development. If you want a real experience, go there, where the unspoilt life thrives.

Where are your spots to withdraw to, after a market day for instance?

As always, the classic coffee house. It plays a very important role here. Even if they don't actually have good coffee! One orders a cup, gets a glass of water with it, and can then read the papers—undisturbed—for hours on end.

Which coffee house can you most often be found?

In the *Café Prückel*, perhaps. Or in the *Café Korb*. But I'm also a bar frequenter. I like to go to bars... and I don't just drink one cocktail!
Which bars do you hit then?

While the *Loos Bar* isn't a well-kept secret anymore, it's still lovely. The new bar in the Park Hyatt Hotel is also worth a visit.

Is there anything you don't like about Vienna?

The opening times on Sundays are a horror. Everything is closed on Sundays. That's a catastrophe here. Especially for visitors who come to town just for the weekend.

Ah... So what is there to do on a Sunday in Vienna?

Stay in bed—hopefully not alone! Or go out to the countryside. It doesn't take long to get to the vineyards and the nice countryside restaurants. It takes 20 minutes by car, and shows you a whole new side of the region.

Editors' Picks
Wiener Melange

Royal Skull
Would a perfume in any other bottle smell as sweet? Wendy & Jim answer that question with their cold modern twist on a classic idea. The fashion designer duo mix clever china work with naturally sourced fragrances. The skeletons are made out of high quality Royal Viennese porcelain.
• Drop No.01, Wendy & Jim, wendyjim.com

Eyecatcher
Melanie Haarhaus's jewellery has a subtle dark touch—and her handmade designs are wonderfully finished. Whether it's a silver skeleton with golden ornaments or a gallant rose, you'll find it difficult not to stare. Her work is playful yet stunning. Either take it with a smile or drink in the artistic originality.
• By appointment only, melaniehaarhaus.com

Heritage Hats
Counting Yoko Ono and Madonna among its patrons, Mühlbauer is a milliner that provides both quality and variety. According to the season of your choice, all fabrics and styles are assembled expressly for you. Their unique hats manage to mix tradition with a modern touch.
• muehlbauer.at

Books

Night Work
• Thomas Glavinic, 2006

The "Omega Man" scenario set in Austria: Jonas wakes up one morning in modern Vienna to discover he's the only human being around. But where Charlton Heston in the 1971 sci-fi film had to shoot zombies, Jonas must fight solitude and the monsters of his subconscious. A thrilling and profound piece of fiction from the city of Freud.

The Third Man
• Graham Greene, 1950

A murder mystery leads a journey through post-war shadows of ruined palaces. Arrived in Vienna for a job, the protagonist finds himself investigating the sudden death of his host. Greene intentionally wrote the novella for its film adaptation: it was published after Carol Reed's classic movie.

The Man Without Qualities
• Robert Musil, 1930–1942

This seminal modernist novel addresses themes ranging from science and economics to philosophy and psychoanalysis. The plot depicts Viennese society months prior to the beginning of the First World War. Musil spent over 20 years writing his masterpiece. And having failed to achieve commercial success at the time, he died in 1942, broke. The novel was never completed.

Movies

Müllers Büro
• Niki List, Hans Selikovsky, 1986

This comedy set in mid-1980s Vienna plays with film noir and musical clichés. A singing private eye meets a strange character in Vienna's red light district. This trippy movie experience is typical fare from cult filmmaker Niki List.

The Piano Teacher
• Michael Hanneke, 2010

The story of a nymphomaniac piano teacher (Isabelle Huppert) teaching at the Vienna Conservatory, and her relationship with a student. A dark tale in which the juxtaposition of piano and hard mechanical sounds makes for interesting, layered melodies. Listen out for those from Vienna's own Schubert. A classic and disturbing European masterpiece.

1st April 2000
• Wolfgang Liebeneiner, 1952

A political satire depicting a vision of Austria in the year 2000. The nation has to prove it is no longer a nation of war but of peace. On the one hand, there's a submersion into creativity and wonder with spaceships, a "Global Union" and futuristic sets. On the other a firm re-attachment to Austria's authentic heritage is presented. The resulting combination is not to be missed.

Music

10 Years Of Cheap Records
• Various, 2004

The A-team of Vienna's electronic music pioneers: listen to techno by Christopher Just, post-punk by Potuznik, or the voice of Louie Austen who transformed from a bar singer to an underground star. The highlight: "Claire"—one of the most elegant club tracks ever—written by Pulsinger & Tunakan.

20th Century Masterpieces
• Alban Berg Quartet, 2008

A product of the Vienna Academy of Music, they're lauded as one of the finest string quartets of our time. Whether they play Alban Berg, Arnold Schoenberg or Béla Bartók—their execution holds up to their reputation. Immaculately harmonious and tastefully subtle: this is surely one of Vienna's finest cover bands.

Psychoterror
• Drahdiwaberl, 1980

If Frank Zappa had been from Vienna this would likely have been his outfit. Punk meets funk with anarcho-inspired lyrics in Austrian-German and English. Falco made his first stage appearances with this combo, and introduced his future hit "Ganz Wien" to their underground audience.

Where to stay

Ruby Sofie Hotel

The Ruby Sofie Hotel represents a new world of hospitality with a strong movement in the direction of lean luxury. The former Sofiensäle music hall was transformed for the opening of the hotel in 2014, and still boasts an impressive Secessionist-style façade. The hotel combines laidback glamour, traditional materials, and an unconventional approach. The rooms come in three sizes and offer very comfortable beds and a modern interpretation of Austrian style. The library and 24-hour bar feature design classics, vintage pieces, and original items from the building's past.

Where to stay

25hours Hotel

The 25hours Hotel Vienna at MuseumsQuartier might be the most playful place to stay in Vienna. The interior design is largely inspired by the circus, perfectly embodying a surreal and surprising fantasy atmosphere. The combination of material, shapes and colours transforms each of the 217 rooms into an individual retreat. With its excellent location, right in the centre of the museum district, all major attractions are easily reachable. The rooftop bar, in-house restaurant, and food truck at the hotel all offer attractive dining options to help you savour your day.

Where to stay

The Guesthouse

Once a 1950s student hostel, The Guesthouse is now a temple of clean, comfortable interior design. The boutique hotel conversion is outfitted with modern design classics by Wittmann and Kiesler, selected by none other than Sir Terence Conran. Its supreme location in the centre of town, surrounded by classical Viennese architecture and charming small streets, makes The Guesthouse an ideal spot for a picture-perfect city break. Personal service and a down-to-earth atmosphere pay homage to the hotel's humble beginnings, and the famous in-house bakery adds an extra cosy charm. Hard not to feel at home in this relaxed and welcoming spot.

Story

Out of the Dark

Something warm and wet is touching my face; it feels like a dog's tongue. I hear music, a rural version of Falco's "Ganz Wien". I've never heard it sound like this, but the voice is definitely his. I hear children laughing and the gentle voice of a woman.

"Hans, come quick—he woke up!"
The music stops. It sounds like people are leaving. Where am I? I open my eyes and see a man entering the room.

"You've been sleeping for days," he says. "The dog found you in the woods under a fallen tree. We're glad you're alive. Are you hungry?"

Wait a minute—this is not happening. Is that Falco? I must be dead then.

Later that night we are sitting at a giant wooden table in the open kitchen of the farmhouse. He's looking good. Not that he ever looked bad, but now he seems very much in balance with himself.

"Hello, stranger," he greets me, smiling and offering wine and food.

"Falco? What happened? The world thinks you're dead—how is this possible?"

"Please don't call me Falco. That was a marketable international brand; it's history. Just call me Hans."

"Of course... Sorry, Hans, but how come you're alive?"

"Well, that was a deal I made with a friend, a famous ex-racecar driver who was very concerned about my lifestyle in the late '80s. He said: 'Hans, you have to get away from all the bullshit. Why can't you stop messing up and start over? Step out of the light. It's time for you to really get clean and start living a happy life.'"

"But wasn't that why you moved to the Dominican Republic in the first place?"

"Of course, but it didn't work, even the Dominican cab drivers were greeting me with 'Alles klar, Herr Kommissar' and offering me everything that would destroy me. So my friend and I figured out the only way to kill Falco was to let Hans Hoelzl die. We faked the accident, arranged a coffin to be flown back to Vienna, and staged a funeral." He laughs.

"What a big success!"

The door flies open, and the room suddenly fills with happy kids. "Daddy!" they cheer, and jump onto him. He laughs, gives them kisses and asks his wife to take them to bed.

"I'll be right there for a goodnight story." He turns back to me.

"You see, I'm a happy farmer now."

"So, you still make music then?"

"Of course. You never stop being a musician; it just doesn't matter anymore if I earn money with it."

"But, if you were young again, would you do it all the same?" He takes a sip of wine and smiles.

"Maybe I wouldn't live my life to such extremes. On the other hand, I doubt that I would have been able to write songs like 'Ganz Wien' or 'Der Kommissar' in a clean state of mind. It's one thing to push things to the limit and another thing to moralise without experience. I guess that just doesn't work. But why would I even think about that? There's no need to." His beautiful wife enters the room.

"Hans, the children—they're waiting for their story."
"I'll be right there, honey—just a second."
He turns towards me.

"I'm an organic farmer now, and at nights I play music in the barn with my boys from the village. It's like a dream to me. Why the hell would I want to go back? Everybody out there is under constant surveillance. From up here it seems to me that I'm observing robots living a fake life, guided by fake promises. It's all fake down there. I don't miss a thing…"

Back in Vienna, I nevertheless decided to buy some flowers, enter the Straßenbahn and pay his grave a visit. Zentralfriedhof, Abteilung Ehrengräber, Gruppe 40.

Text by Jörg Henning
Illustration by Krista Bursey

Available from LOST iN

Next Issue: Ibiza

WWW.LOSTIN.COM

Index

- Ⓒ Culture
- Ⓕ Food
- Ⓝ Night
- Ⓞ Outdoors
- Ⓢ Shop

1st District

Albertina
Albertinaplatz 1
+43 1 53 483
albertina.at
→ p. 27 Ⓒ

Burggarten
Josefsplatz 1
+43 1 53 39 083
→ p. 27 Ⓞ

Burgtheater
Universitätsring 2
+43 1 51 44 441 40
burgtheater.at
→ p. 17 Ⓒ

Burgtor
Heldenplatz
+43 1 211 140 84
→ p. 17 Ⓒ

Café Bräunerhof
Stallburggasse 2
+43 1 51 30 549
braeunerhof.at
→ p. 40 Ⓕ

Café Central
Herrengasse 14
+43 1 53 33 763
palaisevents.at
→ p. 27 Ⓕ

Café Griensteidl
Michaelerplatz 2
+43 1 535 269 20
cafegriensteidl.at
→ p. 27 Ⓕ

Café Havelka
Dorotheergasse 6
+43 1 51 28 230
hawelka.at
→ p. 27 Ⓕ

Café Korb
Brandstätte 9
+43 1 53 37 215
cafekorb.at
→ p. 40, 55 Ⓕ

Café Museum
Operngasse 7
+43 1 241 006 20
cafemuseum.at
→ p. 18 Ⓕ

Café Prückel
Stubenring 24
+43 1 51 26 115
prueckel.at
→ p. 18, 27, 55 Ⓕ

Elmayer
Bräunerstraße 13
+43 1 51 27 197
elmayer.at
→ p. 42 Ⓒ

Firis
Bauernmarkt 19
+43 1 53 34 275
firis.at, → p. 27 Ⓢ

Heldenplatz
Heldenplatz
→ p. 17 Ⓞ

Joseph Brot
Naglergasse 9
+43 1 71 02 881
joseph.co.at
→ p. 45 Ⓕ

Kunsthistorisches Museum
Maria-Theresien-Platz
+43 1 01 525 240
press.khm.at
→ p. 17 Ⓒ

Leupold Restaurant
Schottengasse 7
+43(0)5 33 93 811 4
leupold.at
→ p. 41 Ⓕ

Little Joe's Gang
Operngasse 34
+43 1 96 67 534
→ p. 47 Ⓢ

Loos American Bar
Kärntner Durchgang 10
+43 1 51 23 283
loosbar.at
→ p. 41, 55 Ⓝ

Motto am Fluss
Schwedenplatz 2
+43 1 25 25 510
motto.at
→ p. 42 Ⓕ Ⓝ

Mühlbauer
Seilergasse 10
+43 1 53 35 269
muehlbauer.at
→ p. 43 Ⓢ

Palmenhaus
Burggarten 1,
+43 1 53 31 033
palmenhaus.at
→ p. 9 Ⓕ Ⓞ

Pearl Bar / Park Hyatt
Am Hof 2
+43 1 227 401 183
vienna.park.hyatt.com
→ p. 55 Ⓝ

Plachutta
Wollzeile 38
+43 1 51 21 577
plachutta.at
→ p. 18 Ⓕ

Restaurant Konstantin Filippou
Dominikanerbastei 17
+43 1 51 22 229
konstantinfilippou.com
→ p. 54 Ⓕ

Roberto American Bar
Bauernmarkt 11-13,
+43 6 769 429 001
robertosbar.com
→ p. 10, 27, 41 Ⓝ

Stephansdom
Stephansplatz 3
+43 1 515 523 054
stephanskirche.at
→ p. 17 Ⓒ

Tian
Himmelpfortgasse 23
+43 1 89 04 665
tian-vienna.com
→ p. 54 Ⓕ

Wien Museum
Karlsplatz 8
+43 1 505 874 70
wienmuseum.at
→ p. 27 Ⓒ

2nd District

Büro Weltausstellung
Praterstr 15
+43 6 76 43 021 91
rendi.at
→ p. 17 Ⓒ

Café Ansari
Praterstrasse 15
+43 1 27 65 102
cafeansari.at
→ p. 17, 21 Ⓕ

Café Einfahrt
Haidgasse 3
+43 1 94 26 886
einfahrt.at
→ p. 23 Ⓕ

Das Engel
Große Pfarrgasse 5
+43 1 2127894
dasengel.at
→ p. 23 Ⓕ

Das Loft
Praterstrasse 1
+43 1 90 61 60
sofitel.com
→ p. 21 Ⓕ Ⓝ

Kaas am Markt
Karmelitermarkt 33
+43 1 69 918 140 601
kaasammarkt.at
→ p. 23 Ⓕ

Karmelitermarkt
→ p. 54 Ⓞ Ⓕ Ⓢ

MAK
Stubenring 5
+43 1 71 13 60
mak.at, → p. 9 Ⓒ

Mochi
Praterstraße 15
+43 1 92 51 380
mochi.at
→ p. 17 Ⓕ

Pratersauna
Waldsteingartenstraße 135
+43 1 72 91 927
pratersauna.tv
→ p. 27 Ⓝ Ⓞ

Skopik und Lohn
Leopoldsgasse 17
+43 1 21 98 977
skopikundlohn.at
→ p. 22, 28 Ⓕ

Song
Praterstraße
+43 1 53 22 858
song.at/shop
→ p. 21, 27, 43 Ⓢ

Supersense
Praterstrasse 70
+43 1 96 90 832
the.supersense.com
→ p. 20 Ⓢ

Ü-Lokal
Obere Augartenstraße 46
+43 1 96 91 013
ue-lokal.at
→ p. 22 Ⓕ

Wurstelprater
Prater 7
+43 1 728 05 16
prater.at
→ p. 28 Ⓞ

Zimmer 37
Karmelitermarkt 37-39
+43 6 99 17 23 73 11
zimmer37.at
→ p. 23 Ⓕ

3rd District

21 Haus
Arsenalweg 1
+43 1 79 55 70
21erhaus.at
→ p. 44 ©

And_I
Andreas Eberharter
Hohlweggasse 11/2
and-i.net
→ p. 45 Ⓢ

Belvedere
Prinz-Eugen-Strasse 27
+43 1 79 55 71 34
belvedere.at
→ p. 17 ©

Café Malipop
Ungargasse 10
→ p. 45 Ⓕ

Restaurant Steirereck
Am Heumarkt 2A
+43 1 71 33 168
steirereck.at
→ p. 54 Ⓕ

4th District

Erna B
Große Neugasse 31
+43 6 63 92 59 61
ernab.at
→ p. 46 Ⓕ

Flo Vintage
Schleifmühlgasse 15
+43 1 58 60 773
flovintage.com
→ p. 27 Ⓢ

Gasthaus Wolf
Grosse Neugasse 20
+43 1 58 11 544
gasthauswolf.at
→ p. 53 Ⓕ

Jutta Pregenzer
Schleifmühlgasse 4
+43 1 58 65 758
pregenzer.com
→ p. 47 Ⓢ

5th District

Elfenkleid
Margaretenstr. 39 3–4
+43 1 20 85 241
elfenkleid.com
→ p. 47 Ⓢ

Feinedinge
Margaretenstraße 35
+43 1 95 40 918
feinedinge.at
→ p. 46 Ⓢ

Motto
Rüdigergasse 1
+43 1 58 70 672 11
motto.at
→ p. 17 Ⓝ

ON
Wehrgasse 8
+43 1 58 54 900
restaurant-on.at
→ p. 46 Ⓕ

Samstag Shop
Margaretenstraße 46
+43 6 99 17 095 882
samstag-shop.com
→ p. 47 Ⓢ

Unikatessen
Margaretenstrasse 45
unikatessen.at
→ p. 47 Ⓢ

6th District

Café Sperl
Gumpendorfer Str. 11
+43 1 58 64 158
cafesperl.at
→ p. 27 Ⓕ

Disco Volante
Gumpendorfer Str. 98
+43 6 64 19 525 45
disco-volante.at
→ p. 10 Ⓕ

Feinkoch
Theobaldgasse 14
+43 6 99 10 04 57 12
feinkoch.org
→ p. 43 Ⓢ

Finkh
Esterhazygasse 12
+43 1 91 38 992
finkh.at → p. 28 Ⓕ

Habari
Theobaldgasse 16
+43 1 58 62 340
habari.at
→ p. 43 Ⓢ

Kuishimbo
Linke Wienzeile 40
+43 6 99 11 94 06 73
→ p. 28 Ⓕ

Naschmarkt
Linke Wienzeile
+43 6 60 49 055 61
wienernaschmarkt.eu
→ p. 17, 23, 28, 53, 54 Ⓕ Ⓞ

Saint Charles Apotheke
Gumpendorfer Str. 30
+43 1 58 61 363
saint.info
→ p. 11 Ⓢ Ⓕ

We Bandits
Theobaldgasse 14
+43 6 99 19 003 895
webandits.tictail.com
→ p. 43 Ⓢ

Zum Gschupftn Ferdl
Windmühlgasse 20
+43 1 96 63 066
zumgschupftnferdl.at
→ p. 11 Ⓕ

7th District

Alt Wiener Gastwirtschaft Schilling
Burggasse 103
+43 1 52 41 775
schilling-wirt.at
→ p. 41 Ⓕ

Kussmaul
Spittelberggasse 12
+43 1 58 77 628 5
kussmaul.at
→ p. 8 Ⓕ

Lena Hoschek Store
Gutenberggasse 17
+43(0)5.03 09 200
shop.lenahoschek.com
→ p. 43 Ⓢ

Leopold Museum
Museumsplatz 1
+43 1 52 57 00
leopoldmuseum.org
→ p. 27 ©

Mumok
Museumsplatz 1
+43 1 52 50 00
mumok.at
→ p. 27 ©

Park
Mondscheingasse 20
+43 1 52 64 414
park.co.at
→ p. 27 Ⓢ

9th District

Grelle Forelle
Spittelauer Lände 12
grelleforelle.com
→ p. 27 Ⓝ

Zum Roten Bären
Berggasse 39
+43 1 31 76 150
→ p. 41 Ⓕ

10th District

Meixner's Gasthaus
Buchengasse 64
+43 1 60 42 710
meixners-gastwirtschaft.at
→ p. 53 Ⓕ

Victor-Adler-Markt
Viktor-Adler-Markt 3–5
+43 1 60 35 306
→ p. 54 Ⓞ

11th District

Friedhof der Namenlosen
+43 6 60 60 03 023
friedhof-der-namenlosen.at, → p. 28 Ⓞ

12th District

Lena Hoschek Atelier
Längenfeldgasse 27
+43(0)5.03 09 100
shop.lenahoschek.com
→ p. 43 Ⓢ

13th District

Schönbrunn Palace
Schönbrunner
Schloßstraße 47
+43 1 81 11 3239
schoenbrunn.at
→ p. 27 Ⓞ

14th District

Steinhofgründe
Baumgartner Höhe 1
+43 1 91 06 00
wienkav.at
→ p. 43 Ⓞ

16th District

Restaurant Villa Auora
Wilhelminenstraße 237
+43 1 48 93 333
→ p. 43 Ⓕ

17th District

Neuwaldegger Bad
Promenadegasse 58
+43 1 48 62 452
→ p. 28 Ⓞ

19th District

Kahlenberg
→ p. 8, 28 Ⓞ

TBA21
Scherzergasse 1A
+43 1 513 985 624
tba21.org, → p. 22 Ⓒ

22nd District

Strandbad Gänsehäufel
Moissigasse 21
+43 1 269 901 60
gaensehaeufel.at
→ p. 42 Ⓞ

23rd District

Wotruba Kirche
Ottillingerplatz 1
+43 1 88 86 147
georgenberg.at
→ p. 7 Ⓞ

Outskirts

Heuriger Spaetrot
Wienerstraße 1
Gumpoldskirchen
+43 6 64 88 43 90 10
heuriger-spaetrot.com
→ p. 53 Ⓕ

ON THE ROAD

The App for the Discerning Traveller

Explore insider recommendations and create your personal itinerary with handpicked locations tailored to your desires. Our selection of experiences ranges from independent boutiques, galleries, neighborhood bars to brand new restaurants. Experience a new city from within.

LOST iN